Heavy Equipment Operations

Level Three

Trainee Guide
Third Edition

PEARSON

Boston   Columbus   Indianapolis   New York   San Francisco   Upper Saddle River
Amsterdam   Cape Town   Dubai   London   Madrid   Milan   Munich   Paris   Montreal   Toronto
Delhi   Mexico City   São Paulo   Sydney   Hong Kong   Seoul   Singapore   Taipei   Tokyo

NCCER

President: Don Whyte
Director of Product Development: Daniele Stacey
Heavy Equipment Operations Project Manager: Patty Bird
Senior Manager: Tim Davis
Quality Assurance Coordinator: Debie Ness

Desktop Publishing Coordinator: James McKay
Permissions Specialist: Megan Casey
Production Specialist: Megan Casey
Editor: Chris Wilson

Writing and development services provided by Topaz Publications, Liverpool, NY

Lead Writer/Project Manager: Tom Burke
Desktop Publisher: Joanne Hart
Art Director: Alison Richmond

Permissions Editors: Toni Burke
Writers: Thomas Burke, Darrel Wilkerson

Pearson Education, Inc.

Editorial Director: Vernon R. Anthony
Executive Editor: Alli Gentile
Editorial Assistant: Douglas Greive
Program Manager: Alexandrina B. Wolf
Operations Supervisor: Deidra M. Skahill
Art Director: Jayne Conte
Director of Marketing: David Gesell
Executive Marketing Manager: Derril Trakalo
Marketing Manager: Brian Hoehl
Marketing Coordinator: Crystal Gonzalez

Composition: NCCER
Printer/Binder: LSC Communications
Cover Printer: LSC Communications
Text Fonts: Palatino and Univers

Credits and acknowledgments for content borrowed from other sources and reproduced, with permission, in this textbook appear at the end of each module.

4 16

Perfect bound   ISBN-13: 978-0-13-340256-8
ISBN-10:   0-13-340256-8

Preface

To the Trainee

Welcome to your third year of training in heavy equipment operations. If you are training under an NCCER Accredited Training Program Sponsor, you have successfully completed *Heavy Equipment Operations Level One* and *Heavy Equipment Operations Level Two*, and are well on your way to more advanced training.

Heavy equipment operators work on a wide variety of projects, including building construction, and on roads, bridges, mining, and timber operations, just to name a few. New construction and infrastructure projects continue to increase the demand for qualified operators. The skills that qualified operators provide are vital for clearing sites, moving materials, or any earthmoving operations.

New with *HEO Level Three*

NCCER is proud to release the newest edition of *Heavy Equipment Operations Level Three* in full color with updates to the curriculum that will engage you and give you the best training possible. In this edition, you will find many changes from the last edition. The layout has changed to better align with the learning objectives. There are also new end-of-section review questions to compliment the module review. The changes to the modules include updates to the latest technology and standards. *Advanced Operational Techniques* has been deleted. The module *Rollers* has been renamed *Compaction Equipment* and covers all types of compaction equipment. The new *Off-Road Dump Trucks* module focuses on the safety, operation, and preventive maintenance of off-road dump trucks. The modules *Backhoes*, *Dozers*, *Excavators*, and *Motor Graders* have all been updated.

We invite you to visit the NCCER website at **www.nccer.org** for information on the latest product releases and training, as well as online versions of the *Cornerstone* magazine and Pearson's NCCER product catalog.

Your feedback is welcome. You may email your comments to **curriculum@nccer.org** or send general comments and inquiries to **info@nccer.org**.

NCCER Standardized Curricula

NCCER is a not-for-profit 501(c)(3) education foundation established in 1996 by the world's largest and most progressive construction companies and national construction associations. It was founded to address the severe workforce shortage facing the industry and to develop a standardized training process and curricula. Today, NCCER is supported by hundreds of leading construction and maintenance companies, manufacturers, and national associations. The NCCER Standardized Curricula was developed by NCCER in partnership with Pearson, the world's largest educational publisher.

Some features of the NCCER Standardized Curricula are as follows:

- An industry-proven record of success
- Curricula developed by the industry for the industry
- National standardization providing portability of learned job skills and educational credits
- Compliance with the Office of Apprenticeship requirements for related classroom training (*CFR 29:29*)
- Well-illustrated, up-to-date, and practical information

NCCER also maintains a National Registry that provides transcripts, certificates, and wallet cards to individuals who have successfully completed a level of training within a craft in NCCER's Curricula. *Training programs must be delivered by an NCCER Accredited Training Sponsor in order to receive these credentials.*

Special Features

In an effort to provide a comprehensive, user-friendly training resource, we have incorporated many different features for your use. Whether you are a visual or hands-on learner, this book will provide you with the proper tools to get started in heavy equipment operations.

Color Illustrations and Photographs

Full-color illustrations and photographs are used throughout each module to provide vivid detail. These figures highlight important concepts from the text and provide clarity for complex instructions. Each figure reference is denoted in the text in *italic type* for easy reference.

Figure 26 Turning to the right.

Introduction

This page is found at the beginning of each module and lists the Objectives, Performance Tasks, Trade Terms, and Required Trainee Materials for that module. The Objectives list the skills and knowledge you will need in order to complete the module successfully. The Performance Tasks give you an opportunity to apply your knowledge to the real-world duties that heavy equipment operators perform. The list of Trade Terms identifies important terms you will need to know by the end of the module. Required Trainee Materials list the materials and supplies needed for the module.

Special Features

Features provide a head start for those entering heavy equipment operations by presenting technical tips and professional practices from operators in a variety of disciplines. These features often include real-life scenarios similar to those you might encounter on the job site.

Weigh Stations

Anyone who travels on the Interstate Highway System has seen the weigh stations at intervals along the highway. Commercial vehicles using the highways are required to pass through these stations. The more modern weigh stations can record the weight while the vehicle is moving through the station. Others require the vehicle to stop on a scale. The primary purpose of these

Notes, Cautions, and Warnings

Safety features are set off from the main text in highlighted boxes and are organized into three categories based on the potential danger of the issue being addressed. Notes simply provide additional information on the topic area. Cautions alert you of a danger that does not present potential injury but may cause damage to equipment. Warnings stress a potentially dangerous situation that may cause injury to you or a co-worker.

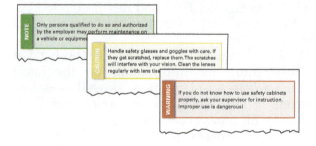

Going Green

Going Green looks at ways to preserve the environment, save energy, and make good choices regarding the health of the planet. Through the introduction of new construction practices and products, you will see how the "greening of America" has already taken root.

GOING GREEN

Hybrid Excavator

As part of an ongoing effort to reduce emissions and improve fuel efficiency, heavy equipment manufacturers are looking for new ways to power their equipment. This Komatsu hybrid excavator uses an electric motor to drive the upper structure. The energy created when the structure brakes and comes to a stop is converted to electricity, which is stored in an electronic device known as an ultra-capacitor. The stored energy is then released and used to drive the swing and to assist the vehicle's engine. Fuel savings can be 20 to 40 percent and the carbon footprint is reduced by a like amount.

Did You Know?

The Did You Know? features offer hints, tips, and other helpful bits of information from the trade.

Did You Know?

Each year in the United States nearly 100 workers are killed and another 20,000 are seriously injured in forklift-related incidents. The most frequent type of accident is a forklift striking a pedestrian. This accounts for 25 percent of all forklift accidents. However, these figures include all forklifts. Rough-terrain models have a lower incidence of accidents, especially in striking pedestrians, since they are generally used outdoors where visibility is better. The tight confines and blind corners of a warehouse significantly increase the accident rate.

Step-by-Step Instructions

Step-by-step instructions are used throughout to guide you through technical procedures and tasks from start to finish. These steps show you not only how to perform a task but also how to do it safely and efficiently.

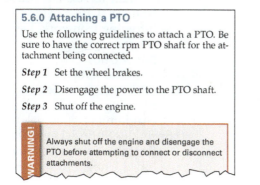

5.6.0 Attaching a PTO

Use the following guidelines to attach a PTO. Be sure to have the correct rpm PTO shaft for the attachment being connected.

Step 1 Set the wheel brakes.

Step 2 Disengage the power to the PTO shaft.

Step 3 Shut off the engine.

WARNING!

Always shut off the engine and disengage the PTO before attempting to connect or disconnect attachments.

Trade Terms

Each module presents a list of Trade Terms that are discussed within the text and defined in the Glossary at the end of the module. These terms are denoted in the text with **bold, blue type** upon their first occurrence. To make searches for key information easier, a comprehensive Glossary of Trade Terms from all modules is located at the back of this book.

Before operating a bulldozer, the operator must understand how to operate the blade and its controls. The blade position can be changed in lift, angle, tilt, and pitch. Changing the position of the blade allows the bulldozer to perform different grading operations. Refer to the O&M manual for each dozer for the location and operation of the blade controls.

The lift control lowers or raises the blade. Lowering the blade allows the operator to change the amount of bite or depth to which the blade will dig into the material. Raising the blade permits the operator to travel, shape slopes, or create stockpiles. The lift lever can also be set to float. The blade adjusts freely to the contour of the ground.

Review Questions

Review Questions are provided to reinforce the knowledge you have gained. This makes them a useful tool for measuring what you have learned.

Review Questions

1. Loaders are grouped into how many main categories?
   a. One
   b. Two
   c. Three
   d. Four

2. What plays a major role in the breakout force and the tipping load abilities of a loader?
   a. The type and size of the engine
   b. The type and size of the loader's frame
   c. The type of lift arms and hydraulic components
   d. The size of the bucket and the arms that lift and tilt it

3. Bucket controls (levers or joystick) are typically located ____.
   a. on the left armrest
   b. on the right armrest
   c. in front of or slightly to the left of the left armrest
   d. in front of or slightly to the right of the right armrest

4. What is the term used for a control position that allows the hydraulic fluid to the lift arm hydraulic cylinders to flow in and out both ends of the cylinders so that the bucket can follow the contour of the ground as the loader moves forward or backward?
   a. Tilt
   b. Hold
   c. Float
   d. Skim

5. What control term used with a loader's arm controls is considered to be a neutral position?
   a. Hold
   b. Float
   c. Stay
   d. Neutral

6. If a loader's diesel engine runs out of fuel, refuel the loader and ____.
   a. restart the engine
   b. clean the injectors
   c. allow the engine to completely cool before restarting the engine
   d. bleed the air out of the fuel lines and injectors before restarting the engine

7. A flashing indicator light means ____.
   a. that the system associated with the indicator needs attention
   b. to recheck the machine system associated with the indicator
   c. to check all machine temperatures and levels
   d. to stop all operations immediately

8. The width of a loader's bucket is normally the same as the ____.
   a. loader's back wheels
   b. loader's front wheels
   c. width of the loader's chassis or frame
   d. length of the loader's chassis or frame

9. A broom attachment can be angled horizontally up to how many degrees on either side?
   a. 10 degrees
   b. 15 degrees
   c. 20 degrees
   d. 30 degrees

10. A grouser is part of the ____.
    a. track
    b. idler
    c. sprocket
    d. ROPS

11. On a track loader, what component(s) keep tension on the tracks to keep them from jumping off?
    a. The idlers
    b. The sprockets
    c. The grousers
    d. The tensioning springs on the axles

NCCER Standardized Curricula

NCCER's training programs comprise more than 80 construction, maintenance, pipeline, and utility areas and include skills assessments, safety training, and management education.

Boilermaking
Cabinetmaking
Carpentry
Concrete Finishing
Construction Craft Laborer
Construction Technology
Core Curriculum:
    Introductory Craft Skills
Drywall
Electrical
Electronic Systems Technician
Heating, Ventilating, and
    Air Conditioning
Heavy Equipment Operations
Highway/Heavy Construction
Hydroblasting
Industrial Coating and Lining
    Application Specialist
Industrial Maintenance
    Electrical and Instrumentation
    Technician
Industrial Maintenance
    Mechanic
Instrumentation
Insulating
Ironworking
Masonry
Millwright
Mobile Crane Operations
Painting
Painting, Industrial
Pipefitting
Pipelayer
Plumbing
Reinforcing Ironwork
Rigging
Scaffolding
Sheet Metal
Signal Person
Site Layout
Sprinkler Fitting
Tower Crane Operator
Welding

Maritime

Maritime Industry Fundamentals
Maritime Pipefitting
Structural Fitter

Green/Sustainable Construction

Building Auditor
Fundamentals of Weatherization
Introduction to Weatherization
Sustainable Construction
    Supervisor
Weatherization Crew Chief
Weatherization Technician
Your Role in the Green
    Environment

Energy

Alternative Energy
Introduction to the Power
    Industry
Introduction to Solar
    Photovoltaics
Introduction to Wind Energy
Power Industry Fundamentals
Power Generation Maintenance
    Electrician
Power Generation I&C
    Maintenance Technician
Power Generation Maintenance
    Mechanic
Power Line Worker
Power Line Worker: Distribution
Power Line Worker: Substation
Power Line Worker:
    Transmission
Solar Photovoltaic Systems
    Installer
Wind Turbine Maintenance
    Technician

Pipeline

Control Center Operations,
    Liquid
Corrosion Control
Electrical and Instrumentation
Field Operations, Liquid
Field Operations, Gas
Maintenance
Mechanical

Safety

Field Safety
Safety Orientation
Safety Technology

Management

Fundamentals of Crew
    Leadership
Project Management
Project Supervision

Supplemental Titles

Applied Construction Math
Careers in Construction
Tools for Success

Spanish Translations

Basic Rigging
    (Principios Básicos de
    Maniobras)
Carpentry Fundamentals
    (Introducción a la
    Carpintería, Nivel Uno)
Carpentry Forms
    (Formas para Carpintería,
    Nivel Trés)
Concrete Finishing, Level One
    (Acabado de Concreto,
    Nivel Uno)
Core Curriculum:
    Introductory Craft Skills
    (Currículo Básico:
    Habilidades Introductorias del
    Oficio)
Drywall, Level One
    (Paneles de Yeso, Nivel Uno)
Electrical, Level One
    (Electricidad, Nivel Uno)
Field Safety
    (Seguridad de Campo)
Insulating, Level One
    (Aislamiento, Nivel Uno)
Ironworking, Level One
    (Herrería, Nivel Uno)
Masonry, Level One
    (Albañilería, Nivel Uno)
Pipefitting, Level One
    (Instalación de Tubería
    Industrial, Nivel Uno)
Reinforcing Ironwork, Level One
    (Herreria de Refuerzo,
    Nivel Uno)
Safety Orientation
    (Orientación de Seguridad)
Scaffolding
    (Andamios)
Sprinkler Fitting, Level One
    (Instalación de Rociadores,
    Nivel Uno)

Acknowledgments

This curriculum was revised as a result of the farsightedness and leadership of the following sponsors:

Bridgerland Applied Technology College
Carolina Bridge Company, Inc.
Caterpillar, Inc.
John Deere

Phillips and Jordan Inc.
Skyview Construction and Engineering, Inc.
Southland Safety, LLC

This curriculum would not exist were it not for the dedication and unselfish energy of those volunteers who served on the Authoring Team. A sincere thanks is extended to the following:

Roger Arnett
Jonathan Goodney
Paul James

Mark Jones
Dan Nickel

Larry Proemsey
Joseph Watts

NCCER Partners

American Fire Sprinkler Association
Associated Builders and Contractors, Inc.
Associated General Contractors of America
Association for Career and Technical Education
Association for Skilled and Technical Sciences
Carolinas AGC, Inc.
Carolinas Electrical Contractors Association
Center for the Improvement of Construction
   Management and Processes
Construction Industry Institute
Construction Users Roundtable
Construction Workforce Development Center
Design Build Institute of America
GSSC – Gulf States Shipbuilders Consortium
Manufacturing Institute
Mason Contractors Association of America
Merit Contractors Association of Canada
NACE International
National Association of Minority Contractors
National Association of Women in Construction
National Insulation Association
National Ready Mixed Concrete Association
National Technical Honor Society
National Utility Contractors Association

NAWIC Education Foundation
North American Technician Excellence
Painting & Decorating Contractors of America
Portland Cement Association
Skills USA
Steel Erectors Association of America
U.S. Army Corps of Engineers
University of Florida, M. E. Rinker School of
   Building Construction
Women Construction Owners & Executives, USA

Contents

Module One
Finishing and Grading

Provides training on common types of equipment and instruments used for finish grading, materials and methods used to stabilize soils and control soil erosion, and finishing and grading methods used for various applications. (Module ID 22307-14; 25 Hours)

Module Two
Compaction Equipment

Provides training on common types of compaction equipment; the primary instruments, controls, and attachments of a roller; safety guidelines associated with compaction equipment; and prestart inspections, preventive maintenance, and proper operating procedures. Factors involved in work activities associated with a roller are also presented. (Module ID 22203-14; 25 Hours)

Module Three
Backhoes

Identifies and describes the common uses, types, components, instruments, controls, and attachments of backhoes. Safety guidelines, prestart inspection procedures, and preventive maintenance requirements are presented. Basic startup and operation are described, and common work activities associated with backhoes are covered. (Module ID 22303-14; 30 Hours)

Module Four
Off-Road Dump Trucks

Identifies and describes the common types, uses, and components of off-road dump trucks. Safety guidelines, prestart inspection procedures, and preventive maintenance requirements are presented. Basic startup, driving maneuvers, loading, and dumping procedures for off-road dump trucks are covered. (Module ID 22310-14; 30 Hours)

Module Five
Dozers

Identifies and describes the common uses, types, and components of dozers. Safety guidelines, prestart inspection procedures, and preventive maintenance requirements are presented. Basic startup and operation are described, and common work activities associated with dozers are covered. (Module ID 22302-14; 30 Hours)

Module Six
Excavators

Identifies and describes the common types, uses, and components of excavators. Safety guidelines, prestart inspection procedures, and preventive maintenance requirements are presented. Basic startup and operation are described, and common work activities associated with excavators are covered. (Module ID 22304-14; 35 Hours)

Module Seven

Motor Graders

Identifies and describes the common uses and types of motor graders. Safety guidelines, pre-start inspection procedures, and preventive maintenance requirements are presented. Basic startup and operation are described, and common work activities associated with motor graders are covered. (Module ID 22305-14; 40 Hours)

Glossary

Index

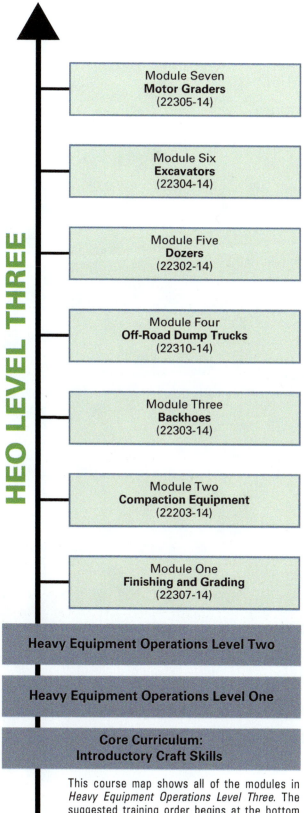

HEO LEVEL THREE

Module Seven
Motor Graders
(22305-14)

Module Six
Excavators
(22304-14)

Module Five
Dozers
(22302-14)

Module Four
Off-Road Dump Trucks
(22310-14)

Module Three
Backhoes
(22303-14)

Module Two
Compaction Equipment
(22203-14)

Module One
Finishing and Grading
(22307-14)

Heavy Equipment Operations Level Two

Heavy Equipment Operations Level One

**Core Curriculum:
Introductory Craft Skills**

This course map shows all of the modules in *Heavy Equipment Operations Level Three*. The suggested training order begins at the bottom and proceeds up. Skill levels increase as you advance on the course map. The local Training Program Sponsor may adjust the training order.

22307-14

Finishing and Grading

OVERVIEW

The final stage in an earthmoving operation is finishing and grading. Equipment operators must be able to select the appropriate machines in order to work efficiently. Some soils must be stabilized before they are graded. Understanding different stabilization methods may be critical for a successful result. In addition to performing various finishing techniques, finish operators must understand how erosion is controlled and how instruments are used in finish grading.

Module One

Trainees with successful module completions may be eligible for credentialing through NCCER's National Registry. To learn more, go to **www.nccer.org** or contact us at **1.888.622.3720**. Our website has information on the latest product releases and training, as well as online versions of our *Cornerstone* magazine and Pearson's product catalog.

Your feedback is welcome. You may email your comments to **curriculum@nccer.org**, send general comments and inquiries to **info@nccer.org**, or fill in the User Update form at the back of this module.

This information is general in nature and intended for training purposes only. Actual performance of activities described in this manual requires compliance with all applicable operating, service, maintenance, and safety procedures under the direction of qualified personnel. References in this manual to patented or proprietary devices do not constitute a recommendation of their use.

Objectives

When you have completed this module, you will be able to do the following:

1. Describe the types of equipment used for finish grading.
   a. Identify equipment used in finish grading.
   b. Describe how laser instruments are used in finish grading.
   c. Describe how a Global Positioning System (GPS) and a robotic total station are used in finish grading.
2. Explain methods used to stabilize soils and control soil erosion.
   a. Identify soil stabilizers and binders.
   b. Describe methods used in the application of soil binders.
   c. Describe methods used to control soil erosion.
3. Describe finish grading methods.
   a. Describe the use of grading specifications.
   b. Explain how finish grade is established for slopes.
   c. Explain how the finish subgrade and base are established.
   d. Describe grading methods used for ditches and trenches.
   e. Describe grading methods used on parking lots, sidewalks, and curbs.

Performance Task

Under the supervision of your instructor, you should be able to do the following:

1. Establish a finish grade after a rough grade has been performed, according to instructions.

Trade Terms

Blue tops
Circle rail
Crown
Detention pond
Geotextile
Gradient
GLONASS
GNSS
Hubs
Infiltration system

Invert
Pozzolan
Red tops
Retention pond
Riprap
Robotic total station
Superelevation
Tassel (whisker)
Virtual reference station (VRS)

Industry Recognized Credentials

If you are training through an NCCER-accredited sponsor, you may be eligible for credentials from NCCER's Registry. The ID number for this module is 22307-14. Note that this module may have been used in other NCCER curricula and may apply to other level completions. Contact NCCER's Registry at 888.622.3720 or go to **www.nccer.org** for more information.

Contents

Topics to be presented in this module include:

1.0.0 Equipment Types...1
  1.1.0 Grading Equipment...1
    1.1.1 Motor Grader...1
    1.1.2 Dozer..2
    1.1.3 Scraper..2
    1.1.4 Telescoping Excavator..3
  1.2.0 Using Laser Instruments...4
    1.2.1 Laser Transmitter..4
    1.2.2 Laser Receivers...5
    1.2.3 Use of Construction Lasers..6
    1.2.4 Laser Instrument Safety...6
    1.2.5 Calibration and Care of Laser Instruments....................7
    1.2.6 Laser-Based Automatic Grade Control Systems............8
  1.3.0 Using a GPS and Robotic Total Stations..8
    1.3.1 GPS-Based Automatic Grade Control Systems.............9
    1.3.2 Robotic Total Stations..10
2.0.0 Soil Stabilization...14
  2.1.0 Soil Stabilizers and Binders...14
    2.1.1 Cement..15
    2.1.2 Lime..15
    2.1.3 Fly Ash..15
    2.1.4 Bitumens...15
    2.1.5 Calcium Chloride...16
    2.1.6 Polymers...16
  2.2.0 Applying Soil Stabilizers..16
    2.2.1 Preparing the Subgrade..16
    2.2.2 Spreading the Binder..17
    2.2.3 Stabilizing...17
  2.3.0 Controlling Soil Erosion...18
    2.3.1 Storm Water Runoff..18
    2.3.2 Geotextiles..18
    2.3.3 Filtration..19
    2.3.4 Separation...19
    2.3.5 Reinforcement...20
3.0.0 Finish Grading Methods..22
  3.1.0 Grading Specifications...22
    3.1.1 Crowns for Roads...24
    3.1.2 Superelevation..25
  3.2.0 Grading Finish Slopes...25
  3.3.0 Finish Subgrade and Base...26
  3.4.0 Ditches and Trenches..27
    3.4.1 Ditches..27
    3.4.2 Trenches...29

3.5.0 Grading Parking Lots, Sidewalks, and Curbs 31

    3.5.1 Parking Lots 31

    3.5.2 Sidewalks and Curbs.. 31

Figures

Figure 1 Motor grader.. 1

Figure 2 Dozer with laser receivers 3

Figure 3 Scraper 3

Figure 4 Excavator ditch cleaning bucket 4

Figure 5 Handheld distance meter.. 4

Figure 6 Rotating beam laser 5

Figure 7 Checking the elevation of the corners of a foundation 5

Figure 8 Laser receiver 6

Figure 9 Caterpillar Accugrade laser grade control system 8

Figure 10 Caterpillar Accugrade laser grade control system display
and control.. 8

Figure 11 GPS grade control system 11

Figure 12 Caterpillar Accugrade GPS grade control system display
and control... 12

Figure 13 Robotic total station 12

Figure 14 Reclaimer/stabilizer mixer machine 15

Figure 15 Scarifier attachment 17

Figure 16 Soil stabilizer.. 18

Figure 17 Retention pond in subdivision 18

Figure 18 Silt fence 19

Figure 19 Silt fence construction 19

Figure 20 Cross-section of a finished roadway.. 23

Figure 21 Finish grade stakes... 24

Figure 22 Crown for a two-lane road... 24

Figure 23 Superelevation on a curve.. 25

Figure 24 Finishing steep slopes 26

Figure 25 Trimming with a dozer 27

Figure 26 Finishing a subgrade with a motor grader................................ 27

Figure 27 Grader cutting a ditch... 27

Figure 28 Cutting a ditch.. 28

Figure 29 Cutting a steep backslope.. 29

Figure 30 Factors determining trench depth.. 29

Figure 31 Laser receiver for backhoes and excavators.............................. 30

Figure 32 Pipe laser... 30

Figure 33 Checking the grade with a laser detector rod 31

Figure 34 Self-trimming curb machine... 31

1.0.0 EQUIPMENT TYPES

Objective

Describe the types of equipment used for finish grading.

a. Identify equipment used in finish grading.
b. Describe how laser instruments are used in finish grading.
c. Describe how a Global Positioning System (GPS) and a robotic total station are used in finish grading.

Trade Terms

GLONASS: A Russian-owned Global Navigation Satellite System operated by the Ministry of Defense of the Russian Federation.

GNSS: An acronym for Global Navigation Satellite System. GNSS is the generic term used to describe a locating or grade control system that uses signals from either GPS or GLONASS satellites.

Riprap: Broken stone, in pieces weighing from 15 to 150 pounds (6.8 to 68 kilograms) each, placed on the ground for protection against the action of water.

Robotic total station: A machine control system that uses infrared signals reflected by a target mounted on the grading equipment to provide data to the equipment's on-board computer about how to adjust the blade.

Virtual reference station (VRS): An imaginary reference station that is established by a computer based on data from numerous GPS receivers and the grading equipment that serves as the source of data for grading control.

Many different pieces of heavy equipment are used for finishing and final grading. The selection of the most appropriate piece of equipment to use will depend on the work that needs to be done. A backslope can be finished with a motor grader up to a certain angle, but it may be necessary to use an excavator or other machine if the motor grader cannot get close enough to the backslope. An excavator would not be used to perform subgrade finishing on a road because the finish would be uneven and the process would not be very efficient. To perform finishing work effectively, the operator must know and understand the characteristics and limitations of each type of heavy equipment. Equipment that is commonly used includes the following:

- Motor grader
- Dozer
- Scraper
- Telescoping excavator

1.1.0 Grading Equipment

During finish grading operations, the operator may work with a grade checker to ensure that the grading is at the correct elevation. The grade checker may use manual or laser instruments. The procedure is basically the same with either instrument.

In addition, technological advances now allow grading operations that are partially or fully automatic. These systems are computer controlled and may use either a laser signal, Global Positioning System (GPS) signals, or optical signals as a reference. These types of systems are extremely accurate and help to save both time and money, but they are expensive to purchase. Contractors that routinely perform large commercial projects are most likely to own this type of equipment.

1.1.1 Motor Grader

A motor grader (*Figure 1*) is usually the first choice for finishing earthwork. It was originally designed for this type of work. It has a wide, sharp blade to cut bumps and high spots while moving the material in front of the blade to fill in holes and low spots. Accurate blade control allows the operator to perform very fine cutting and trimming. Because the blade can be moved outside the

REPRINTED COURTESY OF CATERPILLAR INC.

22307-14_F01.EPS

Figure 1 Motor grader.

motor grader's wheel tracks, it can be positioned very close to curbs and other permanent structures. This eliminates the need for additional hand trimming.

When building subdivisions and industrial layouts, grading can be accelerated by adjusting the blade height to average cuts and fills. This speeds up the grading operation because the operator has longer runs without a grade change.

When fill is needed, the material is dumped and then spread with the motor grader to the proper elevation before being compacted. After compaction, the motor grader will come back and trim the subgrade to the finished elevation.

Motor graders can also be used to finish fairly steep slopes. The wide blade and smooth operation make dressing a slope as easy as grading a subgrade or base course. For slopes greater than 3:1, some additional precautions may be needed to keep the motor grader from tipping. On steep slopes, the motor grader should be operated in first gear at a reduced engine speed, using the accelerator pedal, rather than the speed control, to control the speed of the machine.

When an articulated motor grader is used on a slope, crab steering should be used to keep the back of the motor grader on the downhill side of the cut. Using crab steering will place the frame in an offset position that provides maximum stability. Crab steering should be used any time the machine may tend to slip due to a heavy side load on the blade.

> **WARNING!**
> Evaluate each job carefully. Operating any heavy equipment on a steep slope places the operator at risk of tipping over. Tracked vehicles are more stable on slopes than wheeled vehicles, but are not always available or desirable for use. If you feel that it is not safe to operate the equipment on a slope, you must tell the supervisor.

1.1.2 Dozer

A good dozer operator can perform finish work on several different surfaces. For slopes on embankments, a dozer is a good choice because it will spread and smooth the material, and the tracks can be used to compact the loose soil. A dozer can move across a slope, as well as up and down. Care should be taken to select the right size dozer for a specific job. Too large a machine can cause damage and tear up the surface if it has to make a number of turns on a slope. Moreover, for fine finishing work, the tracks of the dozer will cause unwanted impressions in the surface.

> **WARNING!**
> There is a greater potential for tipping when traveling across a slope. If possible, travel up and down a slope. Do not exceed the manufacturer's recommendations for traveling on inclines.

Dozers can be equipped with automatic control devices that receive a signal from a laser level or GPS and manipulate the equipment's controls to keep the blade at a specified elevation. *Figure 2* shows a dozer equipped with such a leveling device. A mast with the laser receiving unit is attached to the top of the dozer blade. After calibrating the height of the receiving unit, the dozer is free to roam the area and perform grading. The main advantage of this setup is that the dozer can receive a continuous signal once the laser level is set and calibrated properly. This means that the grading operation is being checked for proper elevation continuously rather than only at the grade stakes.

1.1.3 Scraper

Although a scraper (*Figure 3*) is one of the larger pieces of heavy equipment, it can also be used for some finishing work. Scrapers can be used to finish large horizontal areas where tolerances are not as close as those requiring the use of a motor

Ancient Roads

About 2400 years ago, the Romans built a road known as the Appian Way. It was named after Appius Claudius, the official who undertook its construction. When it was completed in 244 BC, it was 362 miles (583 kilometers) long and averaged 20 feet (6.10 meters) wide along its length. The road had a stone foundation and was paved with lava blocks that fit tightly together. Like modern roads, it was crowned to allow water to run off. Parts of the road have been preserved and are used today by automobiles. Given that it was done entirely with human and animal labor, it is a marvelous feat of engineering.

REPRINTED COURTESY OF CATERPILLAR INC.

22307-14_F02.EPS

Figure 2 Dozer with laser receivers.

grader. With its power, weight, and a very sharp blade, the scraper can neatly trim horizontal surfaces and pick up excess material without disturbing the compacted surface underneath.

An elevator scraper is used to pick up windrows of unneeded material made with a motor grader. Using a scraper to do this work saves time in cleaning up the area. It also saves the motor grader from having to push the material off the side of the embankment or to another location.

Scrapers can be used for grading gentle slopes that are long enough to allow for easy maneuvering. When performing this work, a dozer will usually cut a shelf for the scraper to use as a starting point. Since the blade angle on a scraper cannot be adjusted, it must make successive overlapping cuts to trim a slope. Slopes up to 4:1 can be cut and trimmed with the correct size scraper.

There are several operating procedures that affect the scraper's finish grading performance. Good scraper operators always check their equipment before they go to work. A worn cutting edge on the blade will not cut properly. Unequal tire pressure will cause the scraper to lean to one side, causing it to cut unevenly. In addition, when the scraper bowl fills unevenly, it will cause the scraper to lean to one side. To ensure even filling of the bowl, be certain that the scraper is centered on the windrow.

1.1.4 Telescoping Excavator

Telescoping excavators are well-adapted for finish work. This equipment is designed to grade on both horizontal and vertical surfaces. Because the boom works back and forth in one plane, the motion is the same as the scraping motion made by a motor grader or a scraper. The operator can control the angles of both the boom and the bucket to produce a smooth, even finish on horizontal, vertical, and overhead surfaces.

In addition to a general excavation bucket, special attachments, such as a grading blade and a ditch cleaning bucket, can be used with a telescoping excavator for finish grading. Coupled with the excavator's long boom, these attachments make it

22307-14_F03.EPS

Figure 3 Scraper.

possible to finish off areas that cannot be reached with other equipment.

The grading blade attachment is used in the same manner as a motor grader blade. Using the grading blade to finish or trim a slope is easily accomplished by extending the boom and pulling the blade toward the machine in a scraping motion. Tracked excavators provide a stable platform, even while working on a side slope. The track allows the machine to move up and down the slope and perform grading work over the entire work area.

The ditch bucket (*Figure 4*) is designed to provide a long, straight cutting edge that leaves behind a smooth, even surface in soft material. Because of its large size, it can collect large amounts of material as it smoothes and shapes the ditch profile. This reduces the number of cycles required to do a job.

The excavator is also very effective for finishing sloped areas around drainage structures or near waterways. It can finish the slope, help lay any fabric matting for erosion control, and then place riprap on top of the mat.

1.2.0 Using Laser Instruments

Laser instruments are widely used in construction today for site layout and for checking rough and finish grades. Manual instruments rely on the skill of the operator to accurately read the instruments. Laser instruments have digital readouts, so when they are used properly they can provide a greater degree of accuracy. The use of modern construction laser instruments provides many benefits, including the following:

- Ease of use
- Greater productivity
- Increased accuracy with a reduction in measurement errors

Depending on the type of device being used, the laser beam produced may be invisible to the human eye. Because these beams are invisible, they must be detected using an electronic receiver. Typically, a special receiver is used to check grades, because the visibility of the laser light is usually lost in daylight.

1.2.1 Laser Transmitter

Modern laser instruments used for construction layout work are typically battery-operated units that generate a fixed laser beam, a rotating beam, or both. Fixed laser beam instruments can generate either single or multiple beams.

One example of a fixed-beam device is a carpenter's level with a built-in visible beam laser. It can be used as an ordinary carpenter's level or switched on to emit a laser beam, extending the level's reference line, typically between 200 and 300 feet (about 61 to 91 meters). It can be mounted on a tripod or placed on a level surface.

Another example of a single beam device is a handheld distance meter (*Figure 5*) used to measure distances up to about 100 feet (31 meters) without a target, and even farther distances with a reflective target. The auto-leveling laser alignment tool is a good example of a multiple-use laser instrument. It can be used to provide simultaneous plumb, level, and square reference points.

22307-14_F04.EPS

Figure 4 Excavator ditch cleaning bucket.

22307-14_F05.EPS

Figure 5 Handheld distance meter.

Rotating-beam laser level instruments (*Figure 6*) are widely used for establishing grades and leveling over long distances. Both visible-beam and invisible-beam instruments are made in manual, self-leveling, and self-plumbing models. Most self-leveling laser level instruments have a capability that automatically turns off the laser beam whenever the instrument is not leveled. This feature is especially helpful if the instrument is accidentally bumped out of position. Should this happen, the unit automatically stops rotating and shuts off the laser beam. The beam remains off until the instrument is reset by the operator and leveled again. The instrument head can be rotated rapidly or slowly, or it can be stopped and used as a single beam device.

One advantage of using a rotating laser instrument is that it allows one person, instead of two, to perform many site layout operations. After the instrument has been set up and leveled, the laser head can be adjusted to rotate at various speeds, depending on the model. This creates a level, horizontal beam of light that sweeps over a large area so that it can be detected by multiple electronic laser detectors placed at different locations around the job site.

The procedures for determining and establishing elevations with laser levels are similar to those used with optical leveling instruments. To determine elevations, the base of the rod is held to the elevation to be determined. The laser detector is moved up or down on the rod until its display indicates it is centered on the beam. The reading on the rod is then recorded.

An example of this is shown in *Figure 7*, where a rotating laser is being used to check the elevations at the four corners of a simple building foundation.

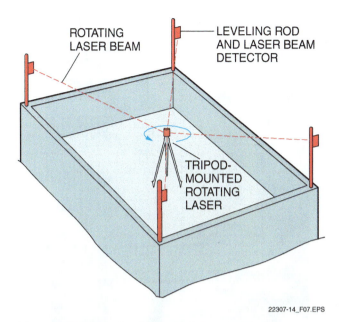

Figure 7 Checking the elevation of the corners of a foundation.

As shown, a rod with a detector attached is placed at each of the four corners where the elevation readings are desired. The rotating laser beam will hit the detector at each location when the detector is in line with the beam.

Similarly, to establish elevations, an electronic detector is attached to a grade rod. The grade rod is moved up or down until the beam indicates that the base of the rod is at grade.

1.2.2 Laser Receivers

A laser beam detector or receiver (*Figure 8*) is used with laser transmitters. It must be used with an invisible beam (infrared) laser instrument such as an electronic level. In bright sunlight, it is also widely used to detect the beam of a visible-light laser instrument.

The laser beam detector is normally mounted on a leveling rod or similar support. However, handheld models are also available. During use, the detector is moved up or down on the rod to intersect the laser beam. A capture window on the beam detector receives the beam from the laser.

The operation of detectors made by different manufacturers is basically the same. Typically, they have an LED visual display that indicates the position of the detector (down, on-center, or up) in relation to the laser beam. Most also have an audible device that gives an indication of beam position. For example, one manufacturer's detector uses a fast beeping tone to indicate that the detector should be moved down. A slow beeping tone means it should be moved up, and a steady continuous tone indicates that the beam is on-center.

Figure 6 Rotating beam laser.

Figure 8 Laser receiver.

Some detectors also have bandwidth and sensitivity adjustments used to adjust the detector for various conditions. These conditions include the degree of ambient light, laser wavelength, focus, power output, and bandwidth tightness required.

1.2.3 Use of Construction Lasers

General procedures for the setup and operation of laser levels to perform horizontal leveling and vertical plumb tasks differ with each manufacturer and model.

Setup and operation of the instrument must be done in accordance with the instructions given in the manufacturer's operator's manual for the specific instrument.

Using a laser level is fairly simple, but mistakes can still be made that will result in measurement errors. These include the following:

- Failure to follow good leveling practices
- Using instruments that are out of calibration
- Setting elevations incorrectly with the rod
- Performing laser measurements near other laser devices, which can cause unwanted signals to be received by the laser detector.

1.2.4 Laser Instrument Safety

Under no circumstances should an unqualified worker operate a laser instrument. Workers must be properly trained, and the training must be documented before a worker can operate a laser beam instrument. Government regulations also require that laser manufacturers provide warnings and cautions on their instruments, and in their literature, regarding the hazards associated with the use of laser instruments. Qualified personnel must adhere to all such manufacturer's warnings and cautions. They must also follow the manufacturer's operating instructions exactly when using a laser instrument.

If a rotating laser level is operated in the rotational mode, the beam will usually not harm the eyes, because a lower power is used. Most lasers used on construction sites are low-power lasers. The laser equipment will have a label on it that indicates the maximum power output, usually less than five milliwatts, which is considered low power.

OSHA 29 CFR Part 1926 provides safety precautions that must be observed when a laser is in use. Manufacturers also provide various warnings and regulations to prevent hazardous situations. Some of these regulations are as follows:

- Prominent warning signs must be posted in the area where a laser is being used. Be alert to all such warning signs.
- Avoid going into areas marked off for laser operation unless it is necessary because of the job.
- Only those workers who have been properly trained are permitted to set up or operate a laser or make any adjustments to it.
- Qualified workers must carry an operator's card at all times when operating laser equipment.
- Always read and follow the manufacturer's operating instructions for the instrument being used.
- Avoid looking directly at the beam. Also avoid looking at any surfaces such as polished metal or a mirror, which can cause the beam to be reflected into the eyes.
- Special hazards are present when the laser is used in a fixed position, such as when it is used for tunnel or sewer work. Never look at the concentrated beam. Never look along the axis from the laser toward the point being sighted because the beam can be reflected from another surface.
- Wear approved safety goggles when the laser has a power output of 5 milliwatts or greater. The laser equipment must have a label indicating the maximum power output.

- Never point a laser beam at anyone. When possible, set the laser up so that it is above or below eye level.
- The laser beam should be turned off, shuttered, or capped when not in use.
- If you suspect you may have a vision problem caused by the laser, such as a persistent after-image, seek immediate medical attention.

1.2.5 Calibration and Care of Laser Instruments

The calibration of a laser instrument can be altered as a result of shock and vibration. Laser instruments should therefore be checked for proper calibration at regular intervals, particularly before starting on a new construction site. Calibration of the instrument should be performed by a qualified person and in accordance with the manufacturer's instructions. When an instrument is calibrated in the field, errors can be introduced because of temperature, humidity, and wind conditions. For this reason, it is recommended that the instrument be checked and calibrated in a controlled environment similar to that in which it will be used.

Although it is designed for construction site work, a laser is a precision instrument requiring the same care as any high-quality leveling device. Observing the following maintenance guidelines will help ensure trouble-free laser operation:

- Always follow the manufacturer's recommended maintenance procedures as directed in the operator's manual for the instrument.
- To prevent moisture and dirt from settling inside the unit, always make sure the laser, accessories, and carrying case are clean and dry before storage. To clean the laser and electronic detectors, wipe them off with a soft, nonabrasive cloth and clear, solvent-free water.
- When not in use, store the laser and its accessories in their original cases.
- Make sure the batteries in the unit and any spare batteries are fully charged to prevent work delays resulting from discharged batteries.
- Do not attempt internal repairs to a laser instrument. Return it to a dealer, manufacturer, or other qualified organization for repair.

A laser level's transmitter beam should be checked for proper calibration at regular intervals, particularly if the instrument is being used for the first time or if the unit has been handled roughly. Severe shock or vibration can cause the instrument to be out of calibration. The laser transmitter needs calibration when the laser beam emitted from one side of the unit is above true level, and the beam emitted from the opposite side is below true level.

When the instrument is correctly calibrated, the beam emitted is horizontal for the 360-degree rotation. If the unit is turned 180 degrees or 90 degrees from its original position, the reading is within the manufacturer's specifications, typically within ±3/32" per 100 feet (±2.38 millimeters per 31 meters) of the original position. Check the laser transmitter calibration as described in the manufacturer's instructions. Manufacturers of different makes and models of instruments recommend checking their instruments in different ways. A general procedure for one method is given here.

> **WARNING!**
>
> Only trained and authorized personnel can operate a laser instrument. Inform all workers within the operating range of the instrument that a laser instrument will be in use. Never look directly into a laser beam or point the beam into the eyes of others. Set up the laser transmitter so that the height of the emitted beam is above or below normal eye level.

Step 1   Set up and level the laser transmitter at a location that has a clear line of sight to an object at least 100 feet (31 meters) away. Attach a laser detector (receiver) unit on a level rod or other calibrated rod and hold it plumb at the 100-foot (31-meter) location.

Step 2   Turn on the laser transmitter and rotate it so that the +X side is aimed at the laser detector.

Step 3   Move the receiver into the beam to get an on-grade reading. Mark and record the elevation, noting that it is for the +X axis.

Step 4   Rotate the laser transmitter 180 degrees so that the –X side is aimed at the laser detector. Mark and record the elevation, noting that it is for the –X axis.

Step 5   Compare the +X and –X elevation readings. If the difference between the readings is less than 3/32 inch (±2.38 millimeters), the X axis is within calibration. If within calibration, go to Step 9. If the difference in readings is more than 3/32 inch (±2.38 millimeters), X-axis calibration is required, as described in Steps 6 through 8.

Step 6   To correct for a calibration error, locate a new mark midway between the +X and –X marks. Move the receiver on the rod until its center-marking notch is aligned with the new midpoint mark.

Step 7 Adjust the X-axis calibration screw to obtain an on-grade laser beam reading at the midpoint line. Most detectors have move-up and move-down indicators that show the direction the beams need to be moved up or down relative to the on grade point.

Step 8 Rotate the laser transmitter 180 degrees back to the original +X face. The on-grade reading should be on or within $\frac{3}{64}$ inch (±1.19 millimeters)of the midpoint line. If not, repeat Steps 3 through 8.

Step 9 Rotate the laser 90 degrees. Repeat Steps 3 through 8 as required to check and adjust the beam emitted from the +Y and –Y sides of the unit.

1.2.6 Laser-Based Automatic Grade Control Systems

Laser-based automatic grade control systems use complicated technology, but their operation is very simple. *Figure 9* shows a typical automatic grade control system. Signals from an off-board laser transmitter are used as the reference. The grader or dozer is outfitted with electronic components that permit a computer processor to determine its exact location on a site and the position of the blade. As the equipment moves throughout the site, the computer determines its exact position and the elevation of the blade. This information is then transmitted to an in-cab display (*Figure 10*) so the operator can make necessary blade adjustments.

The complexity and degree of automation will vary among systems. Some systems use a three-dimensional digitized site plan so the computer processor can calculate the required position of the blade. In a fully automated system, a signal is sent to the grader or dozer and the blade position is adjusted automatically.

1.3.0 Using GPS and Robotic Total Stations

Site layout and grading once were done by manually setting stakes and strings. That method is rapidly falling out of use, especially on larger projects, as technological advances make it possible for grading operations to be partially or fully automated. Using modern technology, entire projects can be completed without setting a single grade stake. Two common types of automatic grade control systems are satellite-based systems and an optical positioning system called a **robotic total station**.

REPRINTED COURTESY OF CATERPILLAR INC.

22307-14_F09.EPS

Figure 9 Caterpillar Accugrade laser grade control system.

Historically, satellite-based systems in general have been referred to as GPS. In reality, GPS is only one of the global satellite navigation systems that are used for machine control. The satellite systems that are in use or in development include the following:

22307-14_F10.EPS

Figure 10 Caterpillar Accugrade laser grade control system display and control.

- *GPS* – a Global Positioning System developed and operated by the United States military and available for use by the public.
- GLONASS – A Russian-owned Global Navigation Satellite System operated by the Ministry of Defense of the Russian Federation.
- *Galileo* – A satellite system under development by the European Union and the European Space Agency.
- *Compass* – A satellite system under development by China.

Signals from both GPS and GLONASS satellites are commonly used by grade control systems. For that reason, GPS and GLONASS are often referred to under the generic term GNSS (Global Navigation Satellite System). Some grade control system manufacturers use GNSS in the names of their equipment. For clarity and familiarity, this training program uses the acronym GPS when describing a satellite-based system.

A robotic total station is similar to a satellite-based system in that it provides three-dimensional (3D) grade control capabilities. However, this type of system uses an electromechanical robot to locate and track the movement of a target mounted on the equipment being controlled. The robot sends infrared signals to the target, which are reflected back to the robot. The robot then sends data via radio signals to the equipment's on-board computer where it can be used to automatically control the grading equipment.

Satellite-based system and robotic total stations can get a job done faster and more accurately than conventional staking. The number of grade checkers on a job site can be greatly reduced. This saves labor costs. It also results in a safer work environment because fewer people are walking around operating equipment. The major disadvantage of these systems is initial cost.

1.3.1 *GPS-Based Automatic Grade Control Systems*

GPS-based automatic grade control systems can deliver very tight tolerances. They operate in a similar fashion to laser-based systems, except the reference signal is from several orbiting satellites. These satellites broadcast radio signals that contain information about satellite locations 24 hours a day. For proper GPS operation, there must be a clear overhead view. Other conditions may affect GPS operation. For example, working underneath high-voltage power lines may create dead zones for satellite coverage and satellite signals may be weaker at certain times of the day.

GPS-based systems are especially helpful on jobs that require a great deal of contouring. These systems compare the blade position to a three-dimensional digitized site plan to determine whether the blade position requires any adjustments.

GPS technology is used in all phases of the job. A surveyor using a GPS receiver in a backpack, can walk across an area in a set pattern to record position information. When all this position information is fed into a computer, it produces a 3-D surface model of the area. The surface characteristics information is then used to lay out a roadway using a computer-aided design (CAD) program. The end product of the design process shows lanes of the road, shoulders, drainage ditches, and other features. An accurate site design is a must when using GPS technology for grading. The CAD program also generates files that are downloaded into the GPS grading system computer mounted in the cab of the dozer or grader.

Before a site is graded, static GPS observations are taken at the site to establish control points. For better grading accuracy, an on-site GPS base station should be used. In this arrangement, satellites transmit radio signals that are picked up by receivers at a base station near the job site and on the grading equipment. The base station makes corrections to the satellite signals to determine its exact position. It then sends corrected data by radio to the grading equipment's on-board computer. The computer uses the satellite signals and the corrected signals from the base station to determine how to position the grading blade to meet the grade specifications in the site plan.

Even tighter grading control can be achieved by using a virtual reference station (VRS). In this type of system, data from numerous GPS receivers is transmitted to a central computing station. The grading equipment's on-board GPS system also signals the computing station of its approximate location. The computing station uses that data to establish an imaginary reference station in the immediate vicinity of the equipment. By doing this, the computing station is able to collect data from numerous receiver locations as if it were sent from a single receiver position (the virtual reference station). The computing station then sends the data of the virtual reference station to the control system of the grading equipment.

There are two modes used when grading with GPS: indicate-only and fully automatic. Indicate-only feeds GPS guidance information to the equipment operator who controls the machine. It requires a minimum of one GPS receiver on the grader and is the less complicated and less costly of the two modes. Without a GPS base station, indicate-only systems cannot deliver the best grading accuracy.

There are several ways to notify the operator that the blade of the dozer or grader is in the correct position. One system uses a set of light-bar position indicators, one for vertical blade location and the other for horizontal blade location. The vertical position light bar has indicators that light up when the blade is on, above, or below grade. The horizontal light bar provides similar guidance for the horizontal position of the blade. An in-cab console can display similar information using changes in color, a numeric readout (plus or minus deviations from grade), or a graphical display where a miniature machine is shown in relation to grade.

Fully automatic systems require a GPS base station and one or two GPS receivers on the grader or a virtual reference station arrangement to deliver the best grading accuracy. The GPS guidance system is fully integrated with the grader's hydraulic system for precise, three-dimensional blade control. Graders can be purchased with a fully automatic system installed, or machines can be retrofitted with aftermarket systems. An override feature allows the operator to always have control of the machine.

Figure 11 shows a typical GPS-based automatic grade control system used on a motor grader. The information can be displayed on an in-cab display (*Figure 12*). The operator can make the required blade adjustments or a signal can be sent to the grader or dozer control system to automatically adjust the blade.

1.3.2 *Robotic Total Stations*

A robotic total station is one of the most accurate 3-D machine control systems available. In some respects, a robotic total station is similar to a machine control system based on GPS. However, a robotic total station uses infrared signals instead of satellite signals, and the robotic station must be in direct line of sight with a prism, or target, mounted on the grading equipment.

A typical robotic total station is shown in *Figure 13*. The system consists of a robotic station, a target that is mounted on top of a mast that is attached to the grading equipment blade, and the on-board computer system that controls the blade movement.

During operation, the robotic station tracks and locates the target and is able to calculate the vertical and horizontal angles of the target. It also sends an infrared signal to the target. The target reflects the signal back to the robot, which enables the robot to determine the distance to the target. The robot then sends the location information via radio signals to the on-board computer of the grading equipment. The on-board computer calculates in 3-D how to adjust the blade to meet the site specifications.

Since robotic total stations do not use satellite signals, they are well suited for use in areas with heavy tree cover, inside buildings, underground, or any other situation where the sky is obstructed. Many companies find that the best method of machine control is to combine GPS and a robotic total station. This way, the advantages of both systems can be utilized.

MOTOR GRADER WITH ONE GPS ANTENNA

MOTOR GRADER WITH TWO GPS ANTENNAS

**GPS BASE STATION WITH ANTENNA
AND RADIO RECEIVER**

22307-14_F11.EPS

Figure 11 GPS grade control system.

REPRINTED COURTESY OF CATERPILLAR INC.

22307-14_F12.EPS

Figure 12 Caterpillar Accugrade GPS grade control system display and control.

Global Positioning System (GPS)

The Global Positioning System was developed by the US Department of Defense in the 1970s for military purposes. In military applications, it is vital to have accurate information about where you are and where the enemy is. A series of satellites in earth orbit constantly transmit signals to GPS receivers on the ground to calculate the latitude, longitude, elevation, and the time at the location of the GPS receiver. Signals from at least three satellites are used. Since the signals travel at a known speed (the speed of light) the time it takes the signal to reach the receiver helps to determine the position of the receiver. GPS technology also has many non-military uses. In 1996, President Bill Clinton issued a directive that effectively made GPS technology available for non-military use.

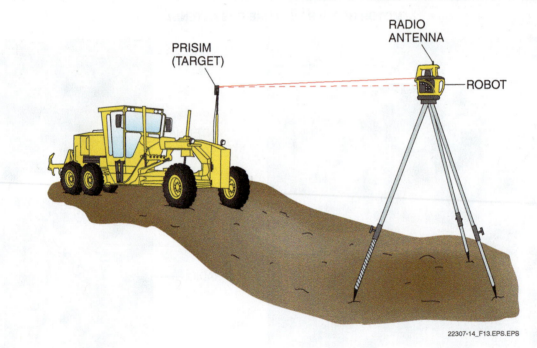

22307-14_F13.EPS.EPS

Figure 13 Robotic total station.

Additional Resources

Moving the Earth, 1988. H.L. Nichols. Greenwich, CT: North Castle Books.

Caterpillar Performance Handbook, Edition 27. A CAT® Publication. Peoria, IL: Caterpillar, Inc.

Soil Stabilization for Pavements Mobilization Construction Engineer Manual (EM 1110-3-137), 1984. Department of the Army, Corps of Engineers.

Basic Equipment Operator, 1994. NAVEDTRA 14081, Morris, John T. (preparer), Naval Education and Training Professional Development and Technology Center.

1.0.0 SECTION REVIEW

1. Adjusting blade height to average cuts and fills results in _____.

   a. better compaction
   b. faster grading
   c. lower engine speed
   d. greater stability

2. Grades are often established over long distances using a(n) _____.

   a. distance meter
   b. spirit level
   c. LED beam receiver
   d. rotating-beam laser

3. GPS-based grade control systems are helpful on jobs that require a lot of _____.

   a. contouring
   b. trenching
   c. soil compaction
   d. fill material

2.0.0 SOIL STABILIZATION

Objectives

Explain methods used to stabilize soils and control soil erosion.
   a. Identify soil stabilizers and binders.
   b. Describe methods used in the application of soil binders.
   c. Describe methods used to control soil erosion.

Performance Task

Establish a finish grade after a rough grade has been performed, according to instructions.

Trade Terms

Crown: A slightly built-up section of the roadway, usually along the center line. The crown provides a slope to the pavement so that water will drain to either side.

Detention pond: A pond that will temporarily hold storm water until it can evaporate or be absorbed.

Geotextile: A synthetic material used to filter water and soil, reinforce soil, or separate to unlike materials.

Infiltration system: Storm water runoff control system that filters contaminants from the runoff before the runoff is allowed to infiltrate soil.

Pozzolan: Originally used to define volcanic ash, but now applies to any similar material that reacts with lime and water to form cement.

Retention pond: A pond that will permanently retain the same level of water.

The stability and loadbearing capacity of a construction site depends on the properties of the soil on which it is built. For a soil to be stable, it must have a high amount of internal friction and cohesiveness. Soils with these two characteristics can bear heavy loads without deforming. When soil mixtures do not have these properties naturally, chemical binders, called soil stabilizers, are often used. Soil stabilizers must be applied before any finish grading is done.

2.1.0 Soil Stabilizers and Binders

There are many advantages to the stabilization of soils, including the following:

- Allows use of existing soil
- Eliminates need for excavating poor soil
- Eliminates the cost of replacing soil
- Reduces cost by mixing in place
- Reduces construction time

Stabilization procedures include the following:

- Determining the logistics of the stabilization job
- Choosing a binder
- Following basic procedures
- Preparing the subgrade
- Spreading the binder
- Stabilizing
- Compacting
- Checking quality control

Sometimes a small amount of binder is added to a soil mixture to change its characteristics, such as particle size or plasticity, but not appreciably increase the strength. This is called soil modification rather than stabilization. The application and mixing process are the same as in stabilization.

When an existing roadway needs to be resurfaced, the deteriorated asphalt surface can be processed and reused as the roadbed. In this case, a reclaimer/stabilizer mixer machine (*Figure 14*) picks up the asphalt and some amount of underlying soil, pulverizes it, and mixes it with an asphaltic binder. This mixture is then laid back down to become the new roadbed. This process not only reuses the old asphalt material, saving disposal costs, but also creates an extremely stable and durable roadbed.

Binders are used to change the characteristics of soil. They are selected based on the particular soil being treated and the desired outcome. The type and amount of stabilizer used depends on the soil mixture's engineering properties and the loadbearing specifications of the project. Common binders for roads include the following:

- Cement
- Lime
- Fly ash
- Bitumen (asphalt and tar)
- Calcium chloride
- Polymers

REPRINTED COURTESY OF CATERPILLAR INC.

22307-14_F14.EPS

Figure 14 Reclaimer/stabilizer mixer machine.

2.1.1 Cement

Cement is a powdered mixture of lime and other minerals. When mixed with water and allowed to cure over time, it will create a hard, durable surface. Cement-stabilized surfaces have greater loadbearing capacity than any other binder. Portland cement is the most common cement in use in the construction industry. It can be used with all soil types as well as with reclaimed asphalt. It is used as a stabilizer by mixing the dry cement and water with the soil that needs to be stabilized. The mixture is then laid back down onto the surface for compaction and paving.

2.1.2 Lime

Quicklime (calcium oxide) and hydrated lime (calcium hydroxide) can be used with any fine-grained soil, but it has the best effect on moderate to high plasticity soils. The lime reacts chemically with the clay to reduce its water-holding capability, thus decreasing swell and increasing stability. The lime needs to be thoroughly mixed with the soil to the proper depth, and then laid back down and compacted. Heavy clay soils may need to be mixed with the lime and allowed to cure for 24 to 48 hours before the final mix.

2.1.3 Fly Ash

When the soil at the construction site has low plasticity, fly ash and lime are often added to increase its stability. Fly ash is a pozzolan and is a byproduct of burning coal. On its own, fly ash does not have any stabilizing properties, but when mixed with lime, they chemically react and act like cement. Fly ash and lime are often mixed with sand or gravel and used as a base for roadways.

2.1.4 Bitumens

Bitumens such as asphalt and tar are mixed with an aggregate and used to surface roadways. Asphaltic binders need to be mixed at the correct temperature to achieve a durable surface. Some are mixed at around 300°F (149°C). These are called hot-mix asphalt. Some asphalts are mixed at a lower temperature; those asphalts are called

---

warm-mix. Others may be mixed at the ambient temperature; they are called cold-mix.

Asphaltic binders include the following:

- Asphalt emulsions
- Cutback asphalt
- Road tars
- Straight asphalt
- Foamed asphalt

2.1.5 Calcium Chloride

Calcium chloride is a salt that increases soil density and lowers the freezing point of water. It helps protect the building foundation or roadway from frost damage during cold weather. Calcium chloride can also be used to settle dust at the construction site.

> **CAUTION**
>
> Calcium chloride is corrosive to metals and attracts moisture. It should be stored away from equipment and tools and protected from moisture.

2.1.6 Polymers

Widespread use of polymers as a soil stabilizer is fairly recent. A simple way to explain a polymer is that it and one or more other materials are bonded together by chemical reaction.

In soil stabilization, polymers bind with soil particles to create a durable surface that supports motorized traffic. Application varies with each product, so it is important to follow the manufacturer's instructions. Most of these products are applied as a thick liquid that dries quickly. It can be a matter of minutes or hours, depending on the weather.

Polymers are often mixed with other binders to strengthen them. Some polymers in a weaker solution can be used to suppress dust and for erosion control.

2.2.0 Applying Soil Stabilizers

During the planning phase of a construction project, the site's soil will be tested and analyzed to determine its properties. Based on the results, engineers and technicians will determine what stabilizing material, if any, needs to be added to the site and the degree of compaction that is needed to complete the project. This information will appear in the project specification.

Planning is key when working on a stabilization job. When planning, always consider the following:

- *Weather conditions* – The temperature should be mild, about 40°F (4.5°C) or above, with little or no rain or freezing weather expected. The initial mixing process should not be done in heavy rain. After the soil is stabilized and compacted, a small amount of rain is not a problem.
- *Binder material supply* – Binder material must be available as needed to avoid delays in application.
- *Water supply* – Water should be clean and available as it is needed. When cement is being used for a binder, optimum moisture content is critical. Lime is usually added to soils already high in moisture, so water is usually not needed for mixing, but may be required for dust control.
- *Compaction equipment* – Equipment should be available for compaction to begin immediately after the mixing and grading process is finished.
- *Trimmer equipment* – After compaction, the surface should be trimmed to accurate grade and **crown** specifications.
- *Curing* – Enough time must be allowed to permit curing. During curing, the surface must be protected as specified to prevent the area from drying out.

2.2.1 Preparing the Subgrade

To prepare a new site for construction, the area is cleared of all vegetation and stripped of top soil. Large objects such as tree stumps or boulders are grubbed to the specified depth. Top soil may be stockpiled for use during the landscaping phase

Polymers

The United States military uses polymers to allow poor soils to support military vehicles and aircraft. In Afghanistan, US Marine helicopters kicked up clouds of damaging fine dust. To defeat the dust problem, helicopter pads were built by scraping away the dust layer, compacting layers of gravel, and then coating them with a polymer. The Marines nicknamed the polymer "rhino snot" because of its appearance. By all accounts, the polymer stabilized the soil and solved the problem.

of the project, or it can be sold. Always check the specification for instructions.

The subgrade is then shaped to specification with a motor grader, grade trimmer, or other suitable equipment. After grading, a stabilizing machine may be used to pulverize the soil and mix it with the binder when soil stabilization is needed. If a stabilizing machine is not available the subgrade surface may be broken up with a scarifier attachment (*Figure 15*) on a motor grader or a discing attachment pulled by a tractor.

2.2.2 Spreading the Binder

The project specification will define the desired soil-to-binder ratio. It is very important that the correct amount of binder is uniformly spread over the area. Dry binder material can be spread with a spreader attachment mounted on a truck. Liquid binders can be spread with a tanker truck. Both can spread measured amounts of the binder.

The binder is then mixed thoroughly with the soil so the mixture is uniform in texture and color. The soil is also pulverized to the desired size and texture with a soil stabilizer machine. A soil stabilizer machine cuts, mixes, and pulverizes soils and other materials so the mixture is uniform in size and texture and the binder is evenly distributed. Some soil stabilizer machines allow the binder to be mixed with the soil in the machine's mixing chamber, saving time.

2.2.3 Stabilizing

Soil stabilizers (*Figure 16*) use blades on rotors or teeth on mandrels for cutting. Many have automatic depth control. Some models have a mixing chamber, which blends materials consistently. The cutting depth is usually to 12 to 15 inches (31 to 38 centimeters). The cutting width is typically 8 feet (about 2.5 meters). Stabilizer features may include the following:

- Mechanical rotor drive
- Microprocessor control of major machine systems
- Four steering modes with automatic rear wheel alignment
- Choice of rotors
- Rollover protection
- Liquid additive and/or water spray systems

REPRINTED COURTESY OF CATERPILLAR INC.

22307-14_F15.EPS

Figure 15 Scarifier attachment.

Dust Control

Dust generated at a construction site is an air pollutant that can cause or aggravate respiratory problems. Most localities require that steps be taken at construction sites to minimize dust generation. Failure to comply with this requirement can result in fines or other penalties.

22307-14_F16.EPS

Figure 16 Soil stabilizer.

Sometimes the machine must make an initial pass to establish proper gradation of the soil, followed by application of the binder, and then make a second pass with the soil stabilizer. Some stabilizer models offer the ability to reverse direction without turning around—a time-saving feature. After the stabilized soil has been compacted, the area may be cut to grade.

2.3.0 Controlling Soil Erosion

It is very important that heavy equipment operators know and follow the specification instructions during the final grading of a site. Finish grading levels are designed to control the flow of runoff and to prevent standing water. Grades left too high can increase runoff and cause erosion. Grades cut too low can cause standing water. Both conditions can result in costly callbacks.

2.3.1 Storm Water Runoff

Before construction, water from rain and snow is absorbed into the ground. After construction, some of the ground is covered. This decreases the area available to absorb water. Excess water is shed to a lower level or stands in puddles until the ground can absorb it. Although standing water is a problem in its own way, runoff causes erosion, so it must be controlled with various engineering methods.

To control storm water runoff and standing water, engineers will study the area to determine the natural flow of runoff. Then they will design methods to lessen the runoff impact. Sometimes engineers will incorporate the use of porous materials into the design of the project. These materials may be used in parking lots and will permit water to quickly and easily pass so that it may be absorbed by the ground. Using porous materials can greatly improve storm water runoff control at a site, but the area will most likely need other methods to fully control runoff.

Other methods that control runoff can be labor-intensive tasks, such as the construction of **retention ponds** (*Figure 17*), **detention ponds**, and **infiltration systems**. Retention ponds are used to hold a certain level of water permanently; when storm water runoff causes the level to rise, the overflow is usually drained to another location. Detention ponds are used to hold storm water runoff temporarily until it is absorbed into the soil. Infiltration systems are designed to filter contaminants from runoff before it is absorbed into the soil.

Other methods are used as well. These include vegetative islands in a paved area to help absorb water or grading ditches to direct and control the flow of water. Often, plants that require a great deal of water to survive are used in these areas. All of these methods will be detailed in the project specification.

2.3.2 Geotextiles

Geotextiles are among the greatest recent advances for the construction industry. Geotextiles are a vast category of synthetic materials that are used to prevent erosion and sedimentation and to improve the strength and durability of soil at a construction site. Geotextiles can be permeable, semi-permeable, or non-permeable. They are used to filter soil from water, separate unlike materials, and reinforce weak areas such as steep slopes.

The use of a geotextile is determined by the material used to make it, the method in which it was

22307-14_F17.EPS

Figure 17 Retention pond in subdivision.

made (woven or non-woven, knit, or grid), and its permeability characteristics. Some geotextiles systems, such as silt fences, are used as a temporary measure (*Figure 18*). Others are used permanently, such as for separating aggregate from soil. Engineers will determine the type of geotextile to use on a job. These instructions will be in the project specification.

When working with geotextiles, check the manufacturers' specifications for storage, use, and handling requirements. The manufacturers' specifications will suggest subgrade preparation, installation, and anchoring methods. Improper handling or installation of a geotextile can cause it to fail during use. This can cause delays during construction and costly repairs after construction.

The appearance of a geotextile can be deceiving. It takes an experienced eye to identify the material. Textiles that are practically identical in appearance can have very different strength and permeability characteristics. One geotextile can be resistant to ultraviolet rays from the sun; others of similar appearance are sensitive to the same rays and must be protected to prevent breakdown. It is very important that you are certain that you are using the correct material. When in doubt, ask your supervisor.

2.3.3 Filtration

Geotextile filtration methods are used whenever there is a need to permit water to flow while filtering soil particles from the water. One of the most often seen uses of geotextile filtration is a silt fence. Silt fences are installed to filter soil particles from runoff when soil has been disturbed. Typically, silt fencing is installed over a broad area to help control sheet flow. Sheet flow is basically a continuous, thin flow of water over a large area, such as a slope. Silt fencing is not designed to be used where a fast flow of deep water occurs, such as in a channel or ditch. For a silt fence to be useful, it must be properly installed (see *Figure 19*). Installation requirements vary from state to state. A geotextile that is ultraviolet-resistant must be used. Its strength and permeability must be adequate for the predicted water flow. The bottom edge of the fabric must be buried in the ground to a uniform depth.

Finally, the soil in the area of the silt fence must be compacted. Always check the project specification for details.

Another use of a geotextile for filtration is around pipes. A trench is lined with a geotextile fabric and a pipe is installed in the trench and covered with gravel. The fabric is then folded over the top of the gravel and the trench is backfilled. This type of installation prevents soil from infiltrating the gravel, but allows water flow. It also separates the gravel and soil.

2.3.4 Separation

Geotextiles are commonly used to prevent two unlike materials from mixing together. Geotextile installation for separation is often similar or identical to that of filtration. In some cases, the method used will serve both purposes.

Figure 18 Silt fence.

22307-14_F18.EPS

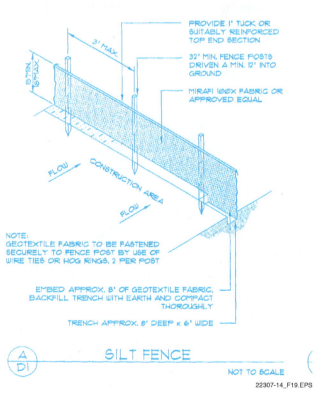

Figure 19 Silt fence construction.

22307-14_F19.EPS

Great effort is often made to be certain that a construction site has the correct mix of large and small soil particles so that it will have adequate strength and durability. Once the effort has been made to get the correct mix, steps must be taken to prevent soil migration that can alter the mixture and weaken the area. For example, the aggregate layer of a roadbed helps to strengthen the surface and make it durable. As vehicles travel over the road, it is subjected to vibrations that over time can move soil and aggregate, altering the mix. This will weaken the road, so a geotextile is used to prevent this type of mixing.

2.3.5 Reinforcement

Geotextiles can be used to temporarily or permanently strengthen the ground at a construction site. This allows construction at sites that would have been rejected because the soil lacked strength. It also saves time and money because geotextiles can be used to strengthen soil that would otherwise need to be replaced with more suitable soil. The reinforcement uses of geotextiles are numerous. Here are a few of the ways geotextiles can be used to improve the strength of a site:

- Stabilize a steep slope and prevent soil slides or creeping
- Reinforce a roadbed or embankment
- Bridge a soft spot in a road
- Stabilize an unpaved area

When used to reinforce roadbeds, geotextiles improve the durability of the surface, decreasing maintenance. They can also reduce the thickness requirement of the aggregate layer, resulting in a substantial decrease in construction costs.

GOING GREEN

Straw Wattles

Straw wattles are flexible tubes filled with straw that are used on construction sites to prevent soil erosion. The tubes are made of a biodegradable open mesh and come in different lengths and diameters. Their flexible nature allows them to conform to any contour. The straw within the tube lets water pass through but blocks any silt or soil. Straw wattles are often supplemented with other methods.

22307-14_SA01.EPS

Additional Resources

Arrest that Fugitive Dust. Roberta Baxter, March/April 2002 Forester Communications, Inc.

National Asphalt Pavement Association (NAPA) web site (**www.hotmix.org**)

Silt Fence Installation Efficacy: Definitive Research Calls for Toughening Specifications and Introducing New Technology. Joel Sprague, P.E., TRI / Environmental, Inc. and Tom Carpenter, Carpenter Erosion Control. (**www.ieca.org**)

Using Lime For Soil Stabilization and Modification, 2001. National Lime Association. (**www.lime.org**)

2.0.0 SECTION REVIEW

1. When heavy clay soil is mixed with lime, it must cure for up to _____.

   a. 2 hours
   b. 12 hours
   c. 20 hours
   d. 48 hours

2. A scarifier attachment is used to _____.

   a. spread binder
   b. mix a binder
   c. prepare the subgrade
   d. inject a stabilizer

3. Using a geotextile to bridge over a soft spot in the road is an example of _____.

   a. separation
   b. filtration
   c. reinforcement
   d. graduation

3.0.0 FINISH GRADING METHODS

Objectives

Describe finish grading methods.
 a. Describe the use of grading specifications.
 b. Explain how finish grade is established for slopes.
 c. Explain how the finish subgrade and base are established.
 d. Describe grading methods used for ditches and trenches.
 e. Describe grading methods used on parking lots, sidewalks, and curbs.

Trade Terms

Blue tops: Stakes that are used to identify the final elevation of the sub-base. The tops are either painted with blue paint or covered with blue plastic tassels.

Circle rail: A ring-shaped component on the motor grader that controls the horizontal position and elevation of the moldboard.

Gradient: The change of elevation per unit length; the slope along a specific line of a road surface, channel, or pipe.

Hubs: Surveying stakes set for reference purposes. Hubs are usually at the edge of, or outside of, the work area.

Invert: The flow line of a pipe. This would be the bottom-most visible surface on the inside of a pipe.

Red tops: Stakes that are used to identify the final elevation of the crushed aggregate base course.

Superelevation: The increased elevation of one side of a curved roadway that allows for banking of the pavement to the inside.

Tassel (whisker): A small plastic colored tag affixed to grade stakes for identification purposes.

Finish grading involves finishing off a part of the earthwork before the application of other material such as crushed stone, pavement, or topsoil. Sometimes it is referred to as trimming or final grading, depending on the activity. Usually there is a specification that defines the tolerance allowed for this work. The tolerance can vary greatly depending on geographic location, type of soil, and type of job.

The ability to perform finish grading comes through experience, practice, and attention to detail. The following are traits of a good operator who knows how to finish earthwork:

- Is aware of the grade changes to the streets, roads, and pads being constructed. When the area being worked gets close to grade, an alert operator will make smaller cuts.
- Is able to understand how the finished job should look.
- Remembers that a road surface is never flat. It will always have a crown, superelevation, elevation changes, or a combination of those features.
- Knows what piece of equipment is best suited for a particular job and will be able to use the most effective attachments to complete the job.

3.1.0 Grading Specifications

On most earthmoving jobs, the plans and specifications require certain elevations or slopes. These elevations and slopes are specified with some amount of tolerance allowed. These tolerances are usually found in the specifications for the project. For the roadway cross-sections shown in *Figure 20* there are separate specifications for each part of the structure.

There will be earthwork specifications, either on the plans or in the specifications book, for the following items:

- Grades for the backslope and inslope (foreslope) of the ditch.
- Grades for finished embankments (below the subgrade). There will also be values specified for the compaction of embankments.
- Grade and compaction requirements for the subgrade, including the shoulder cross slope.
- Grades for the base course, with a required density.

Slope stakes are used as guides when finishing off slopes. For subgrades, hubs and blue tops are placed along the center line and at the edge of the pavement or shoulders as shown in *Figure 21*. The blue tops are driven flush to the required final subgrade.

These finish subgrade stakes have their tops painted blue or have colored plastic tassels (whiskers) covering the tops for easy recognition. Red tops are used in some areas to denote the top of the crushed aggregate base course. In paving, this is also the bottom of the sub-base.

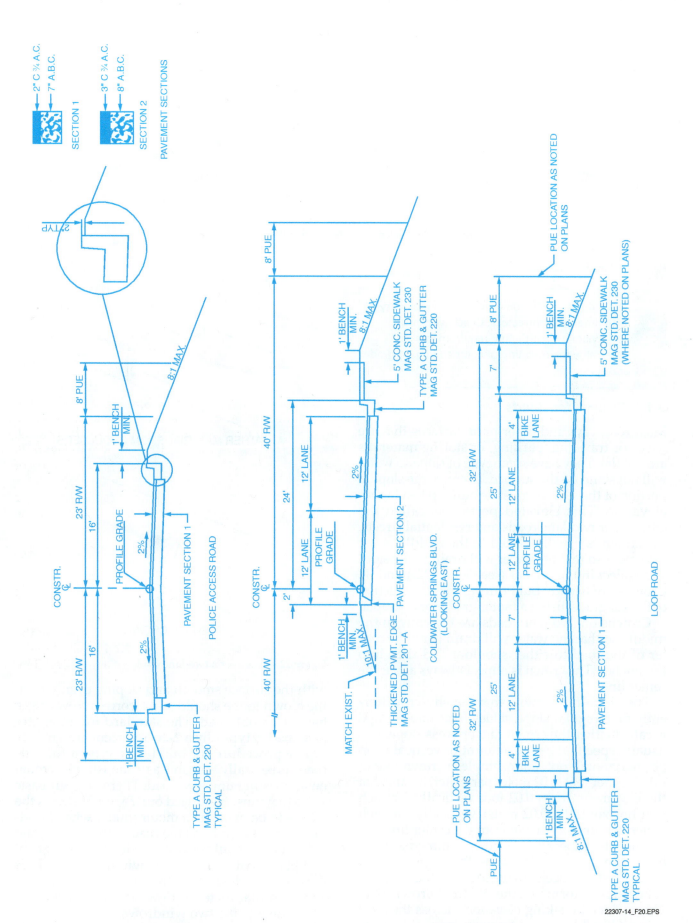

Figure 20 Cross-section of a finished roadway.

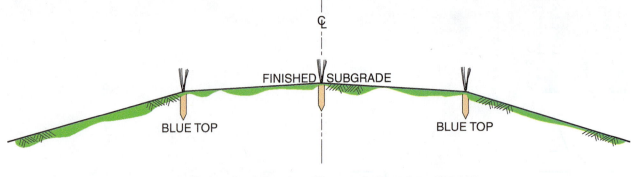

FINISHED SUBGRADE

BLUE TOP

BLUE TOP

SUBGRADE MUST BE EVEN WITH TOPS OF BLUE TOPS

22307-14_F21.EPS

Figure 21 Finish grade stakes.

3.1.1 Crowns for Roads

Most roadways and other hard surfaces that are used for traveling, parking, or storing materials are not flat. They have some type of slope so water will not stand on the surface. To create the slope, a portion of the surface must be built up to a higher elevation. This elevated portion is called the crown. For two-lane roads and residential streets, the crown is usually placed in the middle of the pavement so that the water will drain to the sides and collect in the ditches (see *Figures 22A* and *B*). Crowns are usually built during the construction of the subgrade, using a motor grader or a scraper.

Crowns on two-lane roads are typically in the middle so that the water will drain from the center of the road to either shoulder and then into the ditches. The crown is almost always along the center line.

The slope for crowns may be shown differently than regular slope values. This rate of slope is called either the crown rate or cross slope. It is usually specified as the amount of vertical drop per horizontal foot. For example, a crown rate of 0.02 foot per foot (0.02 meter per meter) means that the pavement drops 0.02 foot vertically for each foot horizontally (0.02 meter vertically for each meter horizontally) away from the center line.

Once the subgrade has been trimmed according to the finish grade stakes, the inspector will make random checks to ensure that the finished subgrade conforms to the planned crown rate. This is done by taking elevations across the road using a hand level and tape measure. When the subgrade is built flush to the tops of the blue tops,

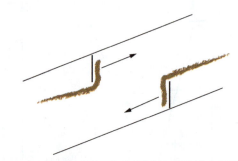

A. GATHER MATERIAL FROM SHOULDERS

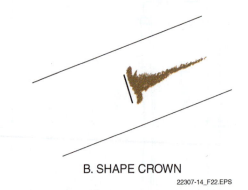

B. SHAPE CROWN

22307-14_F22.EPS

Figure 22 Crown for a two-lane road.

with the surface smooth and sloping evenly from the crown to the shoulder, the correct crown rates have been obtained. The standard for an aggregate road is typically a 2- to 5-percent crown.

The procedure for building a crown for new roads is basically the same as establishing a crown on an existing unpaved road. There are four basic steps that must be carried out. *Figure 22* shows the passes to be made by a motor grader when building a crown on a two-lane road.

Start at a shoulder or one side of the subgrade and blade material into two windrows. Then cut the center of the road with a shallow angle, or no angle, so that material flows from both sides of the blade into the two windrows.

During the next two passes, blade the windrows toward the center of the road. Finally, spread the material to the desired crown.

The steps for performing the work are as follows:

Step 1 Blade the material from the shoulder or sides of the subgrade to the center (*Figure 22A*).

Step 2 Cut the top of the crown with the blade at 0-degree angle or at a slight angle to either side if required (*Figure 22B*). This creates two windrows on the road surface. The activity is performed at a speed of 5 to 7 miles per hour (8 to 11 kilometers per hour).

Step 3 Spread the windrows by putting the blade at an angle of 10 to 25 degrees toward the center. The blade should be held above the level of the undisturbed surface to avoid colliding with any embedded objects. The speed of the operation causes any loose material to be cast out from the blade toward the center line.

Step 4 Spread out any ridge build-up in the center of the road with a straight blade.

This should finish the job. However, you may need to reverse the blade or rework some sections where proper crowning was not obtained.

3.1.2 Superelevation

The process of building up the outside edge of a pavement toward a curve is called superelevating the curve. Superelevation is used on curves in order to give vehicles better traction at a higher rate of speed. Without the superelevation, the side (centrifugal) force exerted on a car going around the curve might be enough to reduce the friction between the tires and the roadway, causing the car to skid sideways or go completely off the road. On the straight sections of a multi-lane road, the travel lanes are typically sloped to the outside edge of the lanes. On a curve, the outside edge of the roadway is elevated (*Figure 23*).

Building roadways that have transitions from a crown in the middle to a superelevation on the outside can be tricky. If the motor grader is to build a smooth grade, the staking must be accurate. If there is an error in the staking or grading that produces a dip on the outside edge, it will be obvious once the pavement has been laid.

The detailed method for providing a transition to a superelevation depends upon design require-

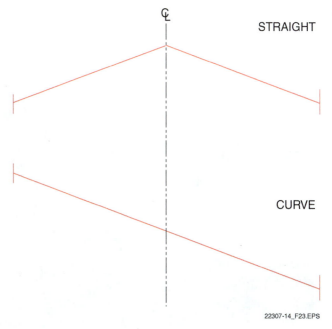

Figure 23 Superelevation on a curve.

ments, which vary from state to state, along with the roadway width and other local factors.

3.2.0 Grading Finish Slopes

All cut and fill slopes around an excavation need to be finished after rough grading has been completed. Tolerances may be very close for areas such as landscaping around buildings, parks, and other public areas. For cut and fill slopes on highways, or for industrial work, the tolerances may not be so tight. These requirements will be recorded in the plans or specifications for the specific job.

In cut areas, it is important to keep the slopes as close as possible. Backslopes are trimmed to grade as the cut proceeds downward to the toe of the slope.

A smooth grade must be established at the bottom of the slope so it can be evenly trimmed. This allows the motor grader to remain level so that the operator can manipulate the blade to get the correct angle for the final slope and make a smooth cut.

If an articulated motor grader is used, the front wheels can be positioned on the slope while the rear wheels are kept level to provide a further reach up the slope (*Figure 24*).

A motor grader cannot be used on fill sections because of the soft material and the danger of tipping. Finishing these areas is usually done by a dozer that can work either across or up and down the slope. An excavator can also do finishing on a fill slope if its reach is far enough. When working on fill slopes, follow the slope stakes or the directions given by the grade setter.

REPRINTED COURTESY OF CATERPILLAR INC.

22307-14_F24.EPS

Figure 24 Finishing steep slopes.

A dozer may also be used to trim slopes, shoulders, and other open areas (*Figure 25*). Operators improve their efficiency if the dozer is equipped with a slope control system. In order to use this method, the blade must be set to level and the control set to zero percent and locked in. Once this is done, the operator can program any percentage of slope required for the cut. The unit will automatically control the cutting edge on the right, left, or both sides of the blade to the selected slope. If too much material is being cut, the operator can override the system and take control manually.

3.3.0 Finish Subgrade and Base

The finishing of the subgrade and base course on a construction project is an important operation. If this surface is not smooth and finished off, the material put on top will reflect any bumps, depressions, or unevenness. *Figure 26* shows just how smooth a motor grader can get a surface. In these situations, the moldboard is positioned forward. This position forces material into low spots. The area in the foreground has been graded smooth, while the area beside the grader still needs some work to eliminate the loose dirt.

A job is ready for subgrade work when the slopes have been trimmed and compacted and the shoulders built and cut off vertically on the road side.

To make sure the correct elevation and slope are obtained, a grade checker usually works with the motor grader or scraper operator to tell the operator to either pull up on the blade or to make a cut. If the grader is not equipped with an automatic grade system, a grade checker will work with the motor grader operator to check the height of the grade relative to the height of the blue top after the blade has passed.

G-Forces

The centrifugal force applied to a vehicle going around a curve is known as the lateral G-force (the G stands for gravity). Race car drivers are very familiar with this term because auto racing is all about managing lateral G-forces. There is a direct relationship between speed and lateral G-force, so the driver has to balance the two in order to get the most speed without losing control of the vehicle when going through a turn. The amount of steering input also has a direct affect on lateral G-forces. Every race track has a line, which is the path through the turns that requires the least steering input. Drivers who consistently hit their marks going through the turns will maintain the highest average lap speed.

Figure 25 Trimming with a dozer.

22307-14_F26.EPS

Figure 26 Finishing a subgrade with a motor grader.

If the motor grader has not taken off enough material, the operator will have to make another pass over this area. If too much has been removed, the stake will have to be reestablished and some extra material brought in on the front of the blade to raise the elevation to the required level.

The objective is to cut enough material to be within the tolerance allowed, but not to cut so much that another pass is needed. Added material will need to be re-compacted. It is important to be sure the entire width needed for the base or pavement has been trimmed.

3.4.0 Ditches and Trenches

Finishing ditches and trenches requires patience and a steady hand on the controls. Cutting ditches is usually done by either a motor grader or an excavator, depending on the area and the type of ditch required. For most highway and other horizontal construction, a motor grader or dozer is the most efficient and effective piece of equipment to use. For trench work, the best equipment is either a backhoe or an excavator, depending on the size of the job.

3.4.1 Ditches

It is very important that water flows through ditches and does not stand in one place. Standing water may saturate the subsurface material underneath the road, preventing proper drainage during the next storm. Standing water also reduces the capacity of a ditch to handle runoff. The next storm could wash out the adjoining roadway or parking lot. Ditches with a **gradient** of 1 to 2 percent are desirable. Anything less than 0.5 percent will not ensure flow and will not drain.

One of the tests of a good motor grader operator is the ability to cut a straight ditch line. This involves cutting the inslope, ditch bottom, and backslope, and then cleaning up the excess material. *Figure 27* shows a motor grader cutting the inslope. The greatest efficiency is achieved when the operator is skillful enough to keep the blade full and to complete the operation with a minimum number of passes. Experienced operators can often cut a ditch with two or three passes.

22307-14_F27.EPS

Figure 27 Grader cutting a ditch.

These operators are familiar with the equipment and have developed a sense for the soil conditions. Initially, it is best to remove the smallest amount of material on the first pass and then gradually increase the depth of the cuts until you are comfortable with your skills.

When cutting a ditch, the first pass is called the marking cut (*Figure 28A*). This is a 3- to 4-inch (76- to 102-millimeter) deep cut made with the toe of the blade. It is used as a guide to help cut a straight ditch. In the second cut, the blade is angled more and positioned over the marking cut (*Figure 28B*). When the second cut is made, more material is deposited onto the road from the heel of the blade. At some point, the cut material needs to be spread toward the center of the road—this is called shoulder pickup (*Figure 28C*). These steps are repeated until the ditch is cut to the desired grade.

Once the backslope is cut down to the required angle, clean up the excess material in the ditch. Back up, reposition the motor grader, and move the material from the ditch bottom and inslope to the shoulder.

At the completion of this operation, there should be a clean V-ditch and with a clearly visible center line. Check that there are no high spots where the flow of water would be restricted or allowed to dam up.

Cutting a flat bottom ditch requires additional steps to the V-ditch procedure, as follows:

Step 1 After cutting the V-ditch to the required alignment, cut another V-ditch on the inslope between the first V-ditch and the shoulder. The blade should be set to move the material out to the top of the shoulder or to the road.

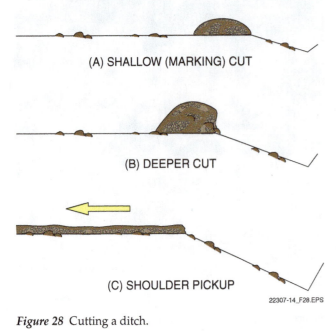

(A) SHALLOW (MARKING) CUT

(B) DEEPER CUT

(C) SHOULDER PICKUP

22307-14_F28.EPS

Figure 28 Cutting a ditch.

Step 2 Make a flat cut in the bottom of the ditch by placing the complete length of the blade in the ditch, with the toe of the blade at the base of the backslope and the heel at the base of the inslope. The angle of the blade will depend on the width of the bottom of the ditch. Begin making a pass along the bottom of the ditch.

Step 3 Make one or two final passes on the inslope to move the material up onto the road or the shoulder and outside the wheels of the grader.

Step 4 The shoulder should now be finished, with any remaining material being windrowed or spread to the center of the road.

The process for cutting a steep backslope requires the operator to move the blade totally out from under the frame (*Figure 29*). Different machines have slightly different approaches to executing this procedure. The following general steps are for cutting backslopes as steep as 90 degrees. Refer to the operator's manual for the machine you are using for the specific procedure:

Step 1 Position the motor grader in the same position as at the start of cutting the inslope.

Step 2 Move the circle rail to the extreme right and shift the moldboard to the right at the same time. This will place the circle rail to the outside of the right side of the machine with the moldboard pointing up the backslope.

Step 3 Position the circle rail at the angle you want to cut into the slope. The moldboard will now be at the desired angle and close to the natural slope.

Step 4 Use the circle rail shift to place the heel of the blade over the center line. Use the blade shift to lower the heel to the desired depth.

Step 5 Place the speed control in the low range and make a pass with the moldboard at the required angle. You will probably have to lean the wheels toward the slope in order to counteract the resistance against the blade.

Step 6 Make a finishing pass to remove material from the ditch bottom and inslope. Then level the windrow created on the shoulder or remove the material from the job site with a loader.

REPRINTED COURTESY OF CATERPILLAR INC.

22307-14_F29.EPS

Figure 29 Cutting a steep backslope.

Another machine used to finish off ditches is the excavator. Its long reach makes it ideal for steep slopes. This machine is also good for shaping around culverts and other irregular objects, and finishing off slopes that a motor grader cannot reach. With the reach of the telescoping boom and the ability to rotate the bucket, grading can be done at any angle above or below the machine.

3.4.2 Trenches

When trenches are dug for laying pipe, the backhoe or excavator operator needs to follow a grade line so that liquid will flow through the pipe correctly. The trench grade line and slope are often established using a laser level instead of a standard level and level rod. The laser level system simply provides a level light beam that can be used to set grade. It is basically the same as checking grade from a string line except that setup is much faster.

The laser level is positioned at one end of the trench and set for direction and slope. It can be set up to transmit a beam of light that is either perfectly horizontal or at a specified slope. The excavator is equipped with a laser receiver attached to the boom. The receiver is adjusted so that the beam from the level is captured at a specified elevation above the grade line of the trench bottom. To dig the trench to the required depth, the operator follows the directions given by the receiver as it senses the beam from the level. A digital display shows the operator whether the equipment is digging too high, too low, or on-grade.

If the excavator is not equipped with a laser receiver, the operator has to follow the directions of a grade checker using a level rod with a receiver attached. The receiver is calibrated for the height of the level plus the depth of the trench. For example, if the level were set up and calibrated at 3 feet (0.9 meters) and the grade hub called for a cut of 7 feet (2.1 meters), the receiver should be clamped to the rod at the 10-foot (3-meter) mark.

Using a laser level eliminates the need for establishing the pipe **invert** (flow line) elevation. Typically, the calculation is taken into account when setting up the level. Laser levels are like any other piece of sophisticated equipment. If used properly for the right application, they speed up production and provide better results. Used incorrectly, the results will be unpredictable.

When grading and grade checking are done manually, remember that the trench must be cut deep enough to accommodate bedding material and the thickness of the pipe. The depth of the undercut will depend on the local conditions and the type of bedding material available. *Figure 30* shows a typical example of how to account for the amount of undercut and pipe thickness to get the correct final excavation depth to the bottom of the trench.

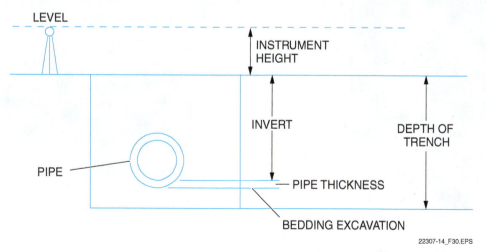

22307-14_F30.EPS

Figure 30 Factors determining trench depth.

The finishing work for a trench usually involves completing the subgrade to the proper slope for the flow line of the pipe to be placed. Skill is required to complete the trimming of the undisturbed material and the placing and trimming of any bedding material for the pipe. This is a difficult operation because it requires working on an area that is below the elevation of the equipment. Unless the trench is very large, you will not be able to place the equipment at the same height as the flow line of the pipe and will have to work from the natural ground level down into a trench bottom that could be as much as 35 feet (11 meters) deep.

Improperly graded trenches can cause serious problems. An uneven grade can make sliding the pipe together difficult, especially when laying large pipe. A poorly compacted trench bottom may settle and cause the pipe to break after the trench has been backfilled and compacted.

Unless you are using a laser leveling system attached to the equipment (*Figure 31*), you will need someone in the trench to guide the work so that the excavation and placement of material is accurate. Some handwork is almost always required for placing bedding material.

If a backhoe or excavator is being used to place the material, the equipment should be placed in-line with the pipe center line so that the material can be placed and smoothed with the bucket instead of by hand. Also, special attachments such as tampers and rollers can be mounted to the end of the hoe to assist in the compaction process.

The bedding material used to fill the trench bottom must have optimum moisture content so it will compact well. If the trench bottom is too wet or too soft, coarse aggregate must be used to provide a firm foundation for the pipe. In this case, the bottom must be cut far enough below the final grade to allow for the extra aggregate. After you have a firm foundation for the pipe, place the bedding material for the pipe to rest upon.

As the pipe is laid onto the bedding material, a pipe laser (*Figure 32*) can be used to ensure the proper slope. Most pipe lasers are small enough to fit inside a 6-inch (152-millimeter) diameter pipe. The depth of the bedding material can be adjusted as needed until the proper slope is achieved.

Most of the actual finish work for pipe placement must be done by crews in the trench using small power or hand tools. The job of the excavator operator is to supply the material from the stockpile to the proper location on the trench bottom so that the bedding crew can spread and compact the material. Sometimes, the bucket of the hoe can be used to help spread the material and rough-grade the bed. To do this, push the dumped pile with the bucket, then scrape the bucket backward to begin forming the saddle where the pipe will lay. This operation requires a good touch on the controls because you must be careful not to dig up material with the bucket and mix the bedding material with the soil underneath.

22307-14_F31.EPS

Figure 31 Laser receiver for backhoes and excavators.

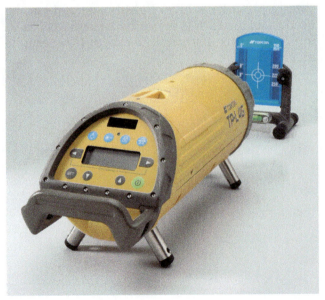

22307-14_F32.EPS

Figure 32 Pipe laser.

3.5.0 Grading Parking Lots, Sidewalks, and Curbs

Parking lots require extra planning to get the finish grading correct the first time. If there is a large parking area or other surface such as an airport apron that drains to swales or curbs, grading must be done precisely. Any low spots will collect water on the finished surface, even though low spots may not be evident by looking at the finished grade.

3.5.1 Parking Lots

The objective is to establish a grid pattern for the area. This is done first on the plans and then transformed to the site. Hubs mark the intersection points of the grid. By figuring out what the elevation should be at each hub, you can see how the entire area will drain. After making any changes on the plans, the hubs can be reset to their correct elevation. They can be set very quickly using a laser and level rod like that shown in *Figure 33*.

After the hubs are set, the motor grader operator can plan the grading pattern so that trimming can be done in an efficient manner without having to push large amounts of dirt from one area to another.

3.5.2 Sidewalks and Curbs

Cutting sidewalk and curb grades takes some practice. First, follow offset stakes as required by the concrete finishing work. If a self-trimming curb machine is used (*Figure 34*), the grade should

22307-14_F33.EPS

Figure 33 Checking the grade with a laser detector rod.

be left high in order to leave the machine enough material for a slight trim. It is important not to undercut the curb grade because the curb machine cannot make fills. If the machine is trimming an undercut curb, it will push loose dirt into low areas, leaving them uncompacted. This will cause the curb to fail.

The curb machine needs room behind the curb, and the grade must be pre-ripped and compacted in hardpan areas. On a large job where a pre-trimmer will precede the concrete machine, leave the grade a bit high. This will increase production substantially. It also eliminates the chattering and override that happens in heavy cuts, leaving a good smooth cut.

Figure 34 Self-trimming curb machine.

22307-14_F34.EPS

Additional Resources

Arrest that Fugitive Dust. Roberta Baxter, March/April 2002 Forester Communications, Inc.

National Asphalt Pavement Association (NAPA) web site (**www.hotmix.org**).

Silt Fence Installation Efficacy: Definitive Research Calls for Toughening Specifications and Introducing New Technology. Joel Sprague, P.E., TRI / Environmental, Inc. and Tom Carpenter, Carpenter Erosion Control. (**www.ieca.org**).

Using Lime For Soil Stabilization and Modification, 2001. National Lime Association. (**www.lime.org**).

3.0.0 SECTION REVIEW

1. A highway pavement is superelevated to _____.

   a. improve high-speed traction
   b. make it easier to grade
   c. slow vehicle speed
   d. improve traffic flow

2. When grading a slope with an articulated motor grader, the rear wheels are _____.

   a. kept sloped
   b. slightly deflated
   c. free-wheeling
   d. kept level

3. The forward position of the moldboard on a motor grader _____.

   a. allows greater ground clearance
   b. improves visibility
   c. forces material into low spots
   d. reduces soil compaction

4. Water will not flow freely in a ditch if the gradient is less than _____.

   a. 10 percent
   b. 4 percent
   c. 2 percent
   d. 0.5 percent

5. When cutting sidewalk and curb grades, always follow the _____.

   a. offset stakes
   b. curb undercut
   c. grid pattern
   d. hub marks

SUMMARY

Many different pieces of heavy equipment are used for finishing and final grading. The correct piece of equipment to use depends on the job. The most common machines for finishing include the motor grader, dozer, scraper, and excavator. Among these, the motor grader is used most often.

Operators will generally have specifications that tell them how much and to what tolerances certain jobs must be finished. These include requirements for grades on slopes, ditches, and embankments, as well as compaction on the embankments, foundations, and other structures that support vertical construction. The specifications are considered part of the contract for the work and must be followed, or the work will have to be torn up and done over.

When building the subgrade on two-lane roads and streets, a crown will have to be built. This slightly increases the height of the center of the road and allows the water to run off to the sides. Some roads also require superelevation on curves to keep vehicles from skidding sideways. Building a subgrade for a superelevated curve requires paying close attention to the staking.

Finish grading on slopes can be done with a motor grader, dozer, or various types of excavators. If the slope can be reached with a motor grader blade, it will do the best job, as long as a level surface can be provided for the wheels. Motor graders should not be used on fill sections where the slope is made of soft material; use dozers instead. If the slope is not too long, a telescoping excavator can also be used.

Grading subgrade and base is a job designed for a motor grader, although it can be done with other equipment such as a dozer or a scraper. Motor graders are best for the fine grading required around structures. The blade of the motor grader can be shifted outside its wheel path and can reach up against structures. Sometimes an area can't be finished with a grader. In this case, the long reach of a telescoping excavator may be the right choice to finish a job.

It is very important that heavy equipment operators know and follow the specification instructions during the final grading of a site. Finished levels are designed to control the flow of runoff, preventing erosion and standing water. Grades left too high can increase the runoff and cause erosion. Grades cut too low can cause standing water. Both conditions can result in the contractor being called back to the site for costly repairs.

Much of the fine grading work performed today is done with heavy equipment that has a laser or GPS leveling system attached. These systems guide the operator in excavating and grading on a continuous basis and are much more accurate than using grading stakes. In some cases, the machine can be set up to automatically perform some grading functions.

1. When using an articulated motor grader to finish steep slopes, use _____.
   a. a ground anchor
   b. crab steering
   c. reverse blading
   d. larger tires

2. How often does a laser leveling system on a dozer check for proper elevation?
   a. Once a minute
   b. Once every 30 seconds
   c. Every 100 feet (30.5 meters)
   d. Continuously

3. When picking up windrows with a scraper, position the scraper _____.
   a. to the right of the windrow
   b. to the left of the windrow
   c. so the bowl is centered on the windrow
   d. with its wheels on top of the windrow

4. What is one of the main things scraper operators should check before they start trimming?
   a. The drive train
   b. Tire pressure
   c. Turn signals
   d. The condition of the bowl

5. A telescoping excavator makes a good trimming machine because _____.
   a. its telescoping boom has an extended reach
   b. its bucket or grading blade can be worked back and forth in a scraping motion
   c. the bucket can turn 360 degrees
   d. several types of buckets are available

6. Binders are used to change the characteristics of soils.
   a. True
   b. False

7. A pozzolan is a byproduct of _____.
   a. oil refining
   b. coal burning
   c. copper mining
   d. agriculture

8. Polymers stabilize soil though _____.
   a. adhesion
   b. compression
   c. chemical reaction
   d. solidification

9. Finish grading levels are designed to _____.
   a. control the flow of runoff and to prevent standing water
   b. increase runoff and cause erosion
   c. cause standing water
   d. help plants and vegetation survive

10. Geotextiles can be used to reduce the depth of aggregate required for a roadway.
    a. True
    b. False

11. A road is crowned because it _____.
    a. helps cars stay on a wet road
    b. makes the road looks better
    c. helps drain water off the road
    d. makes the road stronger

12. When reshaping a crown, the material should be cast _____.
    a. into the ditch
    b. on the center line
    c. in a spoils pile
    d. on the shoulder

13. One of the best machines to use for shaping around culverts and irregular objects is a(n) _____.
    a. motor grader
    b. roller
    c. excavator
    d. scraper

14. Another name for the flow line of a pipe is
    _____.
    a. invert
    b. gradient
    c. alignment
    d. string line

15. What procedure helps the operator determine how a large horizontal surface will drain?
    a. Establishing a grid pattern
    b. Setting bank plugs
    c. Undercutting
    d. Slope staking

Trade Terms Introduced in This Module

Blue tops: Stakes that are used to identify the final elevation of the sub-base. The tops are either painted with blue paint or covered with blue plastic tassels.

Circle rail: A ring-shaped component on the motor grader that controls the horizontal position and elevation of the moldboard.

Crown: A slightly built-up section of the roadway, usually along the center line. The crown provides a slope to the pavement so that water will drain to either side.

Detention pond: A pond that will temporarily hold storm water until it can drain elsewhere.

GLONASS: A Russian-owned Global Navigation Satellite System operated by the Ministry of Defense of the Russian Federation.

GNSS: An acronym for Global Navigation Satellite System. GNSS is the generic term used to describe a locating or grade control system that uses signals from either GPS or GLONASS satellites.

Geotextile: A synthetic material used to filter water and soil, reinforce soil, or separate unlike materials.

Gradient: The change of elevation per unit length; the slope along a specific line of a road surface, channel, or pipe.

Hubs: Surveying stakes set for reference purposes. Hubs are usually at the edge of, or outside of, the work area.

Infiltration system: Storm water runoff control system that filters contaminants from the runoff before the runoff is allowed to infiltrate soil.

Invert: The flow line of a pipe. This would be the bottommost visible surface on the inside of a pipe.

Pozzolan: Originally used for volcanic ash, but now applies to any similar material that reacts with lime and water to form cement.

Red tops: Stakes that are used to identify the final elevation of the crushed aggregate base course.

Retention pond: A pond that will permanently hold the same level of water.

Riprap: Broken stone, in pieces weighing from 15 to 150 pounds (6 to 68 kilograms) each, placed on the ground for protection against the action of water.

Robotic total station: A machine control system that uses infrared signals reflected by a target mounted on the grading equipment to provide data to the equipment's on-board computer about how to adjust the blade.

Superelevation: The increased elevation of one side of a curved roadway that allows for banking of the pavement to the inside.

Tassel (whisker): A small plastic colored tag affixed to grade stakes for identification purposes.

Virtual reference station (VRS): An imaginary reference station that is established by a computer based on data from numerous GPS receivers and the grading equipment that serves as the source of data for grading control.

Additional Resources

This module presents thorough resources for task training. The following resource material is suggested for further study.

Arrest that Fugitive Dust. Roberta Baxter, March/April 2002 Forester Communications, Inc.

Basic Equipment Operator, 1994. NAVEDTRA 14081, Morris, John T. (preparer), Naval Education and Training Professional Development and Technology Center.

Caterpillar Performance Handbook, Edition 27. A CAT® Publication. Peoria, IL: Caterpillar, Inc.

Moving the Earth, 1988. H.L. Nichols. Greenwich, CT: North Castle Books.

National Asphalt Pavement Association (NAPA) web site (**www.hotmix.org**)

Silt Fence Installation Efficacy: Definitive Research Calls for Toughening Specifications and Introducing New Technology. Joel Sprague, P.E., TRI / Environmental, Inc. and Tom Carpenter, Carpenter Erosion Control. (**www.ieca.org**)

Soil Stabilization for Pavements Mobilization Construction Engineer Manual (EM 1110-3-137), 1984. Department of the Army, Corps of Engineers.

Using Lime For Soil Stabilization and Modification, 2001. National Lime Association. (**www.lime.org**)

Figure Credits

Deere & Company, Module opener, Figures 3, 6, 25

Reprinted Courtesy of Caterpillar Inc., Figures 1, 2, 9–12, 14–16, 24, 26, 27, 29

Gradall Industries, Inc., Figure 4

The Stanley Works, Figure 5

Trimble Navigation Limited, Figures 8, 33

Topaz Publications, Inc., Figures 17, 18, SA01

Heavy Equipment Training Academy, Figure 19

Sundt Construction Inc., Figure 20

Courtesy of Topcon Positioning Systems, Inc., Figures 31, 32

Courtesy of Miller Formless Company, Figure 34

Answer	Section Reference	Objective
Section One		
1. b	1.1.1	1a
2. d	1.2.1	1b
3. a	1.3.1	1c
Section Two		
1. d	2.1.2	2a
2. c	2.2.1	2b
3. c	2.3.5	2c
Section Three		
1. a	3.1.2	3a
2. d	3.2.0	3b
3. c	3.3.0	3c
4. d	3.4.1	3d
5. a	3.5.2	3e

NCCER CURRICULA — USER UPDATE

NCCER makes every effort to keep its textbooks up-to-date and free of technical errors. We appreciate your help in this process. If you find an error, a typographical mistake, or an inaccuracy in NCCER's curricula, please fill out this form (or a photocopy), or complete the online form at **www.nccer.org/olf**. Be sure to include the exact module ID number, page number, a detailed description, and your recommended correction. Your input will be brought to the attention of the Authoring Team. Thank you for your assistance.

Instructors – If you have an idea for improving this textbook, or have found that additional materials were necessary to teach this module effectively, please let us know so that we may present your suggestions to the Authoring Team.

NCCER Product Development and Revision

13614 Progress Blvd., Alachua, FL 32615

Email: curriculum@nccer.org
Online: www.nccer.org/olf

❏ Trainee Guide   ❏ Lesson Plans   ❏ Exam   ❏ PowerPoints   Other _____

Craft / Level: _____ Copyright Date: _____

Module ID Number / Title: _____

Section Number(s): _____

Description: _____

_____

_____

_____

_____

Recommended Correction: _____

_____

_____

_____

_____

Your Name: _____

Address: _____

_____

Email: _____ Phone: _____

Compaction Equipment

OVERVIEW

The surfaces on which buildings and roads are built must be able to carry the weight of the structure and its contents or the road and the traffic that will travel on it. Specialized compaction equipment such as rollers and compactors are used to make the underlying soil, asphalt, or gravel as smooth, level, and solid as possible. A skilled operator can recognize different types of compaction equipment and choose the best equipment for the job. Safe operators have a working knowledge of all the equipment's instruments, controls, and attachments, and they understand safety guidelines associated with the use of the equipment.

Module Two

Trainees with successful module completions may be eligible for credentialing through NCCER's National Registry. To learn more, go to www.nccer.org or contact us at **1.888.622.3720**. Our website has information on the latest product releases and training, as well as online versions of our *Cornerstone* magazine and Pearson's product catalog.

Your feedback is welcome. You may email your comments to **curriculum@nccer.org**, send general comments and inquiries to **info@nccer.org**, or fill in the User Update form at the back of this module.

This information is general in nature and intended for training purposes only. Actual performance of activities described in this manual requires compliance with all applicable operating, service, maintenance, and safety procedures under the direction of qualified personnel. References in this manual to patented or proprietary devices do not constitute a recommendation of their use.

Objectives

When you have completed this module, you will be able to do the following:

1. Identify and describe types of compaction equipment.
   a. Describe a pneumatic tire compactor.
   b. Describe a steel-wheel compactor.
   c. Describe a vibratory compactor.
   d. Describe a sheepsfoot compactor.
2. Identify and describe the components, controls, and attachments on a typical compactor.
   a. Describe the features of the operator's cab.
   b. Describe the instruments and indicators.
   c. Describe control systems.
   d. Identify common compactor attachments.
3. Describe safety guidelines and basic preventive maintenance requirements associated with compaction equipment.
   a. Identify specific roller/compactor safety rules.
   b. Identify tire safety rules.
   c. Describe daily inspection and maintenance procedures.
   d. Identify maintenance procedures involved in servicing a compactor.
4. Describe basic procedures for operating a compactor.
   a. Identify steps for preparing, starting up, and shutting down a compactor.
   b. Identify steps for performing basic maneuvers with a compactor.
5. Describe factors involved in work activities associated with a compactor.
   a. Identify factors involved in compaction equipment selection.
   b. Identify factors involved in the method of compaction used.
   c. Identify tests used to check compaction quality.
   d. Describe the process of leveling and compacting soil.
   e. Describe the process for backfilling.
   f. Describe the processes used in compacting cement and asphalt.

Performance Tasks

Under the supervision of your instructor, you should be able to do the following:

1. Complete a proper prestart inspection and maintenance on compaction equipment.
2. Perform the proper startup, warm-up, and shutdown procedures.
3. Execute basic maneuvers with compaction equipment, including forward/backward movement and turning.

Trade Terms

Compaction
Density
Foot
Ground contact pressure (GCP)
Lift
Mat

Pad
Puddling
Sheepsfoot
Settling
Tamping roller
Vibratory roller

Industry Recognized Credentials

If you are training through an NCCER-accredited sponsor, you may be eligible for credentials from NCCER's Registry. The ID number for this module is 22203-14. Note that this module may have been used in other NCCER curricula and may apply to other level completions. Contact NCCER's Registry at 888.622.3720 or go to **www.nccer.org** for more information.

Contents

Topics to be presented in this module include:

1.0.0 Types of Compaction Equipment.. 1
   1.1.0 Pneumatic Tire Roller.. 1
   1.2.0 Steel-Wheel Roller 2
   1.3.0 Vibratory Compactor 4
   1.4.0 Sheepsfoot Roller.. 5
2.0.0 Components and Attachments 7
   2.1.0 Operator's Cab.. 8
   2.2.0 Instruments and Indicators 9
      2.2.1 Engine Coolant Temperature Gauge 9
      2.2.2 Transmission Oil Temperature Gauge................. 9
      2.2.3 Hydraulic Oil Temperature Gauge 10
      2.2.4 Fuel Level Gauge 10
      2.2.5 Digital Display 10
      2.2.6 Indicators 10
   2.3.0 Controls... 11
      2.3.1 Disconnect Switches 11
      2.3.2 Seat and Steering Wheel Adjustment................ 11
      2.3.3 Engine Start Switch 12
      2.3.4 Vehicle Movement Controls 13
   2.4.0 Attachments 13
      2.4.1 Dozer Blade 14
3.0.0 Safety Guidelines and Preventive Maintenance 16
   3.1.0 Specific Roller/Compactor Safety Rules 17
   3.2.0 Tire Safety Rules 17
   3.3.0 Daily Inspection and Maintenance 18
   3.4.0 Servicing a Roller 19
      3.4.1 Preventive Maintenance Records........................ 21
4.0.0 Basic Operation 23
   4.1.0 Preparations, Startup, and Shutdown........................ 23
      4.1.1 Startup.. 24
      4.1.2 Checking Gauges and Indicators........................ 25
      4.1.3 Shutdown.. 25
   4.2.0 Basic Maneuvering.. 25
      4.2.1 Moving Forward 25
      4.2.2 Moving Backward 25
      4.2.3 Steering and Turning 25
      4.2.4 Transporting the Roller 25
5.0.0 Work Activities 27
   5.1.0 Equipment Selection.. 27
   5.2.0 Compaction Method Considerations 27
      5.2.1 Moisture Content... 28
      5.2.2 Layers 28
      5.2.3 Contractual Requirements.................................. 29

Contents (continued)

5.3.0 Checking Quality 29
   5.3.1 Sand Cone Test 29
   5.3.2 Nuclear Testing 29
5.4.0 Leveling and Compacting 30
5.5.0 Backfilling 30
5.6.0 Compacting Cement 31
5.7.0 Compacting Asphalt 31
   5.7.1 Roller Speed 31
   5.7.2 Number of Passes 31
   5.7.3 Rolling Zone 31
   5.7.4 Rolling Pattern 32
   5.7.5 Direction and Mode 32

Figures and Tables

Figure 1   Various types of rollers 3
Figure 2   Pneumatic tire roller 4
Figure 3   Steel-wheel roller 4
Figure 4   Vibratory steel-wheel roller 4
Figure 5   Cutaway of vibratory roller 4
Figure 6   Sheepsfoot roller 5
Figure 7   Towed-type sheepsfoot roller 5
Figure 8   Dropping-weight compactor 6
Figure 9   Roller with dozer blade 7
Figure 10 Operator's cab on a Caterpillar 815F Series 2 8
Figure 11 Operator's cab on an Ingersoll-Rand DD-138 8
Figure 12 Instrument panel for the Caterpillar 825G series roller 9
Figure 13 Typical four-gauge panel for a roller 9
Figure 14 Digital display and warning lights 10
Figure 15 Battery disconnect switch 11
Figure 16 Seat adjustment controls 11
Figure 17 Steering wheel adjustment controls 12
Figure 18 Dual-seat configuration 12
Figure 19 Engine start switch 12
Figure 20 Transmission control 13
Figure 21 Dual joystick controls 13
Figure 22 Vibratory controls 14
Figure 23 Blade control 14
Figure 24 Check pad foot tips 29
Figure 25 Check the engine oil and coolant levels 20
Figure 26 Transmission fluid sight gauge 20
Figure 27 Transmission fluid fill port 20
Figure 28 Hydraulic fluid sight gauge 20
Figure 29 Hydraulic fluid fill port 21
Figure 30 Clean the secondary air filter 21
Figure 31 Check the fuses 21
Figure 32 Parking brake 24
Figure 33 Starting aid switch 24
Figure 34 Roller tie down points 26
Figure 35 Effect of uneven foundation settling 28
Figure 36 Soil density testing equipment 30
Figure 37 Electrical density gauge 30
Figure 38 Combination pneumatic and steel-wheel vibratory roller 31
Figure 39 Typical rolling pattern 32

Table 1   Compaction Selection Chart 28

SECTION ONE

1.0.0 TYPES OF COMPACTION EQUIPMENT

Objective

Identify and describe types of compaction equipment.
   a. Describe a pneumatic tire compactor.
   b. Describe a steel-wheel compactor.
   c. Describe a vibratory compactor.
   d. Describe a sheepsfoot compactor.

Trade Terms

Compaction: Using an engineered process, such as rolling, tamping, or soaking, to reduce the bulk and increase the density of soil.

Density: The ratio of the weight of a substance to its volume.

Ground contact pressure (GCP): The weight of the machine divided by the area in square inches of the ground directly supporting it.

Pad: On a segmented or sheepsfoot roller, the part of the roller that contacts the ground; also called the foot.

Puddling: A process in which water is added to the soil until it is semi-liquid; the soil is then allowed to dry before being vibrated.

Settling: The natural wetting and drying process whereby soil particles become more compact and denser.

Sheepsfoot: A tamping roller with feet expanded at the outer tips.

Tamping roller: One or more steel drums fitted with projecting feet and towed with a box frame.

Vibratory roller: A compacting device that mechanically vibrates the soils while it rolls. It can be self-propelled or towed.

Soil consists of irregular particles; air and water fill the spaces between the particles. When weight is placed on soil, the particles are packed closer together and the air and water are pressed out. This process occurs naturally and is called settling. However, it is a slow and inconsistent process. If an area settles after construction, the structures can fail. For example, if the ground settles unevenly after a building is completed, the foundation can crack. Interior and exterior walls of the building can also crack. Re-

pairing this type of damage is extremely expensive.

The density of the soil is a measure of how closely the particles are packed. Denser soils can bear more weight. Compaction is the engineered process of increasing the density of the soil. It mimics the natural process, but is faster and more consistent. The underlying soil is compacted before construction so that it will not change shape or volume under the weight of the structure, pavement, or traffic.

Soil can be compacted by pressure, vibration, kneading, impact, or a combination of these methods. Different machines with specialized rollers are used to compact soil by using one or more of these methods. For example, a vibratory roller supplies pressure and vibration, while pneumatic rollers supply kneading and pressure. Various compaction techniques are used, depending on site conditions and soil types.

Some fills may be compacted by a process called puddling. This is an enhancement of the natural settling process. Water is added to the area until the soil is semi-liquid. It is then allowed to dry and settle. A small vibrator speeds up the drying process of any puddled fill. Vibration brings excess moisture to the surface. This method is very effective on gravel fills.

Rollers are used for compacting and smoothing materials such as earth, gravel, and asphalt. They compress the materials to the desired density to provide strength so that the specified loads can be carried without causing settling. There are several types of compaction rollers, each used for different types of work.

Early compaction equipment included a horse-drawn steel roller filled with rocks. Various designs have been developed based on these early models. Several companies manufacture rolling equipment in different configurations and sizes. Although there are many different models and makes of equipment, currently there are four basic designs. The four basic types of rollers are shown in *Figure 1* and include the following:

- Pneumatic tire
- Steel-wheel
- Vibratory
- **Sheepsfoot**

Each of these types can be further subdivided into two or more size classifications. For example, pneumatic tire rollers are manufactured in three distinct size categories, based on tire diameter.

Static forces are those produced by the heavy weight of the roller. Dynamic forces are those that use a combination of weight and energy to produce a vibratory or tamping effect on the soil.

Rollers and compactors use both static and dynamic forces to compact the soil and achieve the required soil density.

1.1.0 Pneumatic Tire Roller

Pneumatic tire rollers (*Figure 2*) achieve excellent results in almost every type of compaction. However, some types of wet soils may stick to the tires, limiting their effectiveness. Scrapers are often attached to scrape mud from the tires. The pneumatic tire roller first appeared in the early 1930s as a towed wagon with a rock-filled box. The primary design remains a weight-filled box mounted on two rows of tires. Some models are self-propelled, while others are towed.

The tires are the compaction devices on these rollers. Typically there are two axles with a set of tires on each axle. The tires are smooth with no treads. There are four to five tires on one axle and five or six on the other. Compaction is varied by changing the air pressure in the tires. The normal range is between 60 and 120 psi (413.69 and 827.37 kPa).

Pneumatic tire rollers provide uniform coverage and equal wheel loading. The rear wheels are staggered in relation to the front wheels to roll over the spaces left by the front wheels. This helps achieve uniform coverage. This offset adds a kneading action as the material is pressed toward the spaces between tires. The kneading action aligns the aggregate particles to their most stable positions. Equal wheel loading on uneven terrain is provided by articulated axles.

The tires are designed to provide a specific ground contact pressure (GCP) on the soil. The amount of GCP is controlled by a combination of wheel loading and tire inflation pressure. GCP values were standardized in the United States in 1967. Compactors carry decals that identify their maximum allowable wheel load and the ground contact pressures that can be attained with the various combinations of wheel loadings and tire inflation pressure.

Pneumatic tire rollers are used to compact all types of surfaces. In most cases, the pneumatic-tire roller is followed by finish steel-wheel rolling. When used as an intermediate roller, the pneumatic-tired roller contributes to compaction and particle alignment.

> **NOTE**
>
> Pneumatic tire rollers are used in asphalt road construction. The tendency of the asphalt binder to stick to the rubber tires increases as the temperature of the tires decreases. Heavy mats made of rubber or plywood can be mounted around the tire area to prevent heat loss and keep the tires at or near mat temperature. This will reduce the tendency of the binders to stick to the tires and increase rolling efficiency.

REPRINTED COURTESY OF CATERPILLAR INC.

(A) PNEUMATIC TIRE ROLLER

(B) STEEL-WHEEL ROLLER

REPRINTED COURTESY OF CATERPILLAR INC.

(C) SMOOTH-DRUM, VIBRATING ROLLER

REPRINTED COURTESY OF CATERPILLAR INC.

(D) STEEL-WHEEL (SHEEPSFOOT) ROLLER

22203-14_F01.EPS

Figure 1 Various types of rollers.

1.2.0 Steel-Wheel Roller

Steel-wheel rollers (*Figure 3*) are rollers as well as compactors. Their rolls are machined to provide a smooth, concentric surface. They are designed primarily for pavement course rolling, but are also used for rolling gravel roads, road bases, and some subgrades.

There are three basic types of steel-wheel rollers: three-wheel, two-wheel, and single-wheel. The three-wheel roller has one wide guide roll and two narrow drive rolls. The two-wheel, or tandem, roller has two rolls of equal width. The single-wheel roller has one steel rolling wheel and rubber drive tires. The drive tires are typically mounted on the rear of the machine. There is also a unique version of the tandem roller, called the three-axle tandem, which is similar to the tandem in appearance, but has two guide rolls on a walking beam. The walking beam allows the forces exerted by the rolls to be varied. The three-axle tandem is generally the heaviest of all steel-wheel rollers, but there are different size categories in each style.

Steel-wheel rollers have hydrostatic drive, hydrostatic steering, steering wheels, and a pressure-type fog-spray system for wetting the rolls. A self-contained portability wheel on the 4- to

Figure 2 Pneumatic tire roller.

REPRINTED COURTESY OF CATERPILLAR INC.

22203-14_F03.EPS

Figure 3 Steel-wheel roller.

6-ton (3.63- to 5.44-metric ton) tandems permits towing without the use of a trailer. Hydrostatic drive provides variable speed control, together with dynamic braking action.

1.3.0 Vibratory Compactor

Vibratory compactors (*Figure 4*) combine static weight with a generated, cyclic force. This dynamic type of compaction was developed to provide high densities in granular soils. The vibrations move the particles, and the static pressure forces them into a compact structure. Vibratory compactors work best in granular soils and are ineffective in silt and clay. While vibration is effective at bringing excess moisture to the surface, using a vibratory compactor in sandy soil can sometimes bring so much water to the surface that the finish grade can be affected.

Figure 4 Vibratory steel-wheel roller.

Rollers are manufactured in many styles and sizes. Many vibratory rollers are also available to compact deep-lift asphalt bases, complete with spray systems and smooth tires.

Static rollers depend on the weight of the machine to compress materials. A vibratory roller combines static and dynamic forces to generate higher compaction. The vibration or dynamic force is created by rotating a series of off-balance weights within a steel compacting drum (*Figure 5*). This rotation develops a centrifugal force that is sufficient to lift and drop the steel drum as it turns. The vibrating amplitude is one-half the total lift-and-drop distance.

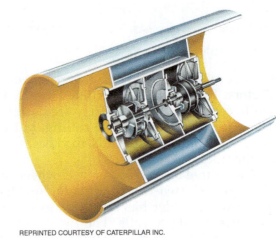

REPRINTED COURTESY OF CATERPILLAR INC.

22203-14_F05.EPS

Figure 5 Cut-away of vibratory roller.

The vibrating amplitude can be changed by adjusting the off-balance weight. The weight can be increased or moved farther from the center axle to increase the amplitude. Decreasing the weight, moving it closer to the center axle, or adding a balancing weight will decrease the vibrating amplitude. Some machines have fluid-filled chambers that can be adjusted to increase the weight or balance the off-balance weight to increase or decrease the vibrating amplitude.

The other measure of vibratory motion is frequency. Frequency is simply the number of times the lift-and-drop cycle is repeated each second. Increasing the speed at which the off-balance weight is rotating will increase the frequency. Decreasing the rotational speed will decrease the frequency. As the machine moves across the surface of the material, the lift-and-drop cycle repeats 28 to 67 times per second. The vibratory motion increases the compacting force up to six times the static weight of the drum assembly.

Most soils vibrate more easily at a specific frequency known as the resonant frequency. In most cases, vibration at this resonant frequency will produce faster compaction. In some vibratory compactors, the frequency can be varied to achieve the best results. Many have a monitoring system that indicates the vibrating amplitude.

1.4.0 Sheepsfoot Roller

The sheepsfoot roller, or **tamping roller** (*Figure 6*), offers fast compaction and high production. Self-propelled units have good maneuverability and pulling power. In addition, a blade can be mounted on the front of a self-propelled compactor to provide grading capabilities.

On a sheepsfoot roller, the steel drum has projecting pads or feet. For that reason, these compactors are sometimes called segmented-pad rollers. The feet are normally 7 to 9 inches (17.78 to 22.86 centimeters) long. There are two basic styles, sheepsfoot and tapered. Cleaner bars can be added to the back of the machine to remove dirt caught between the feet.

Sheepsfoot rollers compact a little at the surface, but provide greater compaction under the feet. In deep or soft fill, the wheel actually rides higher with each pass. This is known as walking out.

Some types of compaction equipment are designed to be used in conjunction with other equipment. For instance, a towed-type sheepsfoot roller (*Figure 7*) can be towed behind a bulldozer. This enables an operator to grade and compact an area in one pass. While many towed-type sheepsfoot rollers have been replaced by large, self-propelled units, smaller towed units still fill a compaction need. With their slower speed and lower cost, towed-type sheepsfoot rollers can provide economical compaction in smaller fill areas.

Another attachment that is used for compacting soil is a dropping-weight compactor (*Figure 8*). It is a type of heavy tamping device that is typically lifted with a crane and then allowed to drop onto the soil being compacted. This process is called impact, or dynamic, compaction. The size and weight of a dropping-weight compactor can vary, but a weight range of 10 to 20 tons (9.07 to 18.14 metric tons) is common. Dropping-weight compactors tend to work best in loose granular soils like sand, or in loose fill material.

REPRINTED COURTESY OF CATERPILLAR INC.

22203-14_F06.EPS

Figure 6 Sheepsfoot roller.

22203-14_F07.EPS

Figure 7 Towed-type sheepsfoot roller.

Sheepsfoot Rollers

The idea for a sheepsfoot roller came about when a contractor observed how a herd of sheep compacted clay soil. The steel projections on the roller drum knead the soil in the same way as sheep feet. Until that time, steel rollers had not done a good job of compacting clay soil.

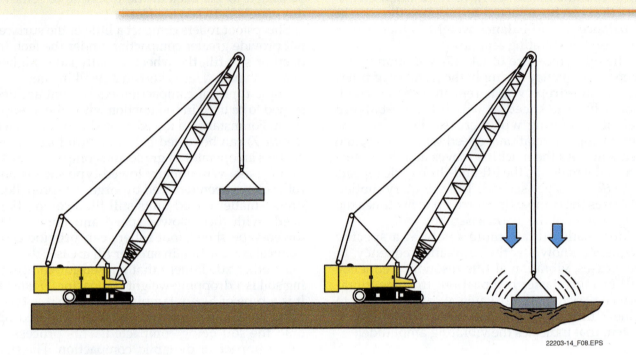

Figure 8 Dropping-weight compactor.

1.0.0 SECTION REVIEW

1. On a pneumatic tire roller, the soil compaction is accomplished by _____.

   a. pneumatic rollers
   b. steel wheels
   c. smooth tires
   d. sheepsfoot compactors

2. A two-wheel, or tandem, roller has _____.

   a. one guide roll and two narrow drive rolls
   b. two rolls of equal width
   c. one steel roller and rubber drive tires
   d. two sets of rubber tires

3. A type of compaction equipment that combines static and dynamic forces is a _____.

   a. static roller
   b. sheepsfoot
   c. tamping roller
   d. vibratory roller

4. A type of compactor in which the roller wheel rides higher in the compacted soil with each pass is a _____.

   a. sheepsfoot roller
   b. tandem roller
   c. walk-in roller
   d. vibratory roller

SECTION TWO

2.0.0 COMPONENTS AND ATTACHMENTS

Objective

Identify and describe the components, controls, and attachments on a typical compactor.

  a. Describe the features of the operator's cab.
  b. Describe the instruments and indicators.
  c. Describe control systems.
  d. Identify common compactor attachments.

Rollers and compactors are designed to perform certain tasks. Because the tasks are similar, the machines designed to complete them are also similar in appearance and operation, regardless of the manufacturer. Optional equipment may be available to improve efficiency or increase the usefulness of the machine, but the basics are the same.

There are more than a dozen companies that manufacture rollers, and describing the operation of each model is impossible. The information presented in this module is largely based on the operation and maintenance requirements of a Caterpillar 815F Series 2 soil compactor. Where significantly different design features occur between manufacturers, those differences will be noted and briefly described. Bomag steel-wheel rollers and Vibromax single-drum rollers are used as comparison machines. The term *roller* will be used to include both rollers and compactors.

It is important to read and understand the operator's manual that applies to the make and model of the equipment being used. If the manual is not available, ask a supervisor for it. No piece of equipment should be operated without understanding how it works and what to do to ensure safety.

The Caterpillar 815F Series 2 soil compactor is a 46,000-pound (20,865-kilogram) diesel-powered articulated compactor with bulldozing capability. The engine is capable of developing 232 horsepower. The transmission has three forward and three reverse gears. *Figure 9* shows the primary components of a Caterpillar roller/compactor with a dozer blade.

A diesel engine provides power to move the roller and drive the hydraulic pumps. On two-wheel smooth rollers and pneumatic rollers, the engine is usually mounted in the center of the machine between the rollers. On articulated rollers, the engine is mounted in the rear section, which also contains the operator's cab. Usually, power is supplied to the rear roller or wheels, which move the machine. However, some units have four-wheel drive. The front or guide roller is used to steer the machine.

Several of the machine systems are hydraulic and one or more of the hydraulic pumps may be interconnected. The hydrostatic drive provides a stepless change of gear ratios and a smooth transition when changing directions. The service brakes may be part of the hydrostatic drive. In this configuration, the forward/neutral/reverse (FNR) lever on the steering column is used for braking. Steering is typically accomplished using two hydraulic rams. On vibratory rollers, the vibration system

ENGINE AIR GAUGE

ENGINE EXHAUST STACK

ENGINE

RADIATOR

OPERATOR'S COMPARTMENT

SHEEPSFOOT ROLLER

CUTTING EDGE

22203-14_F09.EPS

Figure 9 Roller with dozer blade.

may be hydraulically activated. The hydraulic system is also used to control any attachments.

The following provides a brief description of the controls and indicators that are common to most rollers. Consult the instruction manual of the particular make and model being operated for information specific to that machine.

2.1.0 Operator's Cab

The operator's compartment, or cab, is enclosed in a rollover protective structure (ROPS) and/or a falling objects protective structure (FOPS). Most of the machine controls are located in the cab. It is the central hub for roller operations. An operator must understand the controls and instruments before operating a roller. Study the operator's manual to become familiar with the controls and instruments and their functions.

Figure 10 is an overhead view of the interior of the operator's cab on a Caterpillar 815F Series 2. Some switches and controls are located outside of the main cab. However, the controls for normal operations can be reached from the operator's seat. A steering wheel, accelerator, and brakes are used to maneuver the machine. The transmission is controlled with a lever on the left side of the steering wheel. The instrument panel is in front of the steering wheel. The compactor shown has a dozer blade; the joystick control lever to the right of the steering wheel controls that blade. The engine is started with a key switch. An additional

switch panel to control ancillary machine functions is located to the right of the operator's seat.

Figure 11 shows the cab for an Ingersoll-Rand DD-138 steel-wheel vibratory roller. This is a typical open-cab design that provides greater visibility of the leading roller edge. The driver is protected from falling objects by the FOPS/ROPS.

> **NOTE**
>
> Some rollers feature an open operator's platform covered with a ROPS. The open design allows greater visibility in all directions. However, the open design also subjects the operator to greater levels of noise, fumes, heat, and cold. Take appropriate safety precautions to protect yourself from harmful exposures.

22203-14_F11.EPS

Figure 11 Operator's cab on an Ingersoll-Rand DD-138.

TRANSMISSION SPEED SELECT LEVER

LEFT BRAKE PEDAL

INSTRUMENT PANEL

STEERING WHEEL

JOYSTICK BLADE CONTROL

ACCELERATOR PEDAL

RIGHT BRAKE PEDAL

KEY SWITCH

ADDITIONAL SWITCH PANEL

REPRINTED COURTESY OF CATERPILLAR INC.

22203-14_F10.EPS

Figure 10 Operator's cab on a Caterpillar 815F Series 2.

2.2.0 Instruments and Indicators

Operators must pay close attention to the instrument panel when operating a roller. The gauges and lights show the status of the machine's electronically monitored systems. *Figure 12* shows the instrument panel on a Caterpillar model 825GHH. The left-hand panel contains four gauges. In the center is the speedometer and digital readout. The panel on the right contains a series of warning lights.

Figure 13 shows a typical four-gauge panel. These gauges include the water temperature gauge, the transmission oil temperature gauge, the hydraulic oil temperature gauge, and the fuel level gauge. The following sections describe each of these gauges and their functions.

2.2.1 Engine Coolant Temperature Gauge

The engine coolant temperature gauge (*Figure 13*, "A") indicates the temperature of the water or coolant flowing through the cooling system. Refer to the operator's manual to determine the correct operating range for normal roller operations. Temperature gauges normally read left to right, with cold on the left and hot on the right. If the gauge is in the white zone, the coolant temperature is in the normal range. Most gauges have a red segment. If the needle is in the red zone, the coolant temperature is excessive. Some machines may also activate warning lights if the engine overheats.

> **CAUTION**
>
> Operating equipment when temperature gauges are in the red zone may severely damage it. Stop operations, determine the cause of problem, and resolve it before continuing operations.

If the engine temperature gets too high, operation should stop immediately and the problem should be investigated. There are several checks that the operator can perform. First the engine coolant level should be checked. If the level is low, more fluid should be added. The fan belt should be checked to make sure that it is not loose or broken. If necessary, the belt should be replaced. The radiator fins should be checked to ensure that they are not plugged with dirt or debris. If the fins are dirty, they should be cleaned. These are the three primary causes of the engine overheating. If initial troubleshooting fails to resolve the problem, operation should be stopped and the machine should be taken out of service.

REPRINTED COURTESY OF CATERPILLAR INC.

22203-14_F12.EPS

Figure 12 Instrument panel for the Caterpillar 825G series roller.

REPRINTED COURTESY OF CATERPILLAR INC.

22203-14_F13.EPS

Figure 13 Typical four-gauge panel for a roller.

> **WARNING!**
>
> Engine coolant is extremely hot and under pressure. Never remove a radiator cap while the engine is hot. Check the operator's manual and follow the recommended procedure to safely check and fill engine coolant.

2.2.2 Transmission Oil Temperature Gauge

The transmission oil temperature gauge (*Figure 13* "B") indicates the temperature of the oil flowing through the transmission. The normal operating range can be found in the operator's manual. This gauge also reads left to right in increasing temperature. It has a red zone that indicates excessive temperatures. When the weather is colder, the transmission oil should be allowed to warm up sufficiently before the machine is operated.

2.2.3 Hydraulic Oil Temperature Gauge

The hydraulic oil temperature gauge (*Figure 13* "C") indicates the temperature of the oil flowing through the hydraulic system. This gauge also reads left to right in increasing temperature and has a red zone that indicates excessive temperatures. The normal operating range can be found in the operator's manual. The hydraulic oil must be allowed to warm up sufficiently before the machine is operated under a load. The hydraulic system can be cycled under low rpm to warm it up.

> **NOTE**
>
> If the hydraulic system is sluggish, it means the hydraulic oil has not warmed up enough. The machine should be allowed to idle before it is operated under a load.

2.2.4 Fuel Level Gauge

The fuel level gauge (*Figure 13* "D") indicates the amount of fuel in the roller's fuel tank. Some models may have a low fuel warning zone and low fuel warning light. If the machine is running low on fuel, it should be shut down and refueled. Running out of fuel on diesel engine rollers should be avoided because air must be bled from the fuel lines and injectors before the engine can be restarted.

2.2.5 Digital Display

Some machines have a combined digital display. In *Figure 14*, the digital display is on the lower half of the panel. The display can be toggled to show the hour meter, tachometer, speedometer, odometer, or diagnostic codes.

REPRINTED COURTESY OF CATERPILLAR INC.

22203-14_F14.EPS

Figure 14 Digital display and warning lights.

The hour meter indicates the total hours of operation. It shows the period of time the machine has been running. Periodic maintenance is scheduled based on hours of service, which will be covered in more detail later in this module.

The tachometer indicates the engine speed in revolutions per minute (rpm). The speedometer shows the machine's ground speed. Typically they can be set for either miles per hour (mph) or kilometers per hour (kph). Diagnostic codes are shown when the machine is not functioning correctly. They are used by service mechanics to identify problems with the machine's systems.

2.2.6 Indicators

There is a series of warning lights above the display window, as shown in *Figure 14*. When lit, these lights show that the machine systems are not operating under normal conditions and that there is something wrong with the machine. If any of the warning lights start flashing, the machine should be stopped and the problem should be resolved before operations continue. The exception is the parking brake light. If it is lit, the parking brake is activated. However, if it starts flashing the parking brake is not functioning properly.

Typical warning lights show that the following systems are not functioning properly:

- Engine oil pressure
- Parking brake (when flashing)
- Brake oil pressure
- Electrical system
- Low fuel light
- Primary steering
- Supplemental steering

> **CAUTION**
>
> A flashing indicator light requires immediate action by the operator. Stop the machine and investigate the cause. If you continue operations, you may cause serious damage to the equipment. Resolve the problem before continuing operations.

There are often increasing levels of warning alarms on a machine. At the first level, alert indicators will light up. The operator must take action in the near future to correct the problem. At the second level, the alert indicators will flash. The operator must take action immediately to correct the problem and avoid machine damage. At the third level the action indicators will flash and a warning alarm will sound. The operator must immediately shut down the machine to avoid machine damage or operator injury.

2.3.0 Controls

Controls and their locations can vary between makes and models of rollers. Vehicle movement controls on rollers can be similar to those of cars and trucks. A steering wheel is used in combination with throttle and brake foot pedals to control vehicle movement. The dozer blade is controlled with a joystick or levers. Switches and levers activate ancillary and specialty machine functions. Controls described in this module may be different from those of other machines. Operators must review the operator's manual to fully understand the controls for the machine they are using.

2.3.1 Disconnect Switches

Some roller models are equipped with disconnect switches located outside the cab. These switches disconnect critical functions so that the machine cannot be operated. These switches must be activated before the machine can be started. Operators should turn them on before mounting the machine. They offer an additional level of safety and security since unauthorized users usually will not activate these switches and will not be able to operate the equipment.

Some models have a fuel shutoff switch that must be turned to the On position before the machine can be operated. The fuel shutoff switch prevents fuel from flowing from the fuel tank into the supply lines. This prevents unwanted fuel flow during idle periods or when transporting the machine, significantly reducing the potential for fuel leaks.

Alternatively, some machines have a battery disconnect switch, as shown in *Figure 15*. When the battery disconnect switch is turned off, the entire electrical system is disabled. This switch should be turned off when the machine is not in use overnight or longer to prevent a short circuit or active components from draining the battery. Before mounting the machine, check that the switch is in the On position.

Some machines have security panels or vandal guards. These panels can be locked into a no-access position when the machine is not in use. They prevent access to the machine's controls so it cannot be started or operated. Engage any vandal guards when leaving the machine unattended. Move the guard down to the stowed position to operate the machine. Lock the cab door, then secure and lock the engine enclosure. Always engage the security systems when leaving the roller unattended.

REPRINTED COURTESY OF CATERPILLAR INC.

22203-14_F15.EPS

Figure 15 Battery disconnect switch.

2.3.2 Seat and Steering Wheel Adjustment

While seat and steering wheel adjustments are not directly involved in roller operations, correct positioning of these items aids in safe operation. Adjust the seat and steering wheel position and then fasten the seat belt before operating the roller.

Most seats can be moved up or down and forward or backward using a series of levers (*Figure 16*). Some provide an adjustable shock absorber function using springs or hydraulics. The angle of the back of the seat can be set by lifting the seat recline lever. Pulling up on the fore/aft lever allows the seat to be adjusted by sliding it forward and backward. Releasing the lever locks the seat in place. The seat should be adjusted so that the operator's legs are almost straight when the accelerator or brake pedals are fully depressed and the operator's back is flat against the back of the seat. The seat angle can be adjusted to the desired position by lifting the seat cushion angle lever.

22203-14_F16.EPS

Figure 16 Seat adjustment controls.

Releasing the lever locks the seat in position. The stiffness of the seat suspension can be set to different positions by rotating a knob. The correct adjustment is based on the weight of the operator, and an approximate setting is often shown on a scale. The seat height can be raised or lowered by lifting the seat height lever. The armrest can be adjusted by rotating a knob on the side of the armrest.

The steering wheel on some rollers can be adjusted up or down and in or out as desired after the seat is correctly positioned. In *Figure 17*, moving the handle upward unlocks the steering column. The steering column can then be tilted and positioned correctly. Releasing the handle locks the steering column into position. Because seat and steering wheel adjustment devices vary widely for various makes and models, operators should refer to the operator manual for specific instructions.

Many rollers are designed to be driven in both directions. To increase operator comfort and improve visibility, some seats are mounted so that they face the side. Side-mounted seats allow the operator to see equally well in either direction. Other units have dual swivel seats like the ones shown in *Figure 18*. The driver rotates the seat and steering column based on the direction of travel. This allows the operators to see the drum edge and rolling pattern more clearly. Some smaller models feature a seat that can be moved around the operator's platform to access the controls from several different positions.

2.3.3 Engine Start Switch

The ignition switch functions can vary widely between makes and models of rollers. Some only ac-

22203-14_F18.EPS

Figure 18 Dual-seat configuration.

tivate the starter and ignition system. Others may activate fuel pumps, fuel valves, and starting aids. In some cases, the starter and starting aids are engaged by other manual controls. The multi-purpose ignition switch for the Caterpillar 815F series roller is shown in *Figure 19*. This switch is located to the right of the steering wheel on the console.

The engine start switch has the following three positions:

- *Off (1)* – Turning the key to this position stops the engine. It will also disconnect power to electrical circuits in the cab. However, several lights remain active when the key is in the Off position, including the hazard warning light, the interior light, and the parking lights.
- *On (2)* – Turning the key to this position activates all the electrical circuits except the starter motor circuit. When the key is first turned to

22203-14_F17.EPS

Figure 17 Steering wheel adjustment controls.

22203-14_F19.EPS

Figure 19 Engine start switch.

the On position, it may initiate a momentary instrument panel and indicator bulb check.

- *Start (3)* – Turning the key to this position activates the starter and starts the engine. The key should be released when the engine starts. This position is spring-loaded to return to the On position when the key is released. If the engine fails to start, the key must be returned to the off position before the starter can be activated again. To reduce battery load during starting, the ignition switch of some rollers may be configured to shut off power to accessories and lights when the key is in the start position.

> **CAUTION**
>
> To avoid damage to the starter, do not activate it for more than 30 seconds. If the machine does not start, refer to the operator's manual before activating the starter again.

2.3.4 Vehicle Movement Controls

Roller movement is controlled in a manner similar to that of cars and trucks. The throttle and brakes are operated by foot pedals. A steering wheel is used to turn the vehicle. The transmission is controlled with a lever on the right of the steering column. Moving the steering wheel to the left or right steers the vehicle to the left or right. Some steering wheels are equipped with a knob. This gives the operator greater control of the wheel when steering with one hand. This is important when the operator is using one hand to steer and the other to operate the dozer blade with the joystick.

The transmission is controlled by a lever on the left side of the steering column (*Figure 20*). The lever has three positions; forward, neutral, and re-verse. Moving the lever up sets the transmission in forward. The center position is neutral. Moving the lever down sets the transmission in reverse. There are three transmission speeds indicated on the collar. The transmission collar on the control lever is rotated to select the transmission speed. The three speeds can be used for either forward or reverse. Gears should not be skipped when downshifting.

Some rollers have joystick controls. One configuration has two joysticks, one for steering and one for travel control (*Figure 21*). All major machine functions are operated with button switches located at the tips of both joysticks. This allows operators to easily engage and disengage drum vibration and manually control the water-spray system.

On some vibratory rollers, the vibratory action can be adjusted. *Figure 22* shows a typical control for a steel-wheel vibratory roller. The vibrations per minute (vpm) can be varied from 1,225 to 2,050. The high and low amplitude is selected with a switch below the vibration control.

2.4.0 Attachments

Typically, rollers are used only for their primary purpose and do not have interchangeable attachments like forklifts and loaders do. However, there are exceptions. For instance, some models feature interchangeable rollers. The roller can be changed from a smooth, steel-wheel roller to a segmented-pad roller. Generally, roller attachments are limited to water spray units, a dozer blade, or a scarifier.

Water spray units are typically used when rolling asphalt. They can be mounted on steel-wheel or pneumatic-tire rollers but come as standard equipment on many models. They are used when

REPRINTED COURTESY OF CATERPILLAR INC.

22203-14_F20.EPS

Figure 20 Transmission control.

22203-14_F21.EPS

Figure 21 Dual joystick controls.

Figure 22 Vibratory controls.

rolling asphalt to keep the drum moist. Asphalt is less likely to stick to the wet surface of the roller.

2.4.1 Dozer Blade

The blade on a roller can be controlled with levers or with a joystick. The blade can be moved in several directions. The joystick may be used in conjunction with switchers, triggers, or buttons. The four ways to move the blade are as follows:

- *Lift* – Lift lowers or raises the blade. Lowering the blade changes the amount of bite or depth to which the blade will dig into the material. Raising the blade enables operators to travel, shape slopes, or create stockpiles. Most controls have a float position, which permits the blade to adjust freely to the contour of the ground. The float position is commonly used in reverse to smooth the surface.
- *Angle* – Angle adjusts the blade in relation to the direction of travel. When moving a load, the blade should be perpendicular to the line of travel. For filling a ditch, the blade should be angled to permit the load to be pushed off to the side.
- *Tilt* – Tilt changes the angle of the blade relative to the ground. This permits the blade to

cut deeper on one side than on the other. This process is very useful for performing side hill work where the blade tends to hang lower on the downhill side. It is also useful for crowning roads and grading slopes and curves.

- *Pitch* – Pitch is the slope of the blade from top to bottom. The greater the slope, the more the blade tends to dig in. On most machines the blade pitch must be changed manually. Because of the difficulty in controlling the pitch, it is only changed to meet site-specific requirements.

A typical control arrangement is shown in *Figure 23*. The right and left movement of the joystick angles the blade right and left. Backward and forward movement of the joystick raises and lowers the blade. The center position is the float or hold position. The blade will remain in the same position when the controls are in the hold position.

The controls for blade movement on a roller vary. There can be a second lever or joystick to control this movement. Alternatively, the blade could be adjusted with pins that must be moved manually. In a two-lever configuration, moving the lever forward tilts the blade forward. Pulling the lever backward tilts the blade backward. When the lever is released, the lever will return to the center hold position and the blade will stop moving. Operators must always read the operator's manual to understand the controls for the machine they will be operating.

REPRINTED COURTESY OF CATERPILLAR INC.

22203-14_F23.EPS

Figure 23 Blade control.

2.0.0 SECTION REVIEW

1. The operator's cab of a roller is enclosed in a rollover protective structure (ROPS) and/or a(n) _____.

   a. foreign objects repellant structure (FORS)
   b. environmental protective structure (EPS)
   c. weather resistant protective structure (WRPS)
   d. falling objects protective structure (FOPS)

2. If the engine oil pressure warning light on a roller begins to flash, the most suitable operator response is to _____.

   a. rev the engine until pressure builds and the light goes out
   b. stop the machine immediately so the problem can be resolved
   c. do nothing, since a flashing light indicates normal conditions
   d. throttle back on the engine and reset all warning indicators

3. Devices located on the outside of some roller cabs and used to disengage critical functions to prevent machine operation are called _____.

   a. disconnect switches
   b. FNR sensors
   c. multipurpose ignition switches
   d. joystick controls

4. The four ways that a dozer blade attachment on a roller can be moved are _____.

   a. lift, angle, forward, and backward
   b. lift, left, right, and down
   c. lift, angle, tilt, and pitch
   d. lift, tilt, shift, and float

3.0.0 SAFETY GUIDELINES AND PREVENTIVE MAINTENANCE

Objectives

Describe safety guidelines and basic preventive maintenance requirements associated with compaction equipment.
   a. Identify specific roller/compactor safety rules.
   b. Identify tire safety rules.
   c. Describe daily inspection and maintenance procedures.
   d. Identify maintenance procedures involved in servicing a compactor.

Performance Task

Complete a proper prestart inspection and maintenance on a roller.

Trade Terms

Foot: In tamping rollers, one of a number of projections from a cylindrical drum that contact the ground.

Safe equipment operation is the responsibility of the operator. Operators must develop safe working habits and recognize hazardous conditions to protect themselves and others from injury or death. Operators should always be aware of unsafe conditions to protect themselves from injury and the compaction equipment from damage. They should become familiar with the operation and function of all controls and instruments before operating the equipment. They should also read and fully understand the operator's manual.

It is vitally important for everyone on the site to be aware of and follow the safety practices.

Working in the construction industry is hazardous. Working on or around heavy construction equipment adds a layer of hazard. When operating heavy equipment, it is critical for operators to be aware of their surroundings. On a busy construction site, there are many people crossing the path of heavy equipment during the course of a business day. Everyone is concentrating on their jobs, but it is important to remember that safety is the number one consideration.

The first aspect of safe operation is to prepare to work safely. The following are general rules to prepare for safe operation:

- Read and understand the operating manual of the machine being operated.
- Read and understand specific company safety manuals and site safety regulations.
- Inspect the machine for potential safety problems at the beginning of each shift and as possible throughout the day.
- Perform the daily maintenance procedures required by the operator's manual. A well-maintained machine is safer than a poorly maintained machine.
- Be aware of any physical properties at the site, such as overhead wires, that may be a safety hazard.
- Wear appropriate clothing; leave jewelry at home.
- Wear eye and ear protection, a hardhat, and safety shoes.
- Remove or tie down loose tools, tie down safety equipment, and remove or tie down building materials.
- Use the seat belt.
- Read and understand all safety labels or stickers placed on the machine. Keep these signs clean so they can be read. Replace any labels or stickers that are missing or not readable.

The second aspect of safe operations is to avoid dangerous behavior. The following general rules will help prevent risks:

- Do not allow others to ride on the machine.
- Do not drill holes into, or weld anything to, the structure of the machine.
- Never use accessories not specifically designed to be used with the machine.
- Do not disable alarms, lights, or related safety devices and make sure they are working properly.
- Do not smoke while checking machine fluid levels. Some fluids and related vapors are highly explosive. Do not smoke when refueling or when servicing an ether cold weather start system.
- Make sure all shields, guards, and access panels are in place before operating the machine.
- Never drill into or weld to the structure of an ROPS. Weakening of the structure by doing so can result in injury or death of the operator and bystanders.
- Keep objects and hands away from moving fan blades. Contact with the blades will severely cut body parts and cut or throw objects.

- Do not allow oil and grease to accumulate on the machine, as it creates both a fire hazard and a slip-and-fall hazard. Clean spills from the machine immediately and clean with high-pressure water or steam every 1,000 hours or as recommended by the operator's manual.
- Dispose of dangerous fluids and material properly, following all applicable environmental regulations.

Finally, use common sense and advance planning to protect against hazards. Use the following general rules to operate the equipment:

- Stay seated and use the seat belt while operating the machine.
- Wear the appropriate personal protection equipment (PPE) to protect against dust, dirt, and fumes.
- Check the engine coolant level only after the engine has been stopped and the filler cap can be safely removed with bare hands. Vent the pressure and remove the cap slowly.
- When moving the machine to a colder climate, remember to check or replace fluids that may be affected by the temperature change.
- Know the maximum height of the machine.
- Frequent replacement of the same fuse may indicate an electrical problem that should be investigated.
- Always replace fuses with the same type and size.

3.1.0 Specific Roller/Compactor Safety Rules

In addition to the general hazards from operating heavy equipment on a construction site, there are specific hazards related to operating rollers. The following safety rules apply specifically to rollers:

- Know the limits for operating the roller on slopes. Do not turn the machine on a slope or drive across a slope. Only operate up and down slopes.
- Be careful when operating equipment close to the edge of a hill or embankment. Roller compaction can cause the slope to collapse, which can cause the roller to become unstable and overturn.
- Machines that produce vibration can cause the walls of trenches or embankments to collapse. Make sure these walls are properly braced.
- Do not operate the vibrator when the machine is stopped. Excessive vibration can cause the ground to subside and cause the machine to overturn.
- Articulated machines that have a center pivot do not have clearance for personnel in the pivot area. Keep all personnel clear of the pinch area.
- The transmission of some rollers will not hold a parked machine on any grade. Always apply the secondary parking brake and move the transmission control lever to the neutral position after the machine has stopped. Do not use the secondary brake system to slow and then stop the machine. It is to be used to hold the machine in a parked position only. Do not use the machine if it can be moved when the secondary brake system is on.
- Use extreme caution when raising or lowering scrapers. The scrapers are under spring tension. The lower edge of the scraper can become very sharp due to wear.
- Do not operate a water spray system with an empty reservoir. This can damage the pump.

3.2.0 Tire Safety Rules

Tire pressure is an important aspect of roller operation for rubber-tired rollers. Air, nitrogen, or a liquid mixture may be used to inflate rubber tires. Some models require inflation with air and a mixture of calcium chloride and water. Refer to the operator's manual for complete instruction. Remember that most heavy equipment tires are mounted on a split rim and, unlike automobile tires, can explode with devastating force under certain conditions.

Use the following guidelines to be safe when working with rubber tires on a roller:

- Check the tire pressure at the beginning of each shift and at intervals of ten hours thereafter. Inspect each tire for excessive wear and damage to both the tire and rim.
- When inflating the tire, park the machine to position the filler valve at the top of the tire.
- Never inflate a completely deflated tire. Leave that to a trained tire technician.

Roller Rollover

A highway construction worker died after the roller she was operating slipped off the edge of the road surface and rolled over. The worker was compacting a road bed. The earth under the rear tires gave way when she backed up near the edge. She was thrown from the open cab door, pinned under the ROPS, and died from the injury. The machine was equipped with a ROPS, but not a seat belt.

The Bottom Line: Always wear your seat belt. Do not operate a roller that is not equipped with a ROPS and a seat belt.

Source: Electronic Library of Construction Occupational Safety and Health.

- When inflating a tire, never stand directly in front of the tire. Stand to the side of the valve.
- Never expose tire rims to a direct flame or high heat like that from a torch or welding equipment.

3.3.0 Daily Inspection and Maintenance

Routine maintenance includes a regular effort to lubricate and service the equipment. This helps prevent breakdowns and keeps the machine in good working order. Consistent attention to these details keeps the machine operating efficiently, which improves productivity and safety.

These tasks must be performed on a regular basis. Some tasks, like inspection and lubrication, should be done daily. Other tasks must be done weekly, monthly, or annually. Maintenance is scheduled based on the hours of service. The operator must keep track of the hours of service and schedule maintenance. A service log is usually kept with the machine. Some preventive maintenance procedures can be performed easily with the right tools and equipment.

> **CAUTION**
>
> Roller service is normally based on hours of service. A service schedule is contained in the operator's manual. Failure to perform scheduled maintenance could result in damage to the machine.

Maintenance requirements are basically separated into two main categories: as required and scheduled. The scheduled maintenance is further divided into nine cycles:

- Daily (10 hours), or once per shift
- Weekly (50 hours)
- Bi-weekly (100 hours)
- Monthly (250 hours)
- Three months (500 hours)
- Six months (1,000 hours)
- One year (2,000 hours)
- Two years (3,000 hours)
- Four years (6,000 hours)

There is a list of inspections and required service activities in the operator's manual. Make sure all maintenance procedures are performed according to the instruction given in the machine's operating and/or service manual. Before performing any maintenance on the machine, read and understand all safety rules in the operating and/or service manual. Although not all of the maintenance tasks are performed by the operator, it is the operator's responsibility to ensure they have been done.

Before beginning work each day, inspect the machine. Some companies provide a daily inspection checklist. Include all items on the daily or 10-hour maintenance schedule, as well as any as-required maintenance. Complete the inspection and daily

---

Roller Fatality

A construction laborer died after being run over by a roller during highway paving. The worker was repositioning traffic cones placed along the highway. The roller operator made a forward pass on the newly laid asphalt and then reversed the machine. The roller traveled 10 feet (3.05 meters) before the operator sensed something was wrong and stopped the machine. The laborer was discovered lying face down, his head crushed by the roller.

The Bottom Line: Do not back up unless you know that no one is behind you. Look in the direction of travel and know where your blind spots are.

Source: Electronic Library of Construction Occupational Safety and Health.

---

maintenance before starting the engine. This will help prevent a breakdown during operation.

The daily inspection is often called a walk-around. The operator should walk completely around the machine checking various items. Look around and under the machine for leaks, damaged components, or missing bolts or pins. If there is evidence of a leak, locate the source and get it repaired. If leaks are suspected, check fluid levels more frequently. Items to be checked and serviced on a daily inspection include the following:

- Engine precleaner screen for debris
- Engine compartment for debris
- Engine for damaged belts
- Hydraulic system for leaks
- Cooling system for leaks, debris, damaged hoses
- Transmission for leaks
- Fuel tank for water and sediment
- Covers and guards secured in place
- Front and rear differentials for leaks
- Tires for proper inflation and damage
- Blade control linkage for damage and/or wear (if applicable)
- Steps, walkways and handholds for debris and damage
- ROPS for damage
- Windows for cleanliness
- Cab for cleanliness
- Seat belt for damage or wear
- Roller cleaner bars
- Pad **foot** tips (*Figure 24*)
- Lights for damage
- Indicators, gauges, and brakes
- Back-up alarm for proper operation
- Roller and cleaner bar for damage

During the daily inspection, check the fluid levels and add fluids as needed. Make sure that the machine is level to ensure that fluid levels are accurate. The following fluids should be checked:

- Engine oil (*Figure 25*)
- Coolant
- Transmission oil (*Figures 26* and *27*)
- Hydraulic fluid (*Figures 28* and *29*)

> **WARNING!**
>
> Do not check for hydraulic leaks with your bare hands. Use a piece of wood, cardboard, or another solid device. Pressurized fluids can cause severe injuries to unprotected skin. Long-term exposure can cause cancer or other chronic diseases.

Some maintenance is performed as required. This means that it should be done whenever it is needed. As-required maintenance should be

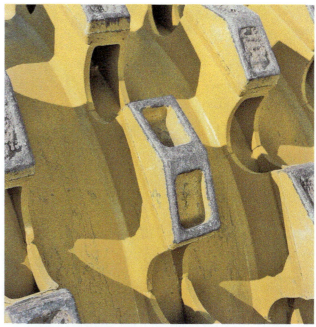

22203-14_F24.EPS

Figure 24 Check pad foot tips.

included in the daily walk-around. These tasks should also be performed during or at the end of the work shift, as needed. Typical as-required maintenance includes the following:

- Clean air intake
- Service primary air filter element
- Service secondary air filter element (*Figure 30*)
- Check fuses (*Figure 31*)
- Check circuit breakers
- Inspect windshield wipers
- Fill windshield washer fluid
- Inspect cutting edge and end bits
- Inspect tamping tips
- Inspect cleaner bar
- Inspect oil filter

The operator's manual usually has detailed instructions for performing periodic maintenance. If an operator finds any problems with the machine but is not authorized to fix it, he or she should inform the foreman or field mechanic before operating the machine.

3.4.0 Servicing a Roller

A roller should be serviced by a trained mechanic. When servicing a roller, follow the manufacturer's recommendations and service chart. Any special servicing for a particular piece of equipment will be highlighted in the manual. Normally, the service chart recommends specific intervals, based on hours of run time, for such things as changing oil, filters, and coolant. In addition, some ma-

REPRINTED COURTESY OF CATERPILLAR INC.

22203-14_F25.EPS

Figure 25 Check the engine oil and coolant levels.

REPRINTED COURTESY OF CATERPILLAR INC.

22203-14_F26.EPS

Figure 26 Transmission fluid sight gauge.

chines' systems are monitored by a computer and provide diagnostic information. The computer will display diagnostic codes that provide service information. The codes and required service are explained in the machine's service manual.

Hydraulic fluids should be changed whenever they become dirty or break down due to overheat-

REPRINTED COURTESY OF CATERPILLAR INC.

22203-14_F27.EPS

Figure 27 Transmission fluid fill port.

ing. Continuous and hard operation of the hydraulic system can heat the hydraulic fluid to the boiling point and cause it to break down. Filters should also be replaced during regular servicing.

Before performing maintenance procedures, always complete the following steps:

Step 1   Park the machine on a level surface to ensure that fluid levels are indicated correctly.

Step 2   Lower all equipment to the ground. Operate the controls to relieve hydraulic pressure.

Step 3   Engage the parking brake.

Step 4   Lock the transmission in neutral.

3.4.1 Preventive Maintenance Records

Accurate, up-to-date maintenance records are essential. Each machine should have a record that describes any inspection or service that is to be performed and the corresponding time intervals. Typically, an operator's manual and some sort of inspection sheet are kept with the equipment at all times.

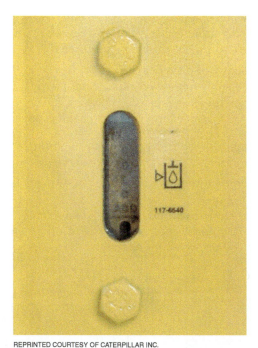

22203-14_F28.EPS

Figure 28 Hydraulic fluid sight gauge.

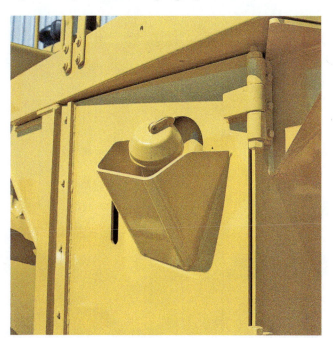

22203-14_F29.EPS

Figure 29 Hydraulic fluid fill port.

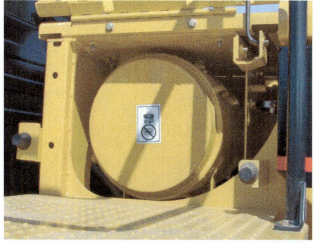

22203-14_F30.EPS

Figure 30 Clean the secondary air filter.

22203-14_F31.EPS

Figure 31 Check the fuses.

Tires

Roller tires can be inflated with air or a liquid mixture. Tires inflated with liquid ballast provide added machine stability. The liquid is a mixture of water and calcium chloride. The latter provides antifreeze protection. No matter what is inside the tires, check the tire pressure daily to make sure they meet the recommended pressure in the operator's manual.

3.0.0 SECTION REVIEW

1. When a roller is being operated on a slope, it is important to _____.

   a. make any necessary turns while on the slope
   b. only operate up and down the slope
   c. only operate across the length of the slope
   d. operate the vibrator only when the roller is stopped

2. When inflating a tire on a pneumatic tire roller, an operator should _____.

   a. make sure the tire is positioned with the filler valve at the bottom
   b. first completely deflate the tire and then re-inflate it
   c. inflate the tire using only air and a water/sodium chloride mixture
   d. make sure the filler valve is at the top and stand to the side of the valve

3. A walk-around inspection of a piece of compaction equipment should be performed _____.

   a. daily or once per shift
   b. weekly or every 50 hours
   c. monthly or every 250 hours
   d. only as needed when a problem is noticed

4. Some rollers have an on-board computer that monitors major systems and provides _____.

   a. recommendations about operating techniques
   b. oil and hydraulic fluid rejuvenation
   c. diagnostic codes for service information
   d. online access for operators

4.0.0 BASIC OPERATION

Objectives

Describe basic procedures for operating a compactor.
  a. Identify steps for starting up and shutting down a compactor.
  b. Identify steps for performing basic maneuvers with a compactor.

Performance Tasks

Perform the proper startup, warm-up, and shutdown procedures.

Carry out basic operations with a roller:

  • Compact an area approximately 20 feet long (6.10 meters) and 10 feet (3.05 meters) wide.

Operation of a roller requires constant attention to the controls and instruments, the rollers, and the surrounding environment. Rollers often work in conjunction with other machines. Operators must be aware of other operations going on around the equipment.

Rollers move slowly; typical speeds are between three and five miles (4.83 and 8.05 kilometers) per hour.

The initial operations that operators must learn include the proper ways to prepare, start up, and shut down the machine and how to move forward and backward.

4.1.0 Preparations, Startup, and Shutdown

Before starting any roller operation, it is important to become familiar with the area of opera-

Over-Compaction

Soil can become too compacted. Over-compaction can occur if the roller makes too many passes over the soil. Excess compaction can have the effect of loosening the soil, so that it may eventually develop cracks, which can affect structures built on the soil.

tions and the soil conditions. Check the area for both vertical and horizontal clearances. Make sure the path is clear of electrical power lines and other obstacles. Make note of soft areas, trenches, embankments, and other unstable areas. Mark any hidden dangers.

Check the machine to make sure that it is adjusted properly for the work to be performed. Where appropriate, check that the following are set correctly:

• Water spray unit
• Tire inflation
• Ballast or other weights

Mount the equipment using the grab rails and foot rests. Always maintain three points of contact when mounting equipment. Keep grab rails and foot rests clear of dirt, mud, grease, ice, and snow. Adjust the seat to a comfortable operating position. The seat should be adjusted to allow full pedal travel with the operator's back against the seat back. This will permit the application of maximum force on the brake pedals. Operators must verify that they can see clearly and reach all the controls. If the machine has dual seats, both seats should be adjusted before the machine is operated.

> **WARNING!**
>
> OSHA requires that approved seat belts and a rollover protective structure (ROPS) be installed on all heavy equipment. Old equipment must be retrofitted. Do not use heavy equipment that is not equipped with these safety devices.

Operator stations vary depending on the manufacturer, size, and age of the equipment. However, all stations have gauges, indicators, switches, levers, and pedals. Gauges show the status of critical items such as water temperature, oil pressure, and fuel level. Indicators alert operators about low oil pressure, engine overheating, and electrical system malfunctions. Switches activate starting aids and turn accessories such as lights on and off. Typical instruments and controls were described previously. Review the operator's manual for the specifics of the machine being operated.

The proper startup and shutdown of an engine is very important. Proper startup lengthens the life of the engine and other components. A slow warm-up is essential for proper operation of the machine while under load. Similarly, the machine must be shut down properly to cool the hot fluids circulating through the system. These fluids must cool so they can cool the metal parts of the engine before it is switched off.

4.1.1 Startup

There may be specific startup procedures for different pieces of equipment, but in general, the startup procedures are done in this sequence:

Step 1 Verify that the transmission control is in neutral.

Step 2 Use a lever or knob to engage the parking brake (*Figure 32, B*), depending on the roller make and model. Some rollers have a parking brake test switch (*Figure 32, A*).

> **NOTE**
>
> When the parking brake is engaged, an indicator light on the dash will light up or flash. If it does not, stop and correct the problem before operating the equipment.

Step 3 Depress the throttle control slightly.

Step 4 Turn the ignition switch to the Start position. The engine should turn over. Do not operate the starter for more than 30 seconds at a time. If the engine fails to start, refer to the operator's manual before cranking again.

Step 5 Warm up the engine for at least 5 minutes. In colder temperatures, warm up the machine for a longer period of time.

Step 6 Make sure all the gauges and instruments are working properly.

Step 7 Shift the transmission into forward and move the gear control to low range.

Step 8 Release the parking brake and depress the service brakes.

Step 9 Check all the controls for proper operation.

Step 10 Check service brakes for proper operation.

Step 11 Check the steering for proper operation.

Step 12 Manipulate the controls to be sure all components are operating properly.

Step 13 Shift the transmission to neutral.

Step 14 Reset the parking brake.

Step 15 Make a final visual check for leaks, unusual noises, or vibrations.

If the machine has a diesel engine, there are special procedures for starting the engine in cold temperatures. Many diesel engines have glow plugs that heat up the engine for ignition. Follow

22203-14_F32.EPS

Figure 32 Parking brake.

the operator's manual for starting the engine in cold temperatures.

Some units are also equipped with a starting aid. Depress the switch (*Figure 33*) while cranking the engine. Depress the switch every two seconds until the engine is running smoothly. Review the operator's manual to fully understand the procedures for using these aids.

As soon as the engine starts, release the key; it should return to the On position. Adjust the engine speed to approximately half throttle. Let the engine warm up to operating temperature before moving the roller.

> **WARNING!**
>
> Ether is used in some older equipment to start cold engines. Ether is a highly flammable gas and should only be used under strict supervision and in accordance with the manufacturer's instructions. Use of ether may void the manufacturer's warranty. Before using any starting aid, check the manufacturer's instructions.

REPRINTED COURTESY OF CATERPILLAR INC.

22203-14_F33.EPS

Figure 33 Starting aid switch.

4.1.2 Checking Gauges and Indicators

Keep the engine speed on idle until the oil pressure registers. The oil pressure light should initially light and then go out. If the oil pressure light does not turn off within 10 seconds, stop the engine, investigate, and correct the problem.

Check the other gauges and indicators to see that the engine is operating normally. On most rollers, the instrument panel will run a self-test when the machine is first started. Check that the coolant temperature, hydraulic oil temperature, and oil pressure indicators are in the normal range. If there are any problems, shut down the machine and investigate, or get a mechanic to look at the problem.

4.1.3 Shutdown

Shutdown should also follow the manufacturer's procedure. Proper shutdown will reduce engine wear and possible damage to the machine.

> **CAUTION**
>
> Rollers tend to be very heavy. Parking on a slope or incline is not advised. If you must park on a slope, park so that the machine is facing uphill. Provide a support behind the machine such as a concrete barrier.

4.2.0 Basic Maneuvering

Basic roller maneuvers include moving forward, moving backward, and turning.

4.2.1 Moving Forward

The first basic maneuver is learning to drive forward. Adjust the seat and fasten the seat belt before operating the machine. To move forward, follow these steps:

Step 1   Before starting to move, use the joystick to raise the blade to clear any obstructions.

Step 2   Depress the service brakes and release the parking brake. Move the shift lever to low forward.

Step 3   Release the service brakes and depress the accelerator pedal to start the roller moving.

Step 4   Drive the machine forward for the best visibility. Steer the machine using a steering wheel.

> **NOTE**
>
> Slow down or brake before changing direction. This will increase the service life of the engine.

4.2.2 Moving Backward

To back up or reverse direction, slowly come to a complete stop. Engage the parking brake. If necessary adjust the seat or switch seats for optimal visibility.

Check to make sure that the travel area is free from pedestrians and obstacles. Depress the service brakes and release the parking brake. Move the shift lever to reverse. Release the service brake and slowly depress the accelerator to begin to move backward.

> **NOTE**
>
> Many rollers use dynamic braking, which uses the engine to slow the machine. To slow the machine, move the transmission lever slowly to the neutral position. The machine will coast to a stop. Once the machine is stopped, move the shift lever to reverse and increase engine speed to move in the desired direction.

4.2.3 Steering and Turning

How a roller is steered depends on the make and model; however, most have a steering wheel. Some roller steering wheels have a knob that allows the wheel to be turned with one hand. This allows the operator to use the other hand to control the joystick to move the blade or operate other controls.

The steering wheel on a roller operates in the same manner as the steering wheel on a car or truck. Moving the wheel to the right turns the roller to the right. Turning the wheel to the left moves the wheels to the left.

4.2.4 Transporting the Roller

Typically a roller is only driven around a site. If it must be transported from one job site to another, it is usually loaded and hauled on a transporter. Transport the roller on a properly equipped trailer or other transport vehicle. Before beginning to load the equipment for transport, make sure the following tasks have been completed:

- Check the operator's manual to determine if the loaded equipment complies with height, width, and weight limitations for over-the-road hauling.

- Check the operator's manual to identify the correct tie down points on the equipment (*Figure 34*).
- Be sure to get the proper permits, if required.

Once the above tasks are completed and the loading plan determined, carry out the following procedures:

Step 1 Position the trailer or transporting vehicle. Always block the wheels of the transporter after it is in position but before loading is started.

Step 2 If equipped, raise the blade slightly. Drive the roller onto the transporter. Whether the roller is facing forward or backward will depend on the recommendation of

22203-14_F34.EPS

Figure 34 Roller tie-down points.

the manufacturer. Most manufacturers recommend backing the roller onto the transporter.

Step 3 Move the transmission lever to neutral, engage the parking brake, turn off the engine, and remove the key.

Step 4 Manipulate the hydraulic controls to relieve any remaining hydraulic pressure.

Step 5 Lock the door to the cab and any access covers. Attach any vandalism protection.

Step 6 Engage the battery disconnect or fuel shutoff switches.

Step 7 Secure the machine with the proper tie-down equipment as specified by the manufacturer. Place chocks at the front and back of all four tires.

Step 8 Cover the exhaust and air intake openings with tape or a plastic cover.

Step 9 Place appropriate flags or markers on the equipment if needed for height and width restrictions.

WARNING!	The machine may shift while in transit if it is not properly tied down. If the machine shifts in transport it could cause personal injury or death. Follow the manufacturer's safety procedures.

Unloading the equipment from the transporter is accomplished by reversing the steps of the loading operation.

4.0.0 SECTION REVIEW

1. Before starting any roller operation, an operator should thoroughly check _____.

   a. work area clearances, obvious hazards, and computer diagnostics
   b. the area of operations, the soil conditions, and the roller setup
   c. soil depth, angle of slope, and the roller setup
   d. area hazards, weather conditions, and roller transport damage

2. To move a roller backward, move the shift lever to reverse, release the brakes, and _____.

   a. pull the joystick trigger
   b. push forward on the hand lever
   c. lower the blade
   d. depress the accelerator

5.0.0 WORK ACTIVITIES

Objectives

Describe factors associated with work activities asociated with a compactor.

a. Identify factors involved in compaction equipment selection.
b. Identify factors involved in the method of compaction used.
c. Identify tests used to check compaction quality.
d. Describe the process of leveling and compacting soil.
e. Describe the process for backfilling.
f. Describe the process used in compacting cement and asphalt.

Trade Terms

Lift: A layer of material that is a specific thickness; the depth of material that is being rolled by the roller.

Mat: Asphalt as it comes out of a spreader box or paving machine in a smooth flat form.

The operation of a roller requires an understanding of the compaction process. This knowledge is essential for operators to be able to select the proper equipment for the specific soil conditions. A basic description of compaction and soils is included in this section.

5.1.0 Equipment Selection

There are four basic types of rollers, as previously described: pneumatic tire rollers, steel-wheel rollers, vibratory compactors, and sheepsfoot, or segmented-pad, rollers. This equipment supplies static and dynamic forces to compact soil, depending on the equipment configuration. Therefore, choosing the right equipment is an important part of the compaction process.

Equipment selection is based on the type of material to be compacted. Some types of compactors are more efficient for compacting a certain soil type while other methods are completely ineffective. *Table 1* compares the effectiveness of several types of compactors on different types of soil.

Soil may be compacted with various types of equipment by pressure, kneading, vibration, impact, or a combination of these methods. The pneumatic tire and segmented-pad rollers supply pressure with some kneading. Vibratory compactors supply both pressure and vibration.

The weight of a smooth steel-wheel roller is applied along a straight line across the direction of travel. A high spot may carry the full weight of the roller, while a low spot may be bridged over and receive no compaction. With adequate weight, the soil is squeezed from high to low spots, equalizing irregularities in spreading and producing a smooth surface. However, high spots that are too hard to yield will support the roller and prevent compaction.

Segmented-pad rollers compact mostly with their feet from the bottom up. As the soil is compacted, the roller rises and walks out of the ground. These rollers are good on fine-grained plastic soils and should not be used on granular, noncohesive soils.

Pneumatic tire rollers are suitable for use in any type of soil, but the weight and tire pressure must be proper for the soil type. Results are affected by the shape of the tires, their air pressure, and the total wheel or axle load. (Results are not affected by tire pressure alone, as is often assumed.) Very heavy units of 50 tons (45.36 metric tons) or more may be effective at compressing rock fills.

Vibration is most effective in sand and gravelly soils, but may increase the effectiveness of a roller in any soil. It is particularly effective in bringing excess moisture to the surface.

Trenches and other small areas may be compacted by impact or by jump rammers, vibratory plate tamps, or trench rollers. Gravel fills may be compacted by puddling.

5.2.0 Compaction Method Considerations

Since a building or roadway is only as stable as the ground on which it is built, soil compaction is one of the most important jobs on any project. Fill that is not properly compacted will settle naturally over time, giving the effect of shrinking. Clay soils that are improperly compacted may absorb moisture and swell. In both cases, the resulting movement of the ground can damage roadways and structures, requiring costly repairs.

Uneven compaction across a fill area can cause the greatest damage. An entire project that settles evenly may cause very little harm, but when part of a foundation settles more than another part, it can severely damage the building foundation (see *Figure 35*). Uneven settling of a roadbed can ruin its entire surface.

Table 1 Compaction Selection Chart

Materials					
		Vibrating Sheepsfoot Rammer	**Static Sheepsfoot Grid Roller Scraper**	**Vibrating Plate Compactor Vibrating Roller Vibrating Sheepsfoot**	**Scraper Rubber-Tired Roller Loader Grid Roller**
	Lift Thickness	Impact	Pressure (with kneading)	Vibration	Kneading (with Pressure)
Gravel	12+ in (30.48+ cm)	Poor	No	Good	Very Good
Sand	10+/– in (25.40 cm)	Poor	No	Excellent	Good
Silt	6+/– in (15.24 cm)	Good	Good	Poor	Excellent
Clay	6+/– in (15.24 cm)	Excellent	Very Good	No	Good

22203-14_T01.EPS

The following should be considered prior to the compaction of any area:

- Moisture content
- Layers
- Contractual requirements

5.2.1 Moisture Content

Moisture content is the single most important variable in soil compaction. Moisture acts as a lubricant, allowing soil particles to easily slide across one another to fill in air gaps. In soil that contains too little moisture, the particles resist sliding. In soils that contain too much moisture, the particles float and permit the gaps to be filled with water. Water cannot be compressed; when it drains, it will leave air gaps. Since both conditions are unacceptable, a great deal of effort goes into determining the correct amount of moisture for each mixture. This amount is called the optimal moisture content.

Soil is commonly sprayed with water as it is being placed in a fill so that it may be compacted to the specified density. Often, however, soil from a cut contains too much moisture to be compacted properly. In this case, a grader or other equipment may need to be used to turn the soil to allow it to dry out. Then when the moisture content is acceptable, the area can be graded and compacted to specification. However, in naturally wet locations or wet weather, the soil may never dry out enough to meet compaction specifications.

5.2.2 Layers

Fill is typically placed in layers, or **lifts**, with the contract specifying the maximum thickness of each lift. As each lift is placed, it is compacted to the desired density, and another lift is added until the specified depth is reached. Common lift thicknesses are 4, 6, 8, and 10 inches (10.16, 15.24, 20.32, and 25.40 centimeters). Compacting equipment is

22203-14_F35.EPS

Figure 35 Effect of uneven foundation settling.

Soil Testing

The hand test is one method to determine if the soil has sufficient moisture to be properly compacted. Pick up a handful of soil, squeeze it in your hand, and then drop it. The soil should easily form a ball. It should break into a couple of pieces when dropped. If the soil is powdery and does not retain its shape, it is too dry. If the soil shatters when dropped, it is too dry. If the soil leaves traces of moisture on your fingers when squeezed, it is too wet for compaction. If it stays in one piece when dropped, it is also too wet.

designed for particular soil types and lift thicknesses, so it is important to check the manufacturer's specifications when selecting equipment. For example, a vibratory smooth drum roller works well for coarse rock fills, but not on cohesive clay, which needs a sheepsfoot drum.

5.2.3 Contractual Requirements

Compaction is usually specifically addressed in the construction specifications. It is important to know and follow these requirements during compaction. Violating contractual specifications places the contractor at risk for legal liability should something go wrong on the project due to compaction errors.

5.3.0 Checking Quality

Each compacted layer must meet design specifications to ensure that the final product will stand up under projected loads. It must also meet smoothness requirements. Each stabilized layer may be checked for the following:

- Compaction or density
- Thickness
- Mix uniformity
- Gradation
- Loadbearing capacity
- Moisture content
- Binder content

The amount of compaction needed to bring the soil to the engineers' requirements is really just a measure of the density of the soil. There are tests that will determine if the compaction performed meets the specifications of the designers.

Since it is unlikely that a machine operator will be required to do these tests, only a brief overview of the tests will be provided in the following paragraphs. The main types of tests in use today are the sand cone test and nuclear testing.

5.3.1 Sand Cone Test

This is the oldest testing method. Many engineers consider this the most accurate testing method. A sand cone test basically involves digging a round hole with a volume of one-tenth of a cubic foot (0.003 cubic meters) in the area being tested and then determining the dry weight of the soil that was removed. The hole is then filled with a calibrated amount of sand using a jar and cone device (*Figure 36*). The dry weight of the soil removed from the hole is divided by the volume of sand needed to fill the hole to determine the density of the compacted soil in pounds per cubic foot (kilograms per cubic meter). This result is compared to the theoretical maximum density to determine the relative density of the soil that was just compacted.

5.3.2 Nuclear Testing

Nuclear testing involves either placing the testing machine on the ground to get a reading or inserting a probe attached to the machine into a small hole drilled into the soil. Either method sends impulses into the soil that are reflected back to the device and recorded. Denser soils absorb more impulses. The more a soil is compacted the fewer impulses are returned. The lab technician then creates a moisture density curve of the site in the same way as the sand cone test.

> **WARNING!**
>
> Nuclear equipment should only be used by trained personnel. Exposure to nuclear materials can cause serious illness. Testing must be done by qualified personnel.

Nuclear devices are being replaced with other types of equipment to avoid the risks posed by nuclear materials. The electrical density gauge (EDG) measures the physical properties of compacted soils used in road beds and foundations (*Figure 37*). This device is battery operated and

Density Check

On-board density meters are available for many machines. These meters measure the density of the soil and provide the information to the operator through a display panel. The sensors measure the reaction of the drum while it is compacting. As the soil stiffens, less energy is transmitted to the ground and more is reflected back into the equipment. The operator receives feedback from instruments that show if compaction has reached the desired level. These readings can be as accurate as other testing methods.

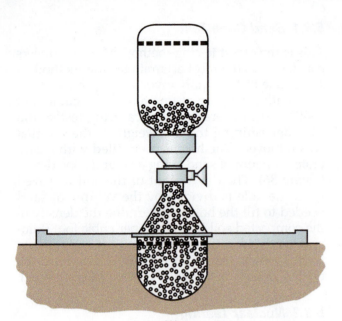

(A)

(B)

22203-14_F36.EPS

Figure 36 Soil density testing equipment.

can determine the wet and dry density, gravimetric moisture content, and percent compaction. The kit includes a console, four electrodes, a hammer, soil sensor and cables, template, temperature probe, and battery charger.

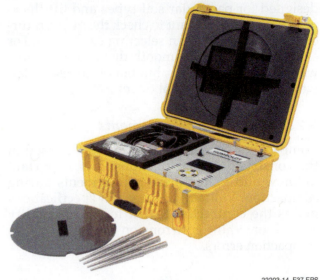

22203-14_F37.EPS

Figure 37 Electrical density gauge.

5.4.0 Leveling and Compacting

A roller equipped with a dozer blade can be used to level and compact an area with a single machine. The dozer blade works to skim soils from high spots and fill in lower areas. The soil is compacted with each pass.

Lower the blade to ground level. Place the transmission in second gear. If the engine slows, lower it to first gear. Doze soils to the desired fill height. When the end of the row is reached, reverse the direction of the machine and place the blade control in the float position. Straddle the previous tracks made with the compactor. Repeat the forward and backward pattern for three or four passes or until the desired compaction is achieved.

5.5.0 Backfilling

Backfilling around culverts or structures must be done carefully so that they are not damaged. The soil should be placed against the structure or pipe in layers, taking care not to concentrate the weight of the fill in one place. Also, be careful not to push against the structure with the material or the blade. To reduce pressure on the material and provide better visibility, raise the blade when approaching the structure.

When backfilling a trench that has pipe or conduit in it, backfill must be placed evenly on both sides of the pipe. Material is usually placed in layers, depending on the compaction required around the pipe. Make sure there is enough fill on top of the pipe or any buried structure before driving the roller over it. Pipe is typically backfilled with a jumping tramp or trench roller that can fit between the pipe and the trench walls.

5.6.0 Compacting Cement

When cement is used, the compaction must be done immediately after mixing because the cement in the mixture will begin to set. A cement mixture also requires that a vibratory roller compaction machine be used to obtain the needed results.

When compacting a lime-stabilized area, the compaction must be done when the moisture content has dropped to a point within the specified range. This may be done immediately or after a short time.

The number of passes made by the compactor is important to the amount of compaction achieved. The number of passes will depend on the moisture content of the material, the layer thickness, the compactor type, and the degree of compaction called for in the specifications. With proper compaction, a final curing stage may be eliminated before the next course is applied.

5.7.0 Compacting Asphalt

It is important that newly placed hot-mix asphalt (HMA) be compacted to a hard, smooth surface. Proper compaction of the asphalt increases its strength and durability, while decreasing deterioration and aging.

As with soil, compacting asphalt compresses the material so that particles in the mix are forced closer together and air voids are reduced. This reduces the overall volume of the asphalt and improves its strength.

Asphalt pavement is usually compacted with static steel-wheel rollers, vibratory steel-wheel rollers, or pneumatic tire rollers. A combination roller with a vibratory drum and pneumatic tires is also used (*Figure 38*).

Factors related to the design of the mix that affect compaction are aggregate properties and cement type. Paving factors that affect the roller's efficiency include the temperature of the mix and the uniformity of the pavement layer. The HMA must be within a specified range in order to achieve optimum compaction or density. Air temperature, surface temperature and layer thickness will also affect the mix temperature. Compaction variables that can be controlled during the process and that have an effect on the level of density include the following:

- Roller speed
- Number of passes
- Rolling zone
- Rolling pattern
- Direction and mode

22203-14_F38.EPS

Figure 38 Combination pneumatic and steel-wheel vibratory roller.

5.7.1 Roller Speed

The more quickly a roller passes over a particular point in the new surface, the less time the weight of the roller spends on any particular point in that surface. This means that as roller speed increases, the amount of compaction decreases. The roller speed that is used will depend on factors such as the thickness of the pavement and the position of the roller in the paving operation. Typical roller speeds range from two to seven miles (3.22 to 11.27 kilometers per hour) per hour.

5.7.2 Number of Passes

A certain number of passes must be made with a roller to adequately compact the asphalt. The exact number of passes needed depends on factors such as the properties of the HMA and the type of roller being used. Usually, a test strip is created at the beginning of a paving operation so that various rollers and patterns can be tested to determine the proper number of passes needed for the job.

5.7.3 Rolling Zone

Obtaining the proper amount of compaction depends on the temperature of the mix being rolled. A basic rule of thumb is that the mix should be above 175°F (79.44°C). Compaction is achieved with hot temperatures, but if the mix is too hot it may become unstable. This means that as long as the mixture is stable enough, the front of the rolling zone should be as close to the paver as possible.

5.7.4 Rolling Pattern

Roller passes must be distributed evenly over the width and length of the **mat**. Proper rolling patterns help ensure that compaction is adequate over the entire surface and not concentrated in certain areas. All too often, the center of the paver lane (the area between the wheel paths) receives greater compaction than the wheel paths and edges do.

The pattern to be used on a specific project should be determined in the beginning by considering the following factors:

- Type of mix
- Type of roller
- Desired density
- Layer thickness
- Temperature

During the rolling process, density can be checked with a nuclear density gauge. If necessary, adjustments can be made then. A typical rolling pattern is shown in *Figure 39*.

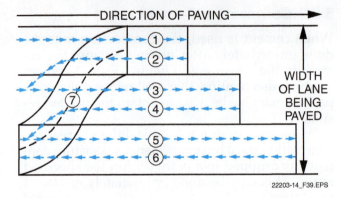

Figure 39 Typical rolling pattern.

5.7.5 Direction and Mode

For vibratory rollers, direction of travel and mode of operation are two other factors that an operator can control. The direction of travel is the orientation of the roller drums to the paver. The mode of operation is either full vibratory mode with both drums vibrating, or combination mode with only one drum vibrating. The direction and mode used generally depends on the type of roller and the type of mixture.

1. A high spot in the soil that carries the full weight of the roll or a low spot that is bridged over and receives no compaction is most likely to occur with a _____.

   a. pneumatic tire roller
   b. smooth steel-wheel roller
   c. sheepsfoot roller
   d. segmented-pad roller

2. Three factors that must be considered prior to a compaction job are _____.

   a. moisture content, sand cone rating, and loadbearing capacity
   b. gradation, binder content, and contractual obligations
   c. soil thickness, shear rating, and mix uniformity
   d. moisture content, layer thickness, and contractual requirements

3. The oldest and possibly the most accurate method for determining soil density is the _____.

   a. hand and ball test
   b. electronic meter test
   c. sand cone test
   d. nuclear test

4. One way to level and compact an area using a single machine is to use a _____.

   a. roller equipped with a dozer blade
   b. segmented-pad on a smooth steel-wheel roller
   c. dropping-weight roller
   d. vibratory compactor

5. When backfilling a trench that has pipe in it, the fill material is usually _____.

   a. mounded high and left to settle on its own
   b. pumped in as a slurry to envelop the pipe
   c. added in layers and carefully compacted
   d. added dry and not compacted at all

6. If cement is used as a soil stabilizer, the compaction must be done _____.

   a. after the cement has set
   b. after about 48 hours
   c. once the HMA drops to 10 percent
   d. immediately after mixing

7. The higher the temperature of hot-mix asphalt, the _____.

   a. less likely it can be compacted to a suitable density
   b. greater the density that can be achieved with compaction
   c. more likely it will shear and stick to a compaction roller
   d. longer the time that must be allowed before compaction

SUMMARY

Different types of compaction equipment are used for compacting and smoothing materials such as earth, gravel, cement, and asphalt. They use static and dynamic force to compress the material to the desired density to provide strength so that the specified loads can be carried without settlement.

Among the more common types of compaction equipment are pneumatic tire rollers, steel-wheel rollers, vibratory compactors, and sheepsfoot rollers. Each piece of equipment is used for specific types of work.

On self-propelled equipment, vehicle movement is controlled with the steering wheel, accelerator, and brake pedals. If the roller is equipped with a blade, it is controlled with levers or a joystick, as well as switches. Operator manuals provide a great deal of information to enable operators to become familiar with the machine they are using.

Safety considerations when operating a roller include keeping the roller in good working condition, obeying all safety rules, being aware of other people and equipment in the area, and not taking chances. Perform inspections and maintenance daily to keep the roller in good working order.

The type of roller selected depends on the material being compacted. Some rollers are more effective with certain types of soils, while others are primarily used for asphalt. Operators must understand the properties of soil that contribute to compaction. Moisture content is one of several factors that contribute to good compaction. Once the soil is compacted, it must be tested to ensure that sufficient density has been achieved.

1. Soil consists of irregular particles, and the spaces between those particles are filled by air and _____.

   a. sand
   b. bacteria
   c. water
   d. dirt

2. A measure of how closely soil particles are packed together is called _____.

   a. psi
   b. plasticity index
   c. compaction
   d. density

3. How many basic roller designs are there?

   a. Two
   b. Four
   c. Six
   d. Ten

4. The compaction from a pneumatic tire roller occurs because of _____.

   a. weight and movement
   b. only tire pressure
   c. vibratory forces
   d. wheel loading and tire pressure

5. On a steel-wheel roller, variable speed control and dynamic braking action are made possible by the _____.

   a. hydrostatic drive
   b. carbon fiber rollers
   c. vibration resistance
   d. tamping blades

6. Vibratory compactors are ineffective in _____.

   a. granular soils
   b. clay
   c. sand
   d. wet areas

7. Typically, the length of the feet on segmented-pad rollers is _____.

   a. 1 to 3 inches (2.54 to 7.62 centimeters)
   b. 3 to 5 inches (7.62 to 12.70 centimeters)
   c. 7 to 9 inches (17.78 to 22.86 centimeters)
   d. 12 inches (30.48 centimeters) or more

8. On a self-propelled roller, a ROPS is most likely to be found over the _____.

   a. instrument panel
   b. exterior switch compartment
   c. digital display cluster
   d. operator's cab

9. In order to allow the hydraulic oil to warm up sufficiently before a roller is operated under load, the hydraulic system should be _____.

   a. isolated temporarily
   b. cycled under low rpm
   c. heated using glow plugs
   d. pressurized with an onboard circulator

10. Roller movement on a self-propelled compactor is normally controlled with _____.

    a. joysticks and triggers
    b. levers and knobs
    c. a steering wheel and switches
    d. a steering wheel and pedals

11. The fuel shutoff switch on a roller physically prevents fuel from flowing from the _____.

    a. fuel tank into the supply lines
    b. supply lines into the filter
    c. gas pump into the fuel tank
    d. filter into the injectors

12. Water spray units are typically used for rolling _____.

    a. sand
    b. gravel
    c. asphalt
    d. clay

13. The intended use of the secondary brake system on a roller is to _____.

    a. dampen the speed of the roller during operation
    b. hold the roller in a parked position
    c. protect the primary brake system from overuse
    d. control the speed of the roller attachment

14. Unlike automobile tires, heavy equipment tires can explode with devastating force because they _____.

    a. contain highly flammable additives
    b. are filled with calcium chloride
    c. are mounted on a split rim
    d. are filled to extremely high pressures

15. To ensure that fluid levels are being accurately indicated during a daily inspection, operators should start by _____.

    a. zeroing all gauges on the instrument panel
    b. activating the disconnect switches
    c. changing all the fluids beforehand
    d. parking the machine on a level surface

16. Continuous and hard operation of a roller's hydraulic system can heat the hydraulic fluid to the boiling point and cause it to _____.

    a. break down
    b. condense
    c. mat
    d. liquefy

17. Before starting any roller operation, it is important to become familiar with the area of operations and the _____.

    a. prior rolling pattern
    b. soil conditions
    c. sand cone results
    d. soil HMA

18. To help warm up the engine before ignition, many diesel engines have devices called

    _____.

    a. hydrostatic transmission
    b. heater coils
    c. glow plugs
    d. starter switches

19. Before moving forward with a roller, an operator should first _____.

    a. disengage the parking brake, start the engine, and depress the accelerator
    b. move the shift lever forward, engage the brake, and raise the blade
    c. lower the blade, disengage the parking brake, and deactivate the engine shutoff
    d. adjust the seat, fasten the seatbelt, and raise to blade to clear obstructions

20. Whether a roller is facing forward or backward on a transport trailer will depend on the recommendation of the _____.

    a. truck driver
    b. roller manufacturer
    c. state highway department
    d. National Transportation Safety Board

21. The segmented-pad roller on a sheepsfoot compacts soil by applying pressure and some _____.

    a. kneading
    b. vibration
    c. impacting
    d. cohesion

22. A soil compaction method that is particularly effective at bringing excess moisture to the surface is _____.

    a. kneading
    b. tamping
    c. vibration
    d. puddling

23. The density of a compacted soil layer is commonly checked using the sand cone test or _____.

    a. nuclear testing
    b. the method only test
    c. plasticity testing
    d. the method and result test

24. When an area is being leveled, soil can be skimmed from high spots to fill lower areas with the use of a _____.

    a. segmented pad
    b. ballast mat
    c. wheel scraper
    d. dozer blade

25. When backfilling a culvert trench, be careful not to _____.

    a. cover the pipe
    b. damage the culvert
    c. over-compact the soil
    d. let in any moisture

26. In order to obtain the needed results from a cement mixture, it is necessary to use _____.

    a. impact compaction
    b. kneading
    c. vibratory roller compaction
    d. tamping

27. As roller speed increases, the amount of compaction achieved with each pass _____.

    a. decreases
    b. increases significantly
    c. increases slightly
    d. is not affected

28. Test strips are performed at the beginning of asphalt compaction to help determine the _____.

    a. mix properties of the asphalt
    b. natural mat settling properties
    c. amount of interparticle friction
    d. proper number of passes needed

29. In HMA compaction, the front of the rolling zone should be _____.

    a. in front of the paving zone
    b. at least 10 feet (3.05 meters) behind the paver
    c. as close to the paver as possible
    d. as far away from the paver as possible

30. Proper rolling patterns help ensure that _____.

    a. each section is only compacted once
    b. compaction is adequate over the entire surface
    c. compaction is concentrated in certain areas
    d. the roller does not have to backtrack

Trade Terms Introduced in This Module

Compaction: Using an engineered process, such as rolling, tamping, or soaking, to reduce the bulk and increase the density of soil.

Density: The ratio of the weight of a substance to its volume.

Foot: In tamping rollers, one of a number of projections from a cylindrical drum that contact the ground.

Ground contact pressure (GCP): The weight of the machine divided by the area in square inches of the ground directly supporting it.

Lift: A layer of material that is a specific thickness; the depth of material that is being rolled by the roller.

Mat: Asphalt as it comes out of a spreader box or paving machine in a smooth, flat form.

Pad: On a segmented or sheepsfoot roller, the part of the roller that contacts the ground; also called the foot.

Puddling: A process in which water is added to the soil until it is semi-liquid; the soil is then allowed to dry before being vibrated.

Sheepsfoot: A tamping roller with feet expanded at the outer tips.

Settling: The natural wetting and drying process whereby soil particles become more compact and denser.

Tamping roller: One or more steel drums fitted with projecting feet and towed with a box frame.

Vibratory roller: A compacting device that mechanically vibrates the soil while it rolls. It can be self-propelled or towed.

Figure Credits

Section Review Answers

Answer	Section Reference	Objective
Section One		
1. c	1.1.0	1a
2. b	1.2.0	1b
3. d	1.3.0	1c
4. a	1.4.0	1d
Section Two		
1. d	2.1.0	2a
2. b	2.2.6	2b
3. a	2.3.1	2c
4. c	2.4.1	2d
Section Three		
1. b	3.1.0	3a
2. d	3.2.0	3b
3. a	3.3.0	3c
4. c	3.4.0	3d
Section Four		
1. b	4.1.0	4a
2. d	4.2.2	4b
Section Five		
1. b	5.1.0	5a
2. d	5.2.0	5b
3. c	5.3.1	5c
4. a	5.4.0	5d
5. c	5.5.0	5e
6. d	5.6.0	5f
7. b	5.7.0	5g

NCCER CURRICULA — USER UPDATE

NCCER makes every effort to keep its textbooks up-to-date and free of technical errors. We appreciate your help in this process. If you find an error, a typographical mistake, or an inaccuracy in NCCER's curricula, please fill out this form (or a photocopy), or complete the online form at **www.nccer.org/olf**. Be sure to include the exact module ID number, page number, a detailed description, and your recommended correction. Your input will be brought to the attention of the Authoring Team. Thank you for your assistance.

Instructors – If you have an idea for improving this textbook, or have found that additional materials were necessary to teach this module effectively, please let us know so that we may present your suggestions to the Authoring Team.

NCCER Product Development and Revision

13614 Progress Blvd., Alachua, FL 32615

Email: curriculum@nccer.org
Online: www.nccer.org/olf

❏ Trainee Guide   ❏ Lesson Plans   ❏ Exam   ❏ PowerPoints   Other _____

Craft / Level: _____   Copyright Date: _____

Module ID Number / Title: _____

Section Number(s): _____

Description: _____

_____

_____

_____

_____

Recommended Correction: _____

_____

_____

_____

_____

Your Name: _____

Address: _____

_____

Email: _____   Phone: _____

22303-14

Backhoes

OVERVIEW

There are several choices of equipment for excavation work. The choice mainly depends on the job to be done. However, one of the most popular items of equipment on a job site is the rubber-tire tractor-mounted backhoe. Operators of a backhoe need to be aware that they are operating a potentially dangerous piece of equipment. They should know all the parts of the equipment and the operating procedures for the particular job. Since safety is a main job requirement, operators should know how to conduct daily preventive maintenance on the machine and practice safety at all times.

Module Three

Trainees with successful module completions may be eligible for credentialing through NCCER's National Registry. To learn more, go to **www.nccer.org** or contact us at **1.888.622.3720**. Our website has information on the latest product releases and training, as well as online versions of our *Cornerstone* magazine and Pearson's product catalog.

Your feedback is welcome. You may email your comments to **curriculum@nccer.org**, send general comments and inquiries to **info@nccer.org**, or fill in the User Update form at the back of this module.

This information is general in nature and intended for training purposes only. Actual performance of activities described in this manual requires compliance with all applicable operating, service, maintenance, and safety procedures under the direction of qualified personnel. References in this manual to patented or proprietary devices do not constitute a recommendation of their use.

Objectives

When you have completed this module, you will be able to do the following:

1. Identify and describe common uses and types of backhoes.
   a. Describe common uses of a backhoe.
   b. Identify types and configurations of backhoes.
2. Identify and describe the components, controls, and attachments on a typical backhoe.
   a. Describe the major parts of a backhoe.
   b. Describe the instruments and controls.
   c. Describe the backhoe controls.
   d. Describe common backhoe attachments.
3. Identify and describe safety, inspection, and service guidelines associated with a backhoe.
   a. Describe rules pertaining to safety.
   b. Describe daily inspection checks.
   c. Describe the servicing requirements for a backhoe.
4. Describe basic operating procedures for a backhoe.
   a. Identify factors for effective backhoe operation.
   b. Identify steps for preparing to work with a backhoe.
   c. Identify steps for performing basic maneuvers with a backhoe.
5. Identify and describe common work activities for a backhoe.
   a. Describe loading from a stockpile with a backhoe.
   b. Describe trenching and loading with a backhoe.
   c. Describe demolition using the hydraulic breaker.
   d. Describe setting pipe using a backhoe.
   e. Describe excavating footings and foundations with a backhoe.
   f. Describe working with a backhoe in confined or unstable areas.
   g. Describe basic procedures for roading and transporting a backhoe.

Performance Tasks

Under the supervision of your instructor, you should be able to do the following:

1. Demonstrate a proper prestart inspection of a backhoe.
2. Perform a proper startup, warm-up, and shutdown of a backhoe.
3. Perform basic backhoe maneuvers, including forward movement, moving in reverse, turning, and operating the front loader bucket.
4. Perform the operation of setting up a backhoe, using stabilizers, and digging with the bucket.
5. Perform an excavation of a trench 20 to 40 feet (6 to 12 meters) long with spoil piles at least 2 feet (0.6 meters) from the edge.

Trade Terms

Bucket
Crowd
Curl
Draw
Extend

Reach
Retract
Rubble
Spoils
Trenching

Industry Recognized Credentials

If you are training through an NCCER-accredited sponsor, you may be eligible for credentials from NCCER's Registry. The ID number for this module is 22303-14. Note that this module may have been used in other NCCER curricula and may apply to other level completions. Contact NCCER's Registry at 888.622.3720 or go to **www.nccer.org** for more information.

Contents ————————————

Topics to be presented in this module include:

1.0.0 Uses and Types of Backhoes... 1
  1.1.0 Common Uses of a Backhoe 1
  1.2.0 Types and Configurations of Backhoes.......................... 1
2.0.0 Identification of Equipment 4
  2.1.0 Major Parts of a Backhoe... 4
    2.1.1 Operator's Cab 5
  2.2.0 Instruments and Controls.. 5
    2.2.1 Voltmeter... 8
    2.2.2 Transmission Oil Temperature Gauge................... 8
    2.2.3 Fuel Level Gauge 8
    2.2.4 Engine Coolant Temperature Gauge 9
    2.2.5 Tachometer... 9
  2.3.0 Backhoe Controls... 9
    2.3.1 Two-Lever Backhoe Controls............................. 10
    2.3.2 Four-Lever Backhoe Controls............................. 11
    2.3.3 Joystick Controls 12
  2.4.0 Backhoe Attachments.. 13
    2.4.1 Ripper 13
    2.4.2 Auger 13
    2.4.3 Street Pads 13
    2.4.4 Hydraulic Breaker 14
    2.4.5 Compactor 14
    2.4.6 Cold Planer... 14
    2.4.7 Buckets 14
3.0.0 Safety, Inspections, and Maintenance 17
  3.1.0 Safety Guidelines.. 17
    3.1.1 Operator Safety 17
    3.1.2 Safety of Co-Workers and the Public.................. 17
    3.1.3 Equipment Safety 18
  3.2.0 Basic Inspections and Preventive Maintenance............ 19
    3.2.1 Daily Inspection Checks 19
  3.3.0 Servicing a Backhoe 20
    3.3.1 Preventive Maintenance Records........................ 21
4.0.0 Basic Operation 23
  4.1.0 Suggestions for Effective Backhoe Operation 23
  4.2.0 Preparing to Work 24
    4.2.1 Startup... 24
    4.2.2 Checking Gauges and Indicators........................ 25
    4.2.3 Shutdown... 26

4.3.0 Basic Maneuvering 26

4.3.1 Positioning the Backhoe 26

4.3.2 Moving Backward 26

4.3.3 Positioning the Stabilizers 27

4.3.4 Setting Up on Level Ground 27

4.3.5 Setting Up on Sloping Terrain 28

4.3.6 Pushing the Backhoe Forward 28

4.3.7 Pulling the Backhoe Backward 29

5.0.0 Work Activities 30

5.1.0 Loading from a Stockpile Using the Backhoe 30

5.2.0 Trenching and Loading 31

5.3.0 Demolition Using the Hydraulic Breaker 32

5.4.0 Setting Pipe 33

5.5.0 Excavating Footings and Foundations 35

5.6.0 Working in Confined or Unstable Areas 36

5.6.1 Digging Between Objects 37

5.6.2 Trench Digging 37

5.6.3 Trenching Next to Walls 38

5.6.4 Working on Unstable Soils 38

5.7.0 Roading and Transporting the Backhoe 39

5.7.1 Roading 39

5.7.2 Transporting 40

Appendix Maintenance Interval Schedule 47

Figures

Figure 1   Backhoe..2
Figure 2   Backhoe digging a narrow ditch...............................2
Figure 3   Excavating arm..3
Figure 4   Basic parts of a typical backhoe 4
Figure 5   Range of motion 4
Figure 6   Cab front position..5
Figure 7   Cab position for backhoe operations........................6
Figure 8   Typical front and side control panels........................7
Figure 9   Vandal cover 8
Figure 10 Instrument panel 9
Figure 11 Stabilizer controls...10
Figure 12 Two-lever backhoe controls 11
Figure 13 Positions for two-lever backhoe controls 11
Figure 14 Standard backhoe control pattern............................12
Figure 15 Alternate backhoe control pattern (excavator pattern) 12
Figure 16 Four-lever backhoe controls 12
Figure 17 Joystick backhoe controls 12
Figure 18 Ripper attachment 13
Figure 19 Auger attachment 14
Figure 20 Street pads..14
Figure 21 Reversible pads 15
Figure 22 Breaker attachment 15
Figure 23 Plate tamper attachment..15
Figure 24 Cold planer attachment...16
Figure 25 Mount using three points of contact.........................18
Figure 26 Establish a safe work zone.....................................18
Figure 27 Check all fluids daily..19
Figure 28 Top off fluids as needed...20
Figure 29 Check the hydraulic fluid level.................................21
Figure 30 Lubricate all pivot points 21
Figure 31 Loading a truck 23
Figure 32 Engine key switch..24
Figure 33 Transmission control..25
Figure 34 Oil pressure light...25
Figure 35 Service brakes 27
Figure 36 Loader joystick control...27
Figure 37 Backhoe with stabilizers lowered.............................28
Figure 38 Loading with the backhoe.......................................31
Figure 39 Secure large pieces with the bucket and stick 31
Figure 40 Position the stick..32
Figure 41 Position the bucket 32
Figure 42 Work the bucket parallel to the ground 32
Figure 43 Finishing...33
Figure 44 Hydraulic breaker 33

Figure 45 Rig a sling to the bucket..34
Figure 46 Setting pipe...34
Figure 47 Backfilling..35
Figure 48 Starting position...35
Figure 49 Position for second cut 36
Figure 50 Digging out a corner 36
Figure 51 Excavating the second half of the foundation 36
Figure 52 Completed excavation 37
Figure 53 Position backhoe to dig between objects......................................38
Figure 54 Excavate the center section...38
Figure 55 Position the backhoe 38
Figure 56 Boom lock. 39
Figure 57 Roading the backhoe/loader 40
Figure 58 Chock trailer wheels 40
Figure 59 Secure the machine on the trailer...40

SECTION ONE

1.0.0 USES AND TYPES OF BACKHOES

Objective

Identify and describe common uses and types of backhoes.
  a. Describe common uses of a backhoe.
  b. Identify types and configurations of backhoes.

Trade Terms

Bucket: A U-shaped closed-end scoop that is attached to the backhoe.

Extend: Move the extendable stick outward.

Spoils: Excavated material from a digging operation.

Trenching: Using a backhoe to dig a long, straight excavation with vertical walls.

The tractor-mounted backhoe is smaller than most other hydraulic excavating equipment. It has a limited swing compared to the swing of the tracked excavator or telescoping excavator, and is less powerful. Nevertheless, it is a useful piece of equipment because it is very mobile and maneuverable.

The backhoe attachment is usually mounted on a rubber-tired two- or four-wheel-drive tractor. It may also be fitted onto a crawler tractor or a truck. The tractor usually has permanent mountings, such as a three-point hitch, to which the backhoe can be attached with pins. Most models, however, are a combination backhoe and loader, such as the one shown in *Figure 1*.

1.1.0 Common Uses of a Backhoe

Although the tractor-mounted backhoe is smaller than the tracked excavator, it is able to do certain

---

What's in a Name?

It is common to think that the backhoe takes its name from the fact that it has a hoe-type attachment at the rear of the tractor. It has been argued that the backhoe is so-called because it draws earth back toward itself, rather than lifting it forward as the excavator does.

---

tasks better. Its compact size allows it to be used in areas that are hard to reach for most excavators. Because it has no tail swing, it can work in small spaces. The rubber tires allow it to cross lawns without causing damage. It can also be driven a short distance to and from jobs on public roads.

Disadvantages of the backhoe include its limited swing and the difficulty in keeping it stable. The backhoe has a swing of only about 200 degrees and may require frequent moving on certain jobs. Because this machine is lightweight compared to its digging power, it is harder to keep stable than an excavator.

The main uses of a backhoe include the following:

- Digging and cleaning narrow ditches (*Figure 2*)
- Excavating smaller areas where it is not practical for other equipment
- Light-duty hoisting
- Loading trucks

1.2.0 Types and Configurations of Backhoes

There are many sizes and models of backhoes. Most have **bucket** sizes for nearly any job. Special round bottom, V-shaped, rock, and trash buckets are also available. A small backhoe/loader combination may be the only equipment needed on a small **trenching** project. A backhoe is also a good piece of equipment for digging and setting manholes. There are several different attachments that can be fitted onto the backhoe, including rippers, augers, and hydraulic breakers. The backhoe/loader also has the front bucket loader and attachments that can be fitted onto the front of the machine. With all of these options, the backhoe/loader becomes a very versatile piece of equipment. It is probably the most common piece of heavy equipment purchased and used by small contractors in North America.

A backhoe uses a hydraulically operated excavating arm that is mounted on the back of a crawler or wheel-type tractor (*Figure 3*). The most commonly used configuration today—the backhoe/loader—combines a tractor with a loader and a backhoe. The backhoe is attached to the back of a rubber-tired tractor with a loader bucket on the front. The tractor is used to maneuver the machine. The backhoe is used for excavation. The loader is used to remove the **spoils** and backfill the excavation. This versatility makes the backhoe/loader very popular for small- and medium-sized jobs that do not need the power of larger excavators.

Figure 1 Backhoe.

The extendable stick, also called extenda-dig option, extenda-hoe, or E-stick, is available on some backhoes. The backhoe in *Figure 3* is using the extendable stick feature. This allows operators to hydraulically **extend** a backhoe boom with a foot pedal in the cab. It gives the operator up to 4 feet (1.2 meters) of additional reach. This is useful when unforeseen obstacles arise and also decreases the number of times the backhoe has to be repositioned in standard trenching applications. Extendable stick backhoes offer enhanced productivity; however, proper maintenance is vital to ensure long life. Keep it well lubricated and check it periodically to ensure that tolerances are correct.

Other backhoe configurations include the following:

- Crawler-type, hydraulic
- Multi-purpose hydraulic
- Trenching machine with backhoe accessory

The backhoe attachment has a bucket attached to a pivoting base. Hydraulic outriggers or stabilizers are hinged on each side of the base. Some backhoe attachments have their own operator's seat. Others have the seat and controls in the cab of the tractor. The tractor seat turns to face the attachment when it is in operation.

22303-14_F02.EPS

Figure 2 Backhoe digging a narrow ditch.

REPRINTED COURTESY OF CATERPILLAR INC.

22303-14_F03.EPS

Figure 3 Excavating arm.

1.0.0 Section Review

1. A backhoe may require frequent moving on certain jobs because it has a swing of only about _____.

   a. 90 degrees
   b. 180 degrees
   c. 200 degrees
   d. 270 degrees

2. The most commonly used backhoe configuration combines an excavating arm and a loader that are mounted onto a _____.

   a. dozer
   b. truck
   c. grader
   d. tractor

2.0.0 IDENTIFICATION OF EQUIPMENT

Objective

Identify and describe the components, controls, and attachments on a typical backhoe.
- a. Describe the major parts of a backhoe.
- b. Describe the instruments and controls.
- c. Describe the backhoe controls.
- d. Describe common backhoe attachments.

Trade Terms

Crowd: The process of forcing the stick into digging, or moving the stick closer to the machine.

Curl: Rotate the bucket.

Reach: Extend the stick away from the cab.

2.1.0 Major Parts of a Backhoe

The basic parts of a typical backhoe are shown in *Figure 4*. The operator sits in the cab that is contained within the rollover protective structure (ROPS). Some operator's cabs are fully enclosed and are known as enclosed rollover protective structures (EROPS). The stabilizers, boom, stick, and bucket are all hydraulically controlled. The two stabilizers at the base of the excavating arm can be raised and lowered. The excavating arm is divided into two sections, known as the boom and stick, which can be raised and lowered. The bucket is attached to the end of the excavating arm, and can be extended and retracted.

The excavation arm can swing approximately 200 degrees side-to-side. This allows the operator to dig and dump spoils to either side of the excavation. The cylinder that operates the boom can be retracted (boom up) for traveling or extended (boom down) for digging. The stick cylinder can be retracted and extended for reaching and crawling during excavation. Finally, the bucket can be extended and retracted for digging and dumping. The range of motion for a typical backhoe/loader is shown in *Figure 5*.

REPRINTED COURTESY OF CATERPILLAR INC.

22303-14_F05.EPS

Figure 5 Range of motion.

DIPPER HYDRAULIC CYLINDER — BOOM — ROLLOVER PROTECTIVE STRUCTURE (ROPS) — OPERATOR'S CAB

BUCKET/DIPPER — DIPPER ARM (STICK) — STABILIZER/OUTRIGGER — LOADER BOOM — FRONT LOADER BUCKET

22303-14_F04.EPS

Figure 4 Basic parts of a typical backhoe.

Backhoes are powered by engines. The engines can be gasoline or diesel. The tractor portion of the backhoe/loader is similar to what was covered in HEO Level One, *Utility Tractors*. Always read the operator's manual to understand the details of your equipment's operation and maintenance requirements.

A loader bucket is attached to the front of a backhoe/loader. The operation of a loader was covered in HEO Level Two, *Loaders*. Review that module if you are unfamiliar with the operation of a loader. This section will focus on backhoe operations and only briefly review the operation of the loader.

2.1.1 Operator's Cab

The operator's cab is the central hub for backhoe and loader operations. An operator must understand the controls and instruments before operating a backhoe. Study the operator's manual to become familiar with the controls and instruments and their functions.

The cab in a typical backhoe/loader is designed so that the operator can face forward or backward. The front position faces the end of the machine with the loader bucket, as shown in *Figure 6*. The operator faces the steering wheel, throttle, brakes, and other controls used to maneuver the machine. The operator controls the loader and maneuvers the machine from this position.

The operator's seat can be turned around to face the backhoe. This position provides easy access to the controls for backhoe operations, as shown in *Figure 7*. Always adjust the seat properly and only operate controls from the designated seat position.

Many of the instruments and controls needed for all operations are contained on a side panel that can be seen from either position. It may take a greater effort to monitor these gauges as they are not directly in front of the operator. The cab is designed for maximum visibility. Most cabs have large windshields that extend to the floor of the cab to allow the operator to see the area around the machine. Backhoe/loaders often work with one or more dump trucks or other equipment. It

REPRINTED COURTESY OF CATERPILLAR INC.

22303-14_F06.EPS

Figure 6 Cab front position.

is important that the operator be able to see the movement of other vehicles that are often positioned to the side of the backhoe/loader. Lights aid vision at night or in poor lighting conditions.

> **WARNING!**
>
> The windshield and lights must be inspected daily and cleaned whenever they become dirty. Equipment damage or injury to other workers can occur if the operator cannot see clearly in all directions.

2.2.0 Instruments and Controls

Location of the backhoe's controls will vary with manufacturer and model. A typical backhoe/loader is maneuvered with the operator facing the loader controls. The controls and instruments on the front control panel are used to move and position the machine. The instruments and controls on the side panels are used to operate the backhoe. *Figure 8* shows the front and side control panels for a typical backhoe. The operator must

---

Where Did It Start?

The invention of the backhoe is credited to Joseph C. Bamford, an Englishman who developed the first model in 1953. He started by adding a hoe attachment to the back of a tractor and then mounted a loading bucket to pick up the spoils. Today, it is one of the most popular items of equipment in construction and agriculture. Although it is generally called a backhoe or backhoe/loader in the United States, in England and on the European continent, it is commonly called a JCB after its inventor. Bamford is the only non-American ever inducted into the USA Construction Industry Hall of Fame.

Figure 7 Cab position for backhoe operations.

understand all the controls of the backhoe before moving the machine.

The basic controls for maneuvering include the following:

- *Parking brake* – Applies the brakes of the tractor when machine is parked.
- *Glow plug switch* – Activates glow plugs for engine preheating in diesel engines.
- *Key switch* – Turns on the electrical system and provides electrical power to the engine.
- *Service brakes* – Applies pressure to rear brakes (drive wheels). Usually there are two pedals, one for each wheel, which allows them to be used for steering. For roading the equipment, the pedals can be locked together to operate as one unit.
- *Transmission direction control* – Operates forward and reverse gears and may control gear selection.
- *Transmission speed control* – Alternate control configuration for gear selection.
- *Foot throttle* – Controls engine speed.
- *Steering wheel* – Controls the direction of the front wheels for steering control.

- *Differential lock* – Locks in the rear differential on both wheels for better traction.

Side panel controls for operating the backhoe include the following:

- *Stabilizer controls* – Raises and lowers stabilizers.
- *Boom lock* – Locks the boom in the retracted position when the boom is not in use.
- *Hand throttle* – Controls engine speed for backhoe operations.

> **WARNING!**
>
> The hand throttle may not be used for forward operation.

All backhoes also have indicator lights and an alert system similar to tractors and loaders. Review the operator's manual and be familiar with all alert and indicator lights before operating the machine.

FRONT PANEL AND CONTROLS

SIDE PANEL AND CONTROLS

22303-14_F08.EPS

Figure 8 Typical front and side control panels.

The primary instrument panel for backhoe operations is the side panel. This panel includes the engine start and glow plug switches used to start the equipment. In some models, a vandal cover (*Figure 9*) slides over the instrument panel and is locked to prevent unauthorized access. On other models, the entire instrument panel, including the engine start switch, rotates and locks so that unauthorized users cannot operate the equipment. *Figure 9* shows the instrument panel in working position (A), the instrument panel in rotation (B), and the vandal cover in the secured position (C).

NOTE

Always secure any vandal covers or other anti-theft protections before leaving the machine unattended.

The side panel contains several light switches, a set of alert indicators, cab comfort controls, and several gauges. The gauges include the voltmeter, transmission oil temperature gauge, fuel level gauge, the engine coolant temperature gauge, and the tachometer (see *Figure 10*). These instruments are explained in the following sections.

NOTE

The callout numbers in the following sections refer to *Figure 10*.

2.2.1 Voltmeter

The voltmeter (1) shows the voltage in the charging system. A needle shows the voltage over a scale with low on the left and higher values on the right. There is a red warning range on both sides, as either too high or too low voltage can cause operational problems. This instrument is optional on some models.

2.2.2 Transmission Oil Temperature Gauge

The transmission oil temperature gauge (2) indicates the temperature of the oil flowing through the transmission. This gauge also reads left to right in increasing temperature. It has a red zone that indicates excessive temperatures.

A

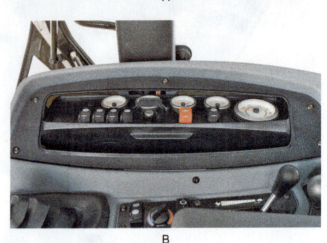

B

C

REPRINTED COURTESY OF CATERPILLAR INC.

22303-14_F09.EPS

Figure 9 Vandal cover.

2.2.3 Fuel Level Gauge

The fuel level gauge (3) indicates the amount of fuel in the fuel tank. On diesel engine loaders, the gauge may contain a low fuel warning zone. Some models have a low fuel warning light.

REPRINTED COURTESY OF CATERPILLAR INC.

22303-14_F10.EPS

Figure 10 Instrument panel.

> **NOTE**
>
> Avoid running out of fuel on diesel engines because the fuel lines and injectors must be bled of air before the engine can be restarted.

2.2.4 Engine Coolant Temperature Gauge

The coolant temperature gauge (4) indicates the temperature of the coolant flowing through the cooling system. Refer to the operator's manual to determine the correct operating range for normal operations. Temperature gauges normally read left to right, with cold on the left and hot on the right. Most gauges have a red zone. If the needle is in the red zone, the coolant temperature is excessive. Some machines may also activate warning lights if the engine overheats.

> **CAUTION**
>
> Operating equipment when temperature gauges are in the red zone may severely damage it. Stop operations, determine the cause of the problem, and resolve it before continuing operations.

2.2.5 Tachometer

The tachometer (5) indicates the engine speed in revolutions per minute (rpm). Most tachometers are marked in hundreds on the meter face and read left to right in an increasing scale. Again, there is a red zone on the high end of the scale that indicates that the engine is overspeeding. On some backhoe/loaders there is also an indicator light that will warn the operator if the engine speed is too high.

2.3.0 Backhoe Controls

The controls for the operation of the backhoe are located behind the operator's seat in a normal wheel-type tractor configuration. The operator swings the seat around to face these controls before beginning operation. In this position, the operator is looking at the controls and the back of the machine where the backhoe is mounted. Before operating the backhoe, the operator must adjust the seat and fasten the seat belt. The controls are designed so that they are within easy reach of operator's seat.

Although there are different control configurations, there are functions common to all backhoes. The controls for normal backhoe operations include the following:

- *Boom control* – Controls the raising and lowering of the boom at the main hinge point.
- *Bucket control* – Controls the **curl** and dump motion of the backhoe bucket.
- *Stick control* – Controls **reach** and **crowd** of the stick.
- *Hand throttle* – Controls the engine speed for backhoe operation. This control may not be used for forward operation.
- *Left and right swing controls* – Control the swing of the boom from side to side.
- *Left-hand stabilizer* – Controls the movement of the left stabilizer.
- *Right-hand stabilizer* – Controls the movement of the right stabilizer.

The stabilizers are controlled hydraulically with two levers or toggle switches (*Figure 11*). The left lever controls the left stabilizer and the right lever controls the right stabilizer. The center position is the default hold position.

> **WARNING!**
>
> The stabilizers help keep the backhoe level and secure. Raising the stabilizers could cause the backhoe to become unstable and fall into the excavation. Use extreme caution when raising the stabilizers. Always set the parking brake before raising the stabilizers when operating on a slope.

The boom, stick, and bucket can be controlled by either levers or joysticks. There are several styles of backhoe controls, including two-lever, three-lever, four-lever, and joystick configurations. The boom swing function can also be controlled with a foot pedal. The controls can be set for different patterns within these control styles. It is very important to review the operator's manual before operating any machine. Do not assume control functions are the same for every piece of equipment.

2.3.1 Two-Lever Backhoe Controls

The operator's cab with two-lever backhoe controls is shown in *Figure 12*. However, all two-lever control systems are not set up in the same way. The backhoe function activated by moving the lever to a certain position may vary from one machine to another, depending on the control configuration. Always read the operator's manual before operating a machine.

Two levers are visible to the operator (*Figure 13*). Each lever has five positions: forward, back, left, right, and center. These positions and their corresponding arm movements are as shown in *Figure 13*. The center position is the hold position for all two-lever control configurations. When the control is in the center position, the movement will stop. The control lever will return to the hold position when it is released.

> **NOTE**
>
> Some two-lever controls are designed to move in a diagonal pattern. If the operator is facing north, the positions would be northeast, southwest, northwest, southeast, and center. These positions are similar to those described, only rotated 45 degrees clockwise.

There are two common configurations for two-lever backhoe controls. In the first, the boom and swing are controlled with one lever, and the stick and bucket are controlled with the other. In the second, one lever controls the stick and boom swing, and the other controls the boom and bucket. On some backhoes, the operator can set the controls for either of these patterns. Some machines may have a different control pattern. Read

the operator's manual to find the control configuration for the machine being operated.

In the standard pattern for two-lever backhoe controls, the boom and swing are controlled with the left lever. The stick and bucket are controlled with the right lever. This control pattern is shown in *Figure 14*. However, this is not the case in all machines.

The control positions and backhoe functions for the left lever are as follows:

- *Forward (A): Lower boom* – The boom will lower when the lever is moved to this position.
- *Back (B): Raise boom* – The boom will raise when the lever is moved to this position.
- *Left (C): Swing left* – The boom will swing to the left when the lever is moved to this position.
- *Right (D): Swing right* – The boom will swing to the right when the lever is moved to this position.

The control positions for the right lever are as follows:

- *Forward (E): Stick out* – The stick will move away from the boom when the lever is moved to this position.
- *Back (F): Stick in* – The stick will crowd toward the boom when the lever is moved to this position.
- *Left (G): Bucket load* – The bucket will curl toward the operator when the lever is moved to this position.
- *Right (H): Bucket dump* – The bucket will open outward when the lever is moved to this position.

One two-lever control pattern is sometimes known as the backhoe pattern, as shown in *Figure 15*. The left lever moves the boom and swings the boom. The right lever controls the stick and bucket.

The control positions and backhoe functions for the left lever are as follows:

- *Forward (A)* – Lower boom
- *Back (B)* – Raise boom
- *Left (C)* – Swing left
- *Right (D)* – Swing right

The control positions for the right lever are as follows:

- *Forward (E)* – Lower boom
- *Back (F)* – Raise boom
- *Left (G)* – Bucket load
- *Right (H)* – Bucket dump

REPRINTED COURTESY OF CATERPILLAR INC.

22303-14_F11.EPS

Figure 11 Stabilizer controls.

22303-14_F12.EPS

Figure 12 Two-lever backhoe controls.

Another two-lever control pattern that may be encountered is the excavator pattern. In this pattern, the controls for the stick and boom are switched. The stick is controlled by the left lever, and the boom is controlled with the right.

2.3.2 Four-Lever Backhoe Controls

Four-lever backhoe controls are shown in *Figure 16*. In this configuration, the levers move forward or away from the operator and backward towards the operator. The center position is the hold position, similar to two-lever controls. When released, the controls move to the hold position and movement stops. A common convention is to number the levers one through four from left to right as seen by the operator at the controls.

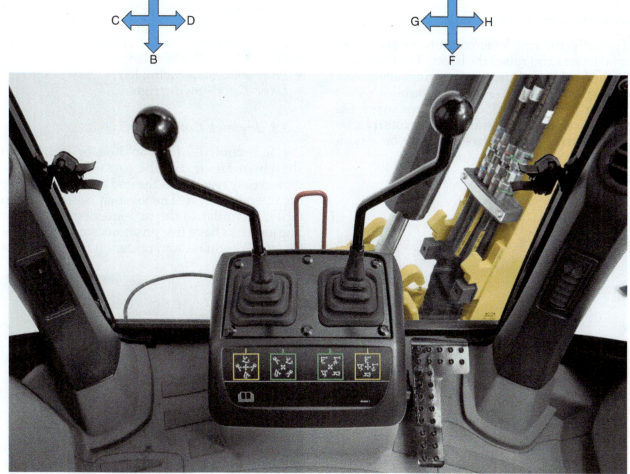

22303-14_F13.EPS

Figure 13 Positions for two-lever backhoe controls.

A – BOOM LOWER   E – STICK OUT
B – BOOM RAISE   F – STICK IN
C – BOOM SWING LEFT   G – BUCKET LOAD
D – BOOM SWING RIGHT   H – BUCKET DUMP

22303-14_F14.EPS

Figure 14 Standard backhoe control pattern.

Typically, the first lever controls the stick. The second lowers and raises the boom. The third controls the bucket and the fourth swings the boom. On some models, boom swing is controlled with a foot pedal and there will only be three levers. On other models, foot pedals are used to control auxiliary functions. The control functions on a typical four-lever system are as follows:

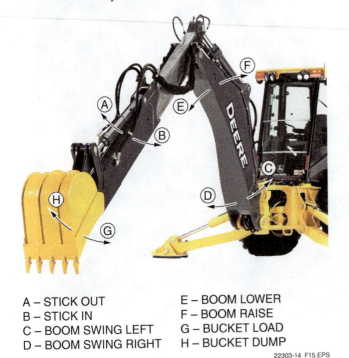

A – STICK OUT   E – BOOM LOWER
B – STICK IN   F – BOOM RAISE
C – BOOM SWING LEFT   G – BUCKET LOAD
D – BOOM SWING RIGHT   H – BUCKET DUMP

22303-14_F15.EPS

Figure 15 Alternate backhoe control pattern (excavator pattern).

22303-14_F16.EPS

Figure 16 Four-lever backhoe controls.

- *Lever 1 forward* – Stick out
- *Lever 1 back* – Stick in
- *Lever 2 forward* – Lower boom
- *Lever 2 back* – Raise boom
- *Lever 3 forward* – Bucket dump
- *Lever 3 back* – Bucket load
- *Lever 4 forward* – Swing left
- *Lever 4 back* – Swing right

2.3.3 Joystick Controls

Joystick controls for backhoe operations are shown in *Figure 17*. The joysticks are located within easy reach near the end of each armrest on the operator's seat. The joystick control configuration is similar to the two-lever configurations. The joysticks have five positions: forward (1), back (2), left (3), right (4), and center (5).

22303-14_F17.EPS

Figure 17 Joystick backhoe controls.

The central position on both joysticks is the hold position. When released, the joystick will return to this position and backhoe movement will stop. In the standard pattern for joystick backhoe controls, the boom and swing are controlled with the left joystick. The stick and bucket are controlled with the right joystick.

The control positions and backhoe functions for the left joystick are as follows:

- *Forward (1)* – Lower boom
- *Back (2)* – Raise boom
- *Left (3)* – Swing left
- *Right (4)* – Swing right

The control positions for the right joystick are as follows:

- *Forward (1)* – Stick out
- *Back (2)* – Stick in
- *Left (3)* – Bucket load
- *Right (4)* – Bucket dump

On some backhoes, the configuration can be changed by adjusting a valve under the rear of the cab. This changes the control pattern to the excavator pattern described previously for two-lever controls. Always read the operator's manual to understand the control configuration before operating a backhoe.

2.4.0 Backhoe Attachments

There are several attachments and accessories available for the backhoe. Each has a specific purpose and should be used in accordance with manufacturer's instructions. Only attachments that are designed for the specific machine should be used. Check the manufacturer's literature for the machine and for the attachment to see if they can be used together before attempting to use a new attachment with a particular machine.

> **WARNING!**
> Using an attachment that is not compatible with the machine can be extremely dangerous. Incompatible attachments may fail and cause severe injury or property damage. Do not alter attachments unless approved by the manufacturer.

Disconnecting the bucket and connecting an attachment must be done correctly. Always read and follow the manufacturer's instructions when connecting and disconnecting attachments.

> **WARNING!**
> An attachment that is not properly mounted on the machine can be extremely dangerous. The attachment or the hydraulics could fail if not connected properly. This can cause severe injury or property damage. Make sure all attachments are properly connected before operating the machine.

2.4.1 Ripper

A one-tooth ripper (*Figure 18*) is a large metal curved blade or tooth. It can be attached to the end of the stick to use on frozen ground, asphalt, tree roots, and hard soils. Different types of teeth can be chosen as accessories. It can be used to cut roots from stumps when clearing an area. Rippers are also used to break up the soil before excavation.

2.4.2 Auger

The auger attachment (*Figure 19*) is a tapered metal spiral blade mounted on a rotating motor. Some models have a variable speed, bi-directional hydraulic motor. This allows the operator to change the rotational speed for varying conditions and change rotational direction for easier removal.

The auger is very useful for boring holes to set posts or pour concrete piers. The auger will always hang vertically because it is hinged at the knuckle point on the stick. This makes centering the drill easier. Various bits are available from 6 to 36 inches (150 to 900 millimeters) in diameter. Only use bits that are compatible with the auger.

2.4.3 Street Pads

The bottoms of the stabilizers are known as pads. They are designed to dig into the soil and not slip

REPRINTED COURTESY OF CATERPILLAR INC.

22303-14_F18.EPS

Figure 18 Ripper attachment.

REPRINTED COURTESY OF CATERPILLAR INC.

22303-14_F19.EPS

Figure 19 Auger attachment.

under pressure. This is useful when the backhoe is operated on dirt. However these pads can make the stabilizer sink into asphalt or chip or crack concrete.

Street pads have rubber soles, as shown in *Figure 20*. They are used when the backhoe sets up on streets or other paved surfaces. They are attached to the bottom of the stabilizers. The rubber sole plate is replaceable.

On some machines the stabilizer pads are reversible (*Figure 21*). One side has rubber for use on pavement. The other side is metal for use on soils.

2.4.4 *Hydraulic Breaker*

A small breaker, or hoe ram, like the one shown in *Figure 22* can be used with the backhoe for demolition work. The breaker is attached to the end of the stick. It can be used on sloping or vertical surfaces. The breaker is used to break up rock, concrete, and other solid materials that are too large for the bucket. The smaller fragments can then be removed with the bucket or loader. These breakers are available in a range of sizes.

REPRINTED COURTESY OF CATERPILLAR INC.

22303-14_F20.EPS

Figure 20 Street pads.

2.4.5 *Compactor*

A compactor is a handy attachment to use when backfilling trenches and other small excavated areas. The two types most commonly used are the vibratory plate and the mechanical roller. Both types accomplish the same thing, which is to compact the soil. *Figure 23* shows a vibratory plate tamper. It attaches to the end of the stick with a specialized fitting and hydraulic attachment kit.

Compactor attachments are available in various sizes with base plate widths from 12 to 36 inches (30.48 to 91.44 centimeters). In large models, the vibratory action cycles at 2,200 per minute to produce a vibratory force of 24,000 pounds (10,886 kilograms).

2.4.6 *Cold Planer*

The cold planer (*Figure 24*) is designed to grind asphalt or concrete paved surfaces to a specific grade. It has a variable-speed hydraulic motor that mills paved surfaces where limited excavation is required. Typical drum width is 18 inches (450 millimeters). The planing depth is adjustable and can dig up to 5 inches (127 millimeters) on each pass.

2.4.7 *Buckets*

Standard buckets are available in many sizes, with widths from 12 inches to 60 inches (305 to 1,524 millimeters). In addition to the standard bucket, there are also different types of buckets available for the backhoe, including high-capacity, heavy-duty, and high-abrasion. High-capacity buckets can handle up to 14 cubic feet (0.4 cubic meters) of material. Heavy-duty and high-abrasion buckets are designed for use in more difficult digging conditions.

22303-14_F21.EPS

Figure 21 Reversible pads.

REPRINTED COURTESY OF CATERPILLAR INC.

22303-14_F22.EPS

Figure 22 Breaker attachment.

REPRINTED COURTESY OF CATERPILLAR INC.

22303-14_F23.EPS

Figure 23 Plate tamper attachment.

Buckets

There are many types of buckets, each designed for a specific task. Examples include the following:

- *Side tip bucket* – These buckets are useful for filling trenches.
- *Ditching bucket* – There are both fixed and tilting models. They are used for cleaning and grading ditches.
- *Multi-purpose bucket* – Can serve as a loader bucket as well as a dozer blade or leveler.

REPRINTED COURTESY OF CATERPILLAR INC.

22303-14_F24.EPS

Figure 24 Cold planer attachment.

3.0.0 SAFETY, INSPECTIONS, AND MAINTENANCE

Objective

Identify and describe safety, inspection, and service guidelines associated with a backhoe.

a. Describe rules pertaining to safety.
b. Describe daily inspection checks.
c. Describe the servicing requirements for a backhoe.

Performance Task

Demonstrate a prestart inspection of a backhoe.

Accidents and equipment breakdowns can be deadly and expensive. Accidents always include the risk of injury to personnel, so operating safely is a critical responsibility. Breakdowns can require expensive repairs that might have been avoided with periodic maintenance. Even minor accidents or breakdowns can delay the completion of a project. The cost and time needed to recover from an accident or breakdown can far exceed the time needed each day to prevent them.

3.1.0 Safety Guidelines

Sometimes it is useful to think of three aspects of safety: keeping yourself, the machine, and others working around you safe. You, as the operator, are responsible for performing your work safely, protecting equipment from damage, and protecting the public and your co-workers from harm. The easiest way to keep a job site safe is to know the equipment and think safety.

3.1.1 Operator Safety

Safety begins with personal responsibility. Safety handbooks and safety training do not prevent accidents, operators do. You can protect yourself and those around you from getting injured on the job. A good operator must be alert and avoid accidents.

One way to avoid accidents is to know and follow your employer's safety rules. Prepare for work by wearing the right clothes and safety equipment. The following are recommended safety procedures for all occasions:

- Wear personal protective clothing as required by site policy.
- Do not wear loose clothing or jewelry, including rings, that could catch on controls or moving parts.
- Never operate equipment under the influence of alcohol or drugs.
- Never smoke while refueling or checking batteries or fluid.
- Always know and use agreed-upon hand signals when working with other personnel.
- Never attempt to search for leaks with your bare hands. Hydraulic and cooling systems operate at high pressure. Fluids under high pressure can cause serious injury.

Backhoes are complex pieces of equipment. Many accidents can happen when performing simple tasks like getting into the cab or parking. Use safety precautions at all times. The following guidelines will help keep you safe when operating a backhoe:

- Wear the seat belt.
- Keep steps, grab irons, and the operator's compartment clean. Mud, oil, or grease on these surfaces is a slip, trip, fall hazard. Clean boots as much as possible before climbing onto the equipment.
- Mount and dismount the equipment carefully using three-point contact and facing the machine (*Figure 25*).
- Never remove protective guards or panels. Replace guards that break or become damaged.
- Keep the windshield, windows, and mirrors clean at all times.
- Always lower the bucket or other attachments to the ground before performing any service or when leaving the backhoe unattended and turn off the machine.
- Observe extreme caution when excavating where there may be overhead power lines and/or underground utilities. Contact the local One-Call number before excavating.

> **WARNING!**
>
> Getting in and out of equipment can be dangerous. Always face the machine and maintain three points of contact when you are mounting and dismounting. That means you should have three out of four of your hands and feet on the equipment. That can be two hands and one foot or one hand and two feet.

3.1.2 Safety of Co-Workers and the Public

Sometimes an operator will work alone in an undeveloped area, but more frequently will work

REPRINTED COURTESY OF CATERPILLAR INC.

22303-14_F25.EPS

Figure 25 Mount using three points of contact.

with other equipment operators and construction workers. As an operator, you are responsible for their safety as well as your own. When working in a public place close to pedestrians or motor vehicles, you are also responsible for their safety. The main safety points when working around other people include the following:

- Walk around the equipment to make sure everyone is clear of the equipment before starting and moving it.
- Maintain a clear view in all directions. Do not carry any equipment or materials that obstruct your view.
- Keep ground personnel in visual contact.
- If you are working in traffic, establish a safe work zone using cones or other barriers (*Figure 26*). Make sure you know the rules and the meaning of all flags, hand signals, signs, and markers used.
- Never allow riders on the backhoe.
- Do not let anyone stand alongside a trench being excavated.

REPRINTED COURTESY OF CATERPILLAR INC.

22303-14_F26.EPS

Figure 26 Establish a safe work zone.

3.1.3 Equipment Safety

Backhoes are designed with many safety features, including guards, canopies, shields, rollover protection, and seat belts. However, they are only effective if operators are aware of them and make sure they are in working order. Backhoes are powerful and expensive pieces of equipment. Make sure the equipment is safe to operate and operated safely.

Use the following guidelines to keep equipment in good working order:

- Perform prestart inspection and lubrication daily (*Figure 27*).
- Look and listen to make sure the equipment is functioning normally. Stop if it is malfunctioning. Correct or report trouble immediately. Remain alert to any change in equipment operation or performance.
- Before beginning work in a new area, take a walk to locate any cliffs, steep banks, holes, power or gas lines, or other obstacles that could cause a hazard to safe operation.
- Always travel with the loader bucket low to the ground and the backhoe bucket retracted and secured.
- Never exceed the manufacturer's limits for speed, lifting, or operating on inclines.
- Always lower the loader bucket and attachments, engage the parking brake, turn off the engine, and secure the controls before leaving the equipment.
- Never park on an incline.
- Check the hinge pins and their securing devices on all components.

REPRINTED COURTESY OF CATERPILLAR INC.

22303-14_F27.EPS

Figure 27 Check all fluids daily.

- Be sure the stabilizers are properly extended before starting operation.
- Use extreme caution when raising the stabilizers. The backhoe can become unstable and tip over.
- Make sure clearance flags, lights, and other required warnings are on the equipment when roading or moving.
- When loading, transporting, and unloading equipment, know and follow the manufacturer's recommendations.
- Always check for overhead power lines.

The basic rule is know the equipment. Know its capabilities and limitations and understand the purpose and use of all gauges and controls. Never operate a machine if it is not in good working order. Some basic safety rules of operation include the following:

- Do not operate the backhoe from any position other than the operator's seat.
- Do not coast. Neutral is for standing only.
- Maintain control when going downhill. Never change gears when going downhill and travel downhill in the same gear as you would travel uphill.
- Whenever possible, avoid obstacles such as rocks, fallen trees, curbs, and ditches.
- If you must cross over an obstacle, reduce speed and approach at an angle to reduce the impact on the equipment and yourself.
- Use caution when undercutting high banks, backfilling new walls, and removing trees.

- If the engine will not start, refer to the manufacturer's instructions.

3.2.0 Basic Inspections and Preventive Maintenance

Preventive maintenance involves an organized effort to regularly perform periodic lubrication and other service work in order to avoid poor performance and breakdowns at critical times. Performing preventive maintenance on the backhoe keeps it operating efficiently and safely and avoids costly failures in the future.

Preventive maintenance of equipment is essential and not that difficult with the right tools and equipment. The leading cause of premature equipment failure is putting things off. Preventive maintenance should become a habit, performed on a regular basis.

3.2.1 Daily Inspection Checks

The first thing to do each day before beginning work is to conduct a daily inspection. This should be done before starting the engine. Inspection will identify any potential problems that could cause a breakdown and will indicate whether the machine can be operated. The equipment should be inspected before, during, and after operation. In addition to inspection, some basic preventive maintenance must be performed daily, including lubricating pivot points and refilling all machine fluids. Before beginning operation, complete the daily inspection and maintenance to ensure that the machine is in good working order.

Maintenance time intervals for most machines are established by the Society of Automotive Engineers (SAE) and have been adopted by most equipment manufacturers. Instructions for preventive maintenance and specified time intervals are usually found in the operator's manual for each piece of equipment. Common time intervals are: 10 hours (daily), 50 hours (weekly), 100 hours, 250 hours, 500 hours, and 1,000 hours. The operator's manual will also include lists of inspections and servicing activities required for each time interval.

The daily inspection is often called a walk-around. The operator should walk completely around the machine checking various items. Items to be checked and serviced on a daily inspection are as follows:

- Inspect the cooling system for leaks and faulty hoses.
- Inspect the engine compartment and remove any debris. Clean access doors.
- Inspect the engine for obvious damage.
- Check the condition and adjustment of drive belts on the engine.
- Inspect tires for damage and replace any missing valve caps.
- Inspect the axles, differentials, wheel brakes, and transmission for leaks.
- Check proper function of braking systems.
- Inspect the hydraulic system for leaks, faulty hoses, and loose clamps.
- Inspect all attachments and the linkage for wear and damage. Check this hardware to make sure there is no damage that would create unsafe operating conditions or cause an equipment breakdown. Make sure the bucket is not cracked or broken. Check to ensure that all attachments are secured properly.
- Inspect and clean steps, walkways, and handholds.
- Inspect the ROPS for damage.
- Inspect the lights and replace any broken bulbs or lenses.
- Inspect the operator's compartment and remove any trash.
- Inspect the windows for visibility and clean them if they are dirty.
- Adjust and clean the mirrors.
- Test the backup alarm and horn. Put equipment in reverse gear and listen for backup alarm.

Some manufacturers require that daily maintenance be performed on specific parts. These parts are usually those that are the most exposed to dirt or dust and may malfunction if not cleaned or serviced. For example, the service manual may recommend lubricating specific pivot points every 10 hours of operation, or always inspecting the air filter pre-cleaner before starting the engine.

Before beginning operation, check the fluid levels and top off any that are low (*Figure 28*). Check all the machine's major functions including the following components:

FLUID LEVEL CHECK/FILL POINTS

REPRINTED COURTESY OF CATERPILLAR INC.

22303-14_F28.EPS

Figure 28 Top off fluids as needed.

- *Battery* – Check the battery cable connections.
- *Crankcase oil* – Check the engine oil level and make sure it is in the safe operating range.
- *Cooling system* – Check the coolant level and make sure it is at the level specified in the operating manual.
- *Fuel level* – Check the fuel level in the fuel tank(s). Do this manually with the aid of the fuel dipstick or marking vial. Do not rely on the fuel gauge at this point. Check the fuel pump sediment bowl if one is fitted on the machine.
- *Hydraulic fluid* – Check the hydraulic fluid level in the reservoir (*Figure 29*).
- *Transmission fluid* – Measure the level of the transmission fluid to make sure it is in the operating range.
- *Pivot points* – Clean and lubricate all pivot points (*Figure 30*).

NOTE

In cold weather it is usually better to lubricate pivot points at the end of a work shift when the grease is warm. Warm the grease gun before using it for better grease penetration.

The operator's manual usually has detailed instructions for performing periodic maintenance. If you find any problems with the machine that you are not authorized to fix, inform the foreman or field mechanic. Get the problem fixed before beginning operations.

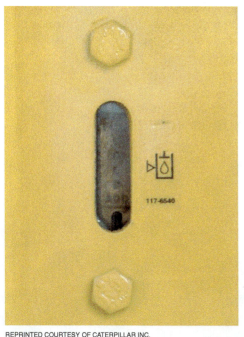

REPRINTED COURTESY OF CATERPILLAR INC.

22303-14_F29.EPS

Figure 29 Check the hydraulic fluid level.

3.3.0 Servicing a Backhoe

When servicing a backhoe follow the manufacturer's recommendations and service chart. Any special servicing for a particular piece of equipment will be highlighted in the manual. Normally, the service chart recommends specific intervals, based on hours of run time, for such things as changing oil, filters, and coolant.

Hydraulic fluids should be changed whenever they become contaminated or break down due to overheating. Continuous and hard operation of the hydraulic system can heat the hydraulic fluid to the boiling point and cause it to break down. Filters should also be replaced during regular servicing.

Before performing maintenance procedures, always complete the following steps:

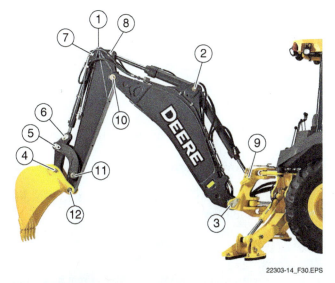

22303-14_F30.EPS

Figure 30 Lubricate all pivot points.

Step 1 Park the machine on a level surface to ensure that fluid levels are indicated correctly.

Step 2 Lower all attachments to the ground. Operate the controls to relieve hydraulic pressure.

Step 3 Engage the parking brake.

Step 4 Lock the transmission in neutral.

3.3.1 Preventive Maintenance Records

Accurate, up-to-date maintenance records are needed in order to know the history of the equipment. Each machine should have a record that describes any inspection or service that is to be performed and the corresponding time intervals. Typically, an operator's manual and some sort of inspection sheet are kept with the equipment at all times. Update maintenance records as part of your daily routine.

3.0.0 Section Review

1. Before leaving a backhoe unattended or performing service on it, it is best to _____.

   a. raise the loader and withdraw the stabilizers for access
   b. lower the bucket or other attachments to the ground
   c. remove whatever attachment is on the backhoe
   d. fully elevate the boom and stick to prevent vandalism

2. The most appropriate time to perform a daily inspection on a backhoe is _____.

   a. immediately after all systems reach normal operating temperatures
   b. as soon as the engine has been started, but before it is warm
   c. midway through the shift or work day
   d. first thing before starting the engine

3. A backhoe's service chart normally recommends specific service intervals for fluid and filter changes based on _____.

   a. hours of run time
   b. mileage
   c. equipment noise
   d. operator judgment

4.0.0 BASIC OPERATION

Objective

Describe basic operating procedures for a backhoe.

   a. Identify factors for effective backhoe operation.

   b. Identify steps for preparing to work with a backhoe.

   c. Identify steps for performing basic maneuvers with a backhoe.

Performance Tasks 1, 2, and 3

Perform a proper startup, warm-up, and shutdown of a backhoe.

Perform basic backhoe maneuvers, including forward movement, turning, moving in reverse, and operating the front loading bucket.

Perform the operation of setting up a backhoe, using stabilizers, and digging with the bucket.

Operation of a backhoe requires constant attention to the controls, instruments, and surrounding environment. Plan work and movements in advance. Be alert to the other operations going on around the machine.

Do not take risks. If there is doubt about the capability of the machine to do some work, stop the equipment and investigate the situation. Discuss it with the foreman or engineer in charge. Whether it is a slope that may be too steep or an area that looks too unstable to work, know the limitations of the equipment. Decide how to do the job before starting. Once you get in the middle of something, it may be too late.

4.1.0 Suggestions for Effective Backhoe Operation

The following are suggestions to improve efficiency when operating a backhoe/loader:

- Observe all safety rules and regulations.
- Keep equipment clean. Make sure the cab is clean so that nothing affects the operation of the controls.
- Calculate and plan operations before starting.

- Set up the work cycle so that it will be as short as possible.
- Spot trucks properly when loading with the bucket.
- Start raising the boom so that the bucket will reach dumping height by the time it arrives at the dump area.
- Work from the top down and take shallow bites to fill the bucket.
- Curl the bucket close to the stick to prevent spilling when loading trucks (*Figure 31*).
- Avoid spinning the wheels of the vehicle, which can cause excessive tire wear and damage.

When operating the loader, maintain good visibility and stability by carrying the bucket low while traveling. Carry the bucket just high enough to clear ground obstacles. When loading trucks, use the wait time to clean and level the work area. Cleanup of spillage around the stockpile smoothes loading cycles and lessens operator fatigue. Maintain traction while loading by not putting excessive down-pressure on the bucket. Excessive down-pressure forces the front wheels to rise up off the ground. When picking up a load, the weight is taken off the front wheels, which makes it harder to get traction.

When operating the loader or the backhoe, control dumping by moving the bucket tilt control lever to the dump position. Repeat this operation until the bucket is empty. When handling dusty material, try to dump the material with the wind to your back. This keeps dust from entering the engine compartment and the operator's cab.

22303-14_F31.EPS

Figure 31 Loading a truck.

Select the correct bucket for backhoe operations. Using the wrong bucket for the task increases wear on the bucket. It also increases the potential to exceed the machine's operating limits. At best, exceeding the machine limits reduces the service life of the machine; at worst, it could damage the machine or cause an accident.

4.2.0 Preparing to Work

Mount the equipment using the grab rails and foot rests. Maintain three points of contact. This means that one hand and two feet or two hands and one foot should be in contact with the machine. Preparing to work involves getting organized in the cab, fastening your seat belt, and starting the machine.

Adjust the seat to a comfortable operating position for both forward and rear operations. The seat should be adjusted to allow full brake pedal travel with your back against the seat back. This will permit maximum force to be applied the pedals. Make sure you can see clearly and can reach all the controls. Adjust the mirrors for maximum visibility from the operator's seat in both directions. Prepare for backhoe operations by adjusting the rear-facing operator's seat and mirrors. Then return the seat to face the front.

> **NOTE**
> Always maintain three points of contact when mounting and dismounting equipment. Keep grab rails and footrests clear of dirt, mud, grease, ice, and snow.

> **WARNING!**
> OSHA requires that approved seat belts and ROPS be installed on heavy equipment. Old equipment must be retrofitted. Do not use heavy equipment that does not have these safety devices.

Operator stations vary, depending on the manufacturer, size, and model of the equipment. However, all stations have gauges, indicators and switches, levers, and pedals. The gauges tell the operator the status of critical items such as water temperature, oil pressure, battery voltage, and fuel level. Indicators alert the operator to low oil pressure, engine overheating, clogged air and oil filters, and electrical system malfunctions. Switches are used to activate the glow plugs, start the engine, and operate accessories such as lights. Typical instruments and controls were described previously. Review the operator's manual so that you know the specifics of the machine you will be operating.

The proper startup and shutdown of an engine is very important. Proper startup lengthens the life of the engine and other components. A slow warm-up is required for proper operation of the machine under load. Similarly, the shutdown of the machine is critical because of all the hot fluids circulating through the system. These fluids must cool so that they can cool the metal parts before the engine is switched off.

4.2.1 Startup

There may be specific startup procedures for the piece of equipment being operated. Check the operator's manual for a specific procedure. In general, the startup procedure should follow this sequence:

Step 1 Be sure all controls are in neutral.

Step 2 Engage the parking brake. Use a lever or knob to engage the parking brake, depending on the backhoe make and model.

> **NOTE**
> When the parking brake is engaged, an indicator light on the dash will light up or flash. If it does not, stop and correct the problem before operating the equipment.

Step 3 Make sure there are no people close to the equipment.

Step 4 Place the ignition switch in the On position (*Figure 32*).

Step 5 Depress the throttle control approximately one-third the total distance.

Step 6 Press the starter button.

REPRINTED COURTESY OF CATERPILLAR INC.

22303-14_F32.EPS

Figure 32 Engine key switch.

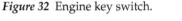

Step 7 Warm up the engine for at least 5 minutes.

Step 8 Check all the gauges and instruments to make sure they are working properly.

Step 9 Shift the gears to low range (*Figure 33*).

Step 10 Release the parking brake and depress the service brakes.

Step 11 Check all the controls for proper operation.

Step 12 Check service brakes for proper operation.

Step 13 Check the steering for proper operation.

Step 14 Manipulate the controls to be sure all components are operating properly.

Step 15 Manipulate controls to extend, set, and retract the stabilizers.

Step 16 Shift the gears to neutral and lock.

Step 17 Reset the brake.

Step 18 Make a final visual check for leaks, unusual noises, or vibrations.

Some machines have special procedures for starting the engine in cold temperatures. Some units are equipped with glow plugs or other starting aids. Review the operator's manual to fully understand the procedures for using these aids.

As soon as the engine starts, release the starter switch and adjust the engine speed to approximately half throttle. Let the engine warm up to operating temperature before moving the backhoe.

Let the machine warm up for a longer period when it is cold. If the temperature is at or slightly above freezing (32°F or 0°C) let the engine warm up for 15 minutes. If the temperature is between 32°F (0°C) and 0°F (–18°C), warm the engine for

REPRINTED COURTESY OF CATERPILLAR INC.

22303-14_F33.EPS

Figure 33 Transmission control.

30 minutes. If the temperature is less than 0°F (–18°C), or hydraulic operations are sluggish, additional time is needed. Follow the manufacturer's procedures for cold starting.

4.2.2 Checking Gauges and Indicators

Keep the engine speed low until the oil pressure registers. The oil pressure light (*Figure 34*) should come on briefly and then go out. If the oil pressure light does not turn off within 10 seconds, stop the engine, investigate, and correct the problem.

Check the other gauges and indicators to see that the engine is operating normally. Check that the water temperature, voltmeter, and oil pressure indicator are in the normal range. If there are any problems, shut down the machine and investigate or get a mechanic to look at the problem.

REPRINTED COURTESY OF CATERPILLAR INC.

22303-14_F34.EPS

Figure 34 Oil pressure light.

4.2.3 Shutdown

Shutdown should also follow a specific procedure. Proper shutdown will reduce engine wear and possible damage to the machine.

Step 1 Find a dry, level spot to park the backhoe. Stop the backhoe by decreasing the engine speed, depressing the clutch, and placing the direction lever in neutral. Depress the service brakes and bring the machine to a full stop (*Figure 35*).

Step 2 Place the transmission in neutral and engage the parking brake. Lock out the controls if the machine has a control lock feature.

Step 3 Lower the bucket so that it rests on the ground. If you have any other attachment, be sure it is also lowered or locked in position.

> **NOTE**
>
> If you must park on an incline, chock the tires.

Step 4 Place the speed control in low idle and let the engine run for approximately five minutes.

> **CAUTION**
>
> Failure to allow the machine to cool down can cause excessive temperatures that could damage the turbocharger.

Step 5 Release hydraulic pressure by moving the control levers until all movement stops.

Step 6 Turn the engine key switch to the Off position and remove the key.

Some machines have disconnect switches or vandal guards for added security. Close the battery disconnect switch and fuel shutoff valve. Close and lock any vandal guards. These controls provide additional safety and deter unauthorized users. Always engage any additional security systems when leaving the backhoe unattended.

4.3.0 Basic Maneuvering

Basic maneuvering of the backhoe involves moving the machine into position and setting the stabilizers before operating the backhoe or other attachments. The backhoe/loader is positioned by driving the machine from the front loader position. Positioning the backhoe on level or sloping ground and positioning the stabilizers must be done carefully to ensure safe backhoe operations.

The backhoe can also be repositioned while operating from the backhoe position.

4.3.1 Positioning the Backhoe

The first basic maneuver is to drive forward while using the bucket loader. Before starting to move, raise the bucket assembly by pulling the boom and bucket control lever or joystick (*Figure 36*). Raise the bucket about 15 inches (38 centimeters) above the ground. This is the travel position.

To move forward, put the shift lever in low forward. Release the parking brake and press the accelerator pedal to start the loader moving. Steering is accomplished by the use of a steering wheel. Stop the machine using the service brakes.

> **NOTE**
>
> Always travel with the bucket low to the ground. This provides better visibility and greater load stability.

Once underway, shift to a higher gear to drive on the road. To shift from a lower to a higher gear, move the shift lever farther forward or rotate the gear selection lever. See the operator's manual for transmission controls for the machine being operated. Remember, high gear is used only for traveling on the road.

> **WARNING!**
>
> Running in high gear can cause the machine to become unstable.

> **NOTE**
>
> Four-wheel drive should not be used when roading the machine.

Some backhoe/loaders have alternative steering patterns including crab steering and circle steering. These patterns are useful when positioning the backhoe/loader in small spaces. If the machine is equipped with these features, follow the manufacturer's instructions for using them. Practice changing steering patterns in an open area before attempting to use them to maneuver in a small space.

4.3.2 Moving Backward

The backhoe is operated from a stationary position with the back of the equipment facing the work area. The operator may need to back the machine into the proper position before starting

---

REPRINTED COURTESY OF CATERPILLAR INC.

22303-14_F35.EPS

Figure 35 Service brakes.

backhoe operations. It can be dangerous to back the equipment up to the edge of an excavation. Use a spotter for safety.

Always come to a complete stop before backing up or reversing direction. Move the shift lever to reverse. Check to make sure that the path is clear before backing up. Slowly apply light pressure to the throttle pedal and begin to move backwards. Watch for hazards while backing, and follow the spotter's signals until the backhoe is in the proper position. Although it is possible to reverse directions while moving, it is not recommended. Changing direction while in motion can cause a sudden jolt to the operator and the load.

4.3.3 Positioning the Stabilizers

Operation of the backhoe can be unstable. Always lower the front bucket and the stabilizers to improve stability. Take care to position the stabilizers and the front bucket so that the weight of the machine is evenly distributed. This creates a firm level platform for working.

> **WARNING!**
>
> Use caution when lowering the stabilizers. Check that no one is in the path of the stabilizers. Use extra precaution when using a spotter to make sure that the spotter is clear.

REPRINTED COURTESY OF CATERPILLAR INC.

22303-14_F36.EPS

Figure 36 Loader joystick control.

First apply the parking brake. Then lower the loader bucket and apply down pressure so that the front wheels are slightly off the ground. Switch seat positions and readjust the seat. Then lower the stabilizers one at a time. The rear wheels should be slightly off the ground. Make sure all personnel are clear of the area before lowering the bucket and stabilizers.

> **NOTE**
>
> Use the appropriate stabilizer pad to prevent damage to the work area. Most stabilizer pads can be reversed.

4.3.4 Setting Up on Level Ground

Setting up on level ground is relatively easy. However, never begin operation until the stabilizers are lowered and adjusted. *Figure 37* shows a backhoe that has been set up properly with the front bucket and the stabilizers in position.

Articulated Backhoes

For articulated backhoes, backing up in a straight line may take additional practice. Steering an articulated machine is about the same as trying to back up a trailer attached to a truck. If you begin to turn too sharply, stop and pull forward. Straighten out the machine before continuing to move backward. Practice in an open area until you are comfortable backing up the machine.

Figure 37 Backhoe with stabilizers lowered.

Step 1 Position the backhoe as near level as possible.

Step 2 Set the parking brake.

Step 3 Lower the loader bucket to the ground. Apply enough downward pressure to take most of the load off the front wheels.

Step 4 Reposition the operator's seat and adjust it.

Step 5 Lower the stabilizers to the ground enough to take most of the pressure off the rear wheels.

Step 6 Adjust the loader bucket and stabilizer controls enough to level the backhoe. After some practice you will get a feel for when the machine is level.

Step 7 If you are not sure about the position of the stabilizers or feel that the machine is not level, shut down the machine and get down from the cab. Look at the setup and position of the stabilizers and the front bucket. If you don't like the position, readjust the stabilizers or reposition the machine.

4.3.5 Setting Up on Sloping Terrain

Setting up on sloping terrain can be more difficult. In many operations, there will be no choice but to set up on uneven areas or sloping grades. For safety, do not operate the backhoe on slopes or hillsides with a slope of more than 15 degrees. Use the following procedure to set up on a slope:

Step 1 Set the parking brake, then lower the loader bucket to the ground just enough to take pressure off the front wheels.

Step 2 Reposition the operator's seat and adjust it.

Step 3 Lower the stabilizer on the downhill side enough to level the backhoe.

Step 4 Lower the opposite stabilizer to the ground.

Step 5 Adjust both stabilizers at the same time to raise the machine enough to take most of the weight off the rear wheels.

Step 6 Adjust the loader bucket to level the backhoe.

Step 7 Begin backhoe operation.

4.3.6 Pushing the Backhoe Forward

When excavating, it is often necessary to reposition the backhoe to continue the cut. It is possible to move the backhoe forward or backward slightly without changing over to front end operation. This is very useful when making slight adjustments to the backhoe position.

For this procedure, the bucket must be firmly positioned either at the bottom of the excavation or along the top edge, whichever is more stable. The loader bucket and stabilizers are raised so that the machine rests on the wheels. The boom and stick controls are used to push the machine to the new location. Once the machine is repositioned, the loader bucket and stabilizers must be lowered and the machine leveled in the new position.

Use the following procedure to push the backhoe forward:

Step 1 Lower the backhoe bucket to the bottom of the excavation.

Step 2 Raise the stabilizers slightly off the ground, high enough to clear any obstacles and to prevent dragging them.

Step 3 Raise the bucket slightly off the ground to avoid obstacles and to prevent dragging it.

Step 4 Check that the front wheels are positioned correctly for the desired travel direction.

Step 5 Release the parking brake.

Step 6 Use the boom and stick controls to maintain downward pressure with the boom control while moving the bucket control away from you. This will push the backhoe forward to the new dig position, keeping the backhoe bucket at the last grade cut in the excavation.

Step 7 After pushing forward, set the parking brake, lower the stabilizers, lower the bucket, and prepare to dig again. Repeat this procedure as needed until the excavation is complete.

> **NOTE**
>
> The controls on specific backhoes may be different than those described in the procedures. Check your operator's manual for information about the controls and limitations of your equipment.

If the bottom of the excavation is soft, position the backhoe bucket along the top edge of the excavation for pushing forward.

4.3.7 Pulling the Backhoe Backward

Pulling the backhoe rearward follows the same procedure as pushing, except that the bucket control is pulled toward the operator. The boom control is used to maintain downward pressure on the bucket. If the bottom of the excavation is soft, position the bucket outside and along the edge of the excavation for pulling rearward.

Working in Mud

The loader bucket can be used to support the front of the machine when working in muddy ground. It will help prevent the front of the machine from sinking into the mud.

4.0.0 Section Review

1. One way to improve operating efficiency with a backhoe is to _____.

    a. avoid extending the stabilizers if the ground is reasonably level
    b. keep the loader attachment elevated for counterweight
    c. work from the top down and take shallow bites to fill the bucket
    d. overfill the bucket if necessary to shorten the job time

2. If a backhoe is started up at a work site where the outside temperature is about 50°F (10°C), the engine should be allowed to warm up for _____.

    a. no more than 1 minute
    b. about 5 minutes
    c. 15 minutes
    d. at least 30 minutes

3. When a backhoe is being driven to position it at a work site, the travel position for the front loader bucket should be _____.

    a. about 15 inches (38 centimeters) above the ground
    b. between 36 and 48 inches (91 and 122 centimeters) above the ground
    c. at least 60 inches (152 centimeters) above the ground
    d. fully curled and secured with the stabilizers

5.0.0 WORK ACTIVITIES

Objective

Identify and describe common work activities for a backhoe.

   a. Describe loading from a stockpile with a backhoe.

   b. Describe trenching and loading with a backhoe.

   c. Describe demolition using the hydraulic breaker.

   d. Describe setting pipe using a backhoe.

   e. Describe excavating footings and foundations with a backhoe.

   f. Describe working with a backhoe in confined or unstable areas.

   g. Describe basic procedures for roading and transporting a backhoe.

Performance Task 5

Perform an excavation of a trench 20 to 40 feet (6 to 12 meters) long with spoil piles at least 2 feet (61 centimeters) from the edge.

Trade Terms

Draw: Move the stick back toward the operator.

Retract: Move the extendable stick in.

Rubble: Fragments of stone, brick, or rock that have broken apart from larger pieces.

The backhoe is chiefly designed for excavation. However, it has several advantages over other types of excavating equipment. For example, it is also good for stripping top soil, making shallow cuts, removing windrows without losing material at the sides, scraping down high embankments, and grubbing smaller objects.

In addition to backhoe operations, there are several different attachments that can be fitted onto the backhoe. The backhoe/loader allows attachments to be fitted onto the front of the machine.

5.1.0 Loading from a Stockpile Using the Backhoe

A dump truck or other type of haul unit can be loaded using either the front loader bucket or the backhoe bucket. Choosing which one to use depends on the amount of space available and the type of hauling equipment being used. The machine must have some room to maneuver when using the loader bucket. However, when loading with the backhoe bucket, the machine stays stationary. In a small space, it can be more efficient to load using the smaller backhoe because the machine does not have to move with each load.

Using the backhoe to load haul units involves excavating the material from the stockpile with the bucket and placing it in the haul unit (*Figure 38*). Take care not to allow material to spill out of the bucket.

Use the following loading techniques when loading a truck with the backhoe:

Step 1   Position the backhoe so that the boom can reach the back of the spoils or stockpile. If you cannot reach the back, pull down the material with a scraping motion from the top of the pile until there is enough material built up to be able to pick it up.

Step 2   Lower the bucket to the pile the same way you would in bucket digging.

Step 3   Use only the crowd cylinder to **draw** or **retract** the stick.

Step 4   Drag the bucket through the material until it is half full.

Step 5   Roll the bucket while retracting the stick until the bucket is full.

Step 6   Raise and swing the bucket up to a height above the sideboard of the truck. Do not move the load over the truck cab.

Sizing the Backhoe

The following is a rule of thumb that can be used to determine the size of the backhoe to use for trenching: the depth of the trench should not be greater than 70 percent of the unit's maximum digging depth. For example, if the trench is to be 10 feet (3 meters) deep, the backhoe should have a maximum digging depth of at least 14 feet (4.3 meters). (Divide the trench depth by 0.70: 10 ÷ 0.70 = 14.28.)

Figure 38 Loading with the backhoe.

Step 7 Move the bucket out over the truck bed while keeping the bucket curled.

Step 8 When the bucket is over the truck bed, uncurl it and release the material into the center of the bed. Load the truck from front to back.

When loading debris or rubble into a dump truck with a backhoe, make sure the chunks of material are clamped well between the bucket and the backhoe arm as shown in *Figure 39*. If it is not clamped well, the rubble can slip out and fall when the hoe arm is raised to dump into the truck bed. As the piece of debris slips free, it will usually fall toward the backhoe. On a rubber-

Figure 39 Secure large pieces with the bucket and stick.

tired backhoe, there is a good chance it will land on the stabilizer, breaking off hydraulic lines or connections.

In some situations, the backhoe may have trouble clearing the side of the truck bed without spilling the load. If necessary, the stabilizers can be lowered in order to raise the tractor. This is done by extending the stabilizers, then lowering the front bucket to help stabilize the machine. Keep in mind, however, that this method affects the stability of the machine.

5.2.0 Trenching and Loading

The backhoe is an excellent machine for trenching and loading. The bucket is well-designed for excavating a narrow trench. Buckets of different sizes can be attached to the backhoe, depending on the size of the trench desired.

The following procedure describes how to carry a straight line while excavating a ditch and loading a haul truck:

Step 1 Position the backhoe in a straight line with markers set by a surveyor or other construction personnel.

Step 2 Position the truck so that it can be loaded over the side, then set the parking brake.

Step 3 Lower the loader bucket to the ground to provide front stabilization for the machine.

Step 4 Lower the stabilizers until the wheels are off the ground and level the machine.

Step 5 Engage the stick control to move the bucket to about 75 percent of the maximum reach, as shown in *Figure 40*.

Step 6 Engage the bucket control to position the cutting edge at about a 30-degree angle to the ground as shown in *Figure 41*. Digging should be done at a slight angle off the vertical to decrease wear on the teeth.

Step 7 Move the boom control to lower the bucket to the ground.

Step 8 Pull the bucket arm control in, and as the bucket enters the material, manipulate the bucket cylinder control so that the cutting edge is working parallel to the ground, as shown in *Figure 42*. Do not force the bucket, as this could damage hydraulics and reduce efficiency.

Step 9 Raise or lower the boom control slightly if the bucket arm slows or stops filling the bucket.

REPRINTED COURTESY OF CATERPILLAR INC.

22303-14_F40.EPS

Figure 40 Position the stick.

Step 10 Pick up the bucket as soon as it is full.

Step 11 Raise the boom control to clear the cut and side of the truck, then swing the bucket over the truck bed and gradually dump the load out of the bucket to ease the strain of added weight on the truck. Keep the bucket well above the side of the truck.

Step 12 Actuate the bucket and bucket arm control levers at the same time for a fast, controlled dump. The truck should be loaded from the far side to the near side.

Step 13 Swing the bucket back to the cut and continue the cycle until the trench is complete or until the truck is loaded.

> **NOTE**
> If you need to wait for an empty truck, use the time to move the backhoe or make adjustments to the stabilizers, loosen dirt, or clean up corners of the cut.

Finishing straight walls takes practice and skill. The best procedure for doing this job well involves the following steps:

Step 1 Finish the farthest wall by moving the stick out while moving the bucket down. Hold the bucket open and vertical, as shown in *Figure 43*.

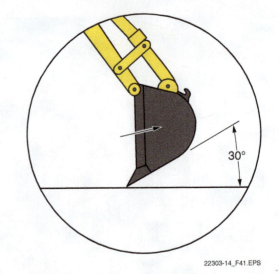

22303-14_F41.EPS

Figure 41 Position the bucket.

Step 2 Finish the closest wall by lifting the bucket up and keeping the cutting edge vertical.

Step 3 In sandy soil, use a platform under the rear tires and stabilizers. The possibility of a cave-in will be lessened by the even distribution of the load over a wider area.

5.3.0 Demolition Using the Hydraulic Breaker

One attachment that comes in very handy on job sites where demolition is required is the hydraulic breaker, or jackhammer (*Figure 44*). This attachment is mounted on the end of the stick in place of the backhoe bucket. It is hydraulically controlled. To use the hydraulic breaker follow these steps:

Step 1 Proceed to the job location.

REPRINTED COURTESY OF CATERPILLAR INC.

22303-14_F42.EPS

Figure 42 Work the bucket parallel to the ground.

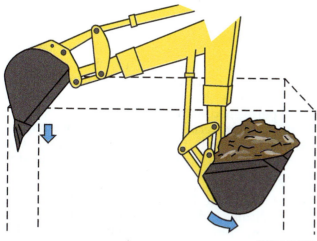

Figure 43 Finishing.

Step 2 Move to the starting point and set up the machine. Usually the areas to be broken will be marked with paint or flagging tape.

Step 3 Lower the loader bucket to the ground enough to take pressure off the front wheels.

Step 4 Reposition the seat to face the backhoe controls.

Step 5 Lower the stabilizers to the ground until most of the rear weight of the backhoe is on the pads and the backhoe is level.

Step 6 Remove the swing lock pin and boom lock pin.

Step 7 Lower the breaker to the ground at the starting point.

Step 8 Use the breaker control to operate the breaker. This control is usually foot-operated, but may be button-operated.

Step 9 When the breaker has been driven into the material, pull the boom up.

Step 10 Move the breaker to the next point to be broken and engage the breaker.

Step 11 After all the material within reach is broken, move the machine by raising the stabilizers and loader bucket. Reposition the machine to the next area. Move far enough away so the breaker can just reach the section already broken.

Step 12 Continue the breaking operations until the job is complete.

REPRINTED COURTESY OF CATERPILLAR INC.

Figure 44 Hydraulic breaker.

5.4.0 Setting Pipe

After excavation of a ditch, the backhoe can be used to do several other tasks that are needed to set pipe in the excavation. This involves placing bedding material for the pipe to rest on, as well as lifting and placing the pipe itself.

To place the bedding material, use the bucket in the following manner:

Step 1 Fill the bucket with the bedding material from a nearby stockpile, truck, or material (bedding) box.

Step 2 Move the bucket into the ditch and position it off the bottom so the bucket will clear the bottom of the ditch when it is uncurled.

Step 3 Cast out the bucket to spread material.

Step 4 Drag the bucket with teeth down through the ditch to help level out material and maintain grade.

Step 5 Remove the bucket from the ditch and fill it with material.

Step 6 Repeat the operation until all the material is placed and leveled.

To place a pipe section into position, use an approved sling or other approved rigging device to attach the pipe. Follow these steps to place the pipe on the bedding material in the ditch:

Step 1 Connect the rigging device to the shackle on the back of the bucket and arrange as shown in *Figure 45*.

Step 2 Lower the boom and center it over the pipe so that the rigging device can be attached.

Step 3 Raise the boom so that the slack is taken out of the rigging device.

Step 4 Continue raising the boom until the pipe clears all obstructions. Be careful not to hit any shoring that may be erected in the trench. Use a spotter or ground personnel to guide the pipe if necessary as shown in *Figure 46*.

Step 5 Lower the pipe so that it will line up with the end of the pipe section already set.

WARNING!

Do not make any sudden movements with the boom or leave the controls while the pipe is attached to the boom because people will be working on the ground nearby and in the trench.

Step 6 Move the bucket control to roll the bucket out and move the pipe in place.

Step 7 Lower the boom slightly to put slack in the sling. The ground crew will disconnect the sling from the pipe.

Step 8 Reposition the backhoe to set the next pipe.

Step 9 Continue this cycle until the job is complete.

Once the pipe has been properly placed, the trench can be backfilled with spoils or select material, as shown in *Figure 47*. This can be done with the front loader bucket to enable the complete job to be performed with one machine.

If the fill material is placed correctly along the side of the trench, the backfill can be done efficiently using the loader bucket in a dozing motion. The seven basic steps to backfilling are as follows:

Step 1 Position the seat to use the loader controls, and fasten the seat belt. Move the machine forward with the bucket level.

Step 2 Adjust the depth of the cut so the load can be moved without stalling the machine.

Step 3 If the engine starts to lug, downshift the transmission or reduce the depth of cut to prevent the engine from stalling.

Step 4 Push the maximum amounts of the load into the ditch without stalling. Leave the initial load in the bucket and use the full bucket to doze material or push soil from the stockpile into the ditch.

Step 5 Lift and level the bucket while backing away from the excavation to prepare for the next pass. Move the load at right angles to the excavation.

Step 6 Dump the load in the bucket as the last load in the excavation.

Step 7 Leave the load bucket spillage for a final lengthwise cleanup pass. Usually this will leave the backfill grade acceptable.

22303-14_F45.EPS

Figure 45 Rig a sling to the bucket.

22303-14_F46.EPS

Figure 46 Setting pipe.

5.5.0 Excavating Footings and Foundations

One of the most common uses of a backhoe besides trenching and ditching is excavating footings and foundations for buildings and other structures. For larger structures, an excavator may be a better choice because of its size. However, a backhoe is the ideal piece of equipment for tight areas and smaller excavations.

To excavate a foundation for a house or other structure, the backhoe will have to be repositioned several times during the operation. This will provide for efficient use of the equipment and produce the proper vertical surface to the walls of the excavation. To excavate a rectangular foundation, follow this approach:

Step 1 Travel to one corner of the excavation and set up the backhoe in the starting position. For example, if you start in the northeast corner, you would move across from east to west, positioning the backhoe so that the fully extended bucket just reaches the foundation wall to the front and right side (*Figure 48*).

Step 2 Lower the front bucket and stabilizers to level the machine.

Step 3 Begin digging on the right side of the area and place the spoils on the outside edge of the excavation. Your reach will be limited by the maximum extension of the boom and the required depth of the excavation.

Step 4 When you have dug all that must be dug from the initial position, move backward and set up in the next area. Position the backhoe so that the bucket will just barely reach the previous cut. Continue to dig placing the spoils on the outside edge (*Figure 49*).

Step 5 Move and set up in the next position until you have reached the south end of the excavation.

Step 6 Turn the backhoe at a right angle to the cut and position it to dig out the southeast corner (*Figure 50*). Make sure you can reach all the way across the face of the cut in front of you.

Step 7 Continue moving across the end of the excavation until reaching the opposite side or west wall.

Step 8 Turn at right angles and start down the west side of the excavation (*Figure 51*).

Step 9 Continue setting up and digging down the west side of the excavation until the opposite end is reached.

Step 10 Turn to the left and clean up across the end.

Step 11 Turn back to the south and continue digging until the excavation is complete (*Figure 52*).

This method will leave a straight, vertical wall, an even floor, and no ramps or uncleared areas.

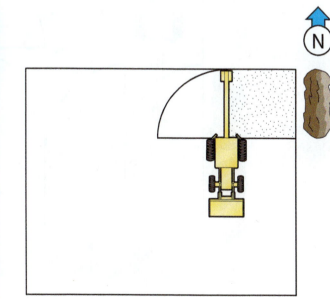

22303-14_F47.EPS

Figure 47 Backfilling.

22303-14_F48.EPS

Figure 48 Starting position.

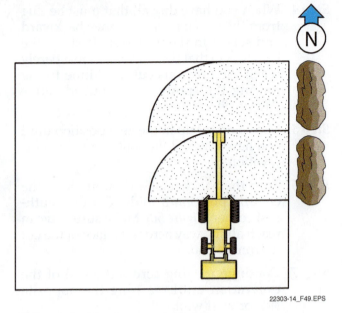

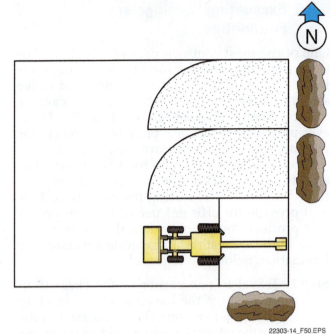

Figure 49 Position for second cut.

Figure 50 Digging out a corner.

5.6.0 Working in Confined or Unstable Areas

There will be times when it will be necessary to work in a confined area such as between two buildings, or up against a wall or fence. This requires careful thought about what has to be done and a plan for carrying it out. Examples of the most common situations are digging between two objects or trenching next to a wall. There also may be times when the backhoe has to be positioned on unstable material. The wheels and stabilizer pads could sink into the ground and create an unstable condition or possible safety hazard, preventing you from doing the work. The following sections explain procedures for working in confined or unstable areas.

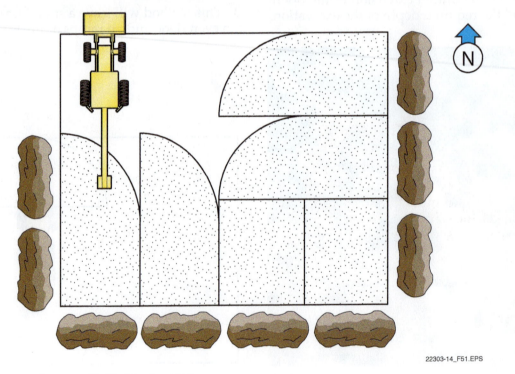

Figure 51 Excavating the second half of the foundation.

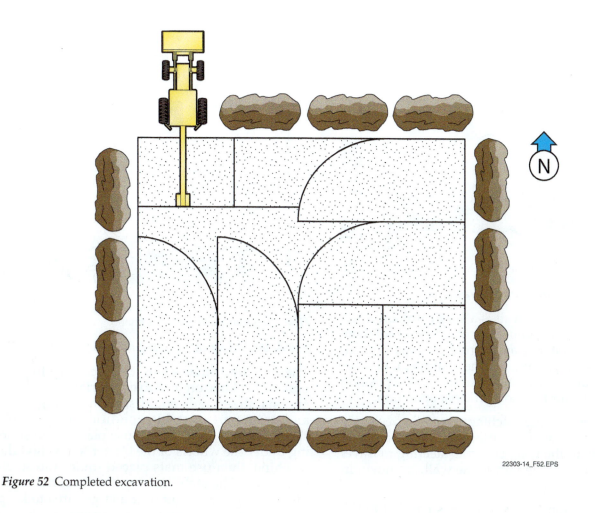

Figure 52 Completed excavation.

5.6.1 Digging Between Objects

Sometimes there is not enough room between the two objects to position the backhoe to dig straight ahead. In these cases, excavate when the boom is in the full swing position. Position the backhoe to the side of the excavation so that the boom is parallel to the excavation when in full swing position, as shown in *Figure 53*.

Step 1 Center the machine over the trench. Lower the stabilizers and excavate in the maximum swing position.

Step 2 Dump the load behind the excavation.

Step 3 Move the machine if necessary to excavate the opposite side.

Step 4 Move the machine back and perpendicular to the excavation as shown in *Figure 54* to excavate a larger area to connect both sides of the excavations.

5.6.2 Trench Digging

Subsurface drain tile is used to lower the water table in an area. These drains are installed to dry out wet basements when a high water table is causing basement water leakage. This system is designed to capture the water that would collect along the building and drain it away. It is also used in agriculture to regulate the water level in fields.

When digging trenches for subsurface drain tile or other pipe, you can backfill as you excavate.

Dig the first trench and stage the spoils. After the tile or pipe has been laid, dig the second trench. As you dig the second trench, backfill the first trench with the spoils from the second trench. Fill the last trench with stockpiled material from the first trench or bring in new material to backfill the final trench. This procedure combines the excavation and backfilling and is more efficient.

Figure 53 Position backhoe to dig between objects.

5.6.3 Trenching Next to Walls

Trenching next to a wall is a very common situation where the backhoe provides a good solution. Excavating material away from an existing wall requires steady hands on the controls. It also requires an awareness of how to set up and position the machine for the most efficient operation.

Step 1 Position the machine at a 45-degree angle to the wall. Position the backhoe parallel to the wall. Use crab steering to travel along the length of the wall, as shown in *Figure 55*.

Step 2 Position the front wheels parallel to the wall in the direction of digging. The machine can be repositioned along the wall by raising the rear of the machine and walking it sideways.

Step 3 Raise the stabilizer closest to the wall while trenching. Lower the other stabilizer for digging stability.

Step 4 Place spoils out of the way and in a location for easy backfilling or removal.

The digging cycle is greatly reduced with this method and it requires less repositioning.

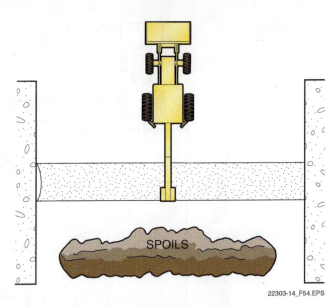

Figure 54 Excavate the center section.

5.6.4 Working on Unstable Soils

There will be times when it is necessary to position the backhoe on unstable material. The wheels and stabilizer pads could sink into the ground and create an unstable condition or possible safety hazard. In this case, use planks and shoring under the wheels of the tractor to position the machine, then use mats placed under the stabilizer pads. The stabilizers will rest on the mats, distributing the weight over a larger area to keep the machine from sinking into the ground during digging operations.

Occasionally, there will be an area of unsuitable material under a layer of firm soil. The firm soil usually bridges the unsuitable material, except for a few isolated areas where the equipment has broken through. For this bridged area, use a backhoe to remove the isolated unsuitable material. The backhoe can load the material into small dump trucks for removal without breaking through the thin bridged area. This reduces the chance of enlarging the unsuitable area before select material can be placed. This method works well in parking

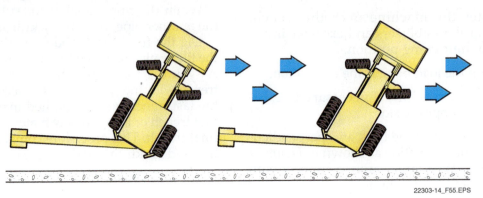

Figure 55 Position the backhoe.

lots where only car traffic will travel over the finished surface.

A backhoe is the best piece of equipment for removing unsuitable material around and above utility lines. If the material must be removed from below the utility line or very close to the top, a worker on the ground with a shovel and a prodding device should direct the backhoe operator.

Use the prodding device ahead of the work to keep the operator from damaging any service lines.

> **WARNING!**
>
> When working under power lines, the minimum clearances must be observed. A spotter should be used.

5.7.0 Roading and Transporting the Backhoe

When backhoe operations in one area are completed, the backhoe must be moved to the new location. For short distances, the backhoe can be driven to the new location. This is known as roading. For longer distances, it is more efficient to load the machine on a trailer and transport it to the new location.

Before moving the backhoe to a new job site, secure the machine for transport. Some machines have a hydraulic lockout switch, fuel shutoff valve, or other lock devices. Many backhoes have a swing pin to prevent the boom from swinging side to side. Backhoes with an extendable stick must be secured with a pin to lock the stick extension. Machines with all-wheel steering should be set in two-wheel steering mode. Follow the procedures in the operator's manual to secure the machine.

All backhoes have a boom lock (*Figure 56*). This is a mechanical latch that secures the backhoe in the closed position. Once the machine is shut down, engage the boom lock. Do not lock the boom for hauling, only use it for roading.

5.7.1 Roading

Backhoes may be driven on public highways. However, it is essential to take adequate safety precautions. Make sure the necessary permits for traveling on a public road have been obtained. Flags must be mounted on the left and right corners of the machine. Lights and flashers should be switched on. Depending on the location, a scout vehicle may also be required. Check the local regulations for roading the machine.

Before driving the backhoe to another work site, raise the loader bucket enough to provide sufficient ground clearance as shown in *Figure 57*. Make sure the boom is locked and the service brakes are locked together. Also, be sure to disengage the four-wheel drive so that the backhoe is roaded in the two-wheel drive mode. Follow the operator's manual for the exact procedures for securing the machine for roading.

Maintain a slow speed at all times in order to remain in complete control of the machine. Be sure to have a slow-moving vehicle sign that can be seen by traffic. Make sure you know the clearance requirements for the machine, both horizontal and vertical.

REPRINTED COURTESY OF CATERPILLAR INC.

22303-14_F56.EPS

Figure 56 Boom lock.

REPRINTED COURTESY OF CATERPILLAR INC.

22303-14_F57.EPS

Figure 57 Roading the backhoe/loader.

If the backhoe becomes disabled, decide whether it can be moved without any further damage. If you need to move it a short distance, it can be towed. Follow the manufacturer's instructions for towing the machine. If the tractor must be moved farther than one mile it should be loaded and transported on a trailer.

5.7.2 Transporting

If the equipment requires moving a long distance, it should be transported on a properly equipped trailer or other transport vehicle. Before loading the equipment for transport, make sure the following tasks have been completed:

- Make sure the loaded equipment complies with local height, width, and weight limitations for over-the-road hauling.
- Check the operator's manual to identify the correct tie-down points on the equipment.
- Be sure to get the proper permits, if required.
- Plan the loading operation so that the loading angle is at a minimum.

Once the above tasks are completed and the loading plan is determined, carry out the following procedures:

Step 1 Position the trailer or transporting vehicle. Always chock the wheels of the transporter before loading the backhoe (*Figure 58*).

Step 2 Place the loader bucket in the travel position and drive the backhoe onto the transporter. Whether the backhoe is facing forward or backward will depend on the recommendation of the manufacturer.

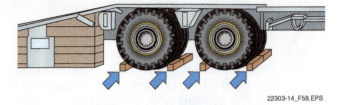

22303-14_F58.EPS

Figure 58 Chock trailer wheels.

Step 3 If the backhoe is articulated, connect the steering frame lock link to hold the front and rear frames together.

Step 4 Release the boom lock and lower both the loader bucket and the backhoe bucket to the floor of the transporter.

Step 5 Move the transmission lever to neutral, and engage the parking brake and the transmission neutral lock. Turn off the engine.

Step 6 Manipulate the bucket and operate all controls to remove any remaining hydraulic pressure.

Step 7 Remove the engine start key and secure vandal guards. Activate any disconnect switches. Place the fuel shutoff valve in the Of position.

Step 8 Lock the door to the cab as well as any access covers. Attach any vandalism protection.

Step 9 Secure the machine with the proper tiedown equipment as specified by the manufacturer (*Figure 59*).

Step 10 Cover the exhaust and air intake openings with tape or a plastic cover to protect the turbocharger.

Step 11 Place appropriate flags or markers on the equipment if needed for height and width restrictions.

Unloading the equipment from the transporter would be the reverse of the loading operation.

22303-14_F59.EPS

Figure 59 Secure the machine on the trailer.

5.0.0 Section Review

1. An advantage of using a backhoe bucket for loading a truck is _____.

   a. the backhoe bucket is larger than the loader bucket
   b. it is easier to maneuver the machine in reverse
   c. the entire backhoe machine can remain stationary
   d. that front loaders are notoriously unstable

2. In order to minimize wear on a backhoe's teeth during trenching, it is best to angle the cutting edge at about a _____.

   a. 10-degree angle to vertical
   b. 30-degree angle to the ground
   c. 60-degree angle to the ground
   d. 90-degree angle to the horizon

3. A hydraulic breaker attachment functions much like a _____.

   a. jackhammer
   b. circuit breaker
   c. check valve
   d. rotary drill

4. When setting pipe in a trench, the backhoe bucket is often used to _____.

   a. remove bedding material from the trench to allow room for the pipe
   b. hammer the piping into place and backfill around it
   c. place bedding material into the trench and lift the pipe into place
   d. position street pads in the trench to cushion the piping sections

5. Using a backhoe to excavate a foundation for a house requires _____.

   a. installing a foundation planer attachment on the front of the backhoe
   b. operating without the use of stabilizers or outriggers
   c. repeatedly drilling along the foundation line with an auger
   d. repositioning the backhoe several times during the operation

6. What can be used under the tractor wheels of a backhoe to help safely position it on unstable soil?

   a. Sand and street pads
   b. Hydraulic breakers
   c. Planks and shoring
   d. Any available bedding material

7. Before roading a backhoe, an operator should raise the loader bucket enough to provide sufficient ground clearance, make sure that the boom is locked, and _____.

   a. carefully back the machine onto the trailer
   b. lock the service brakes together
   c. activate the disconnect switches
   d. hitch the backhoe to the scout vehicle

Summary

The backhoe is a versatile piece of equipment used for excavating and loading. Although it is not as large as other excavating equipment, its size and compactness make it ideal for use in smaller areas and tight places where larger equipment cannot operate.

Backhoes are attachments to wheel-mounted tractors, trucks, and crawler-type tractors that can dig below the ground where the machine is standing. The main uses of a backhoe include the following: digging and cleaning narrow ditches; excavating smaller areas where other equipment cannot work; light-duty hoisting; and loading trucks. There are many sizes and models available. The most common is a combination backhoe/loader, which is a rubber-tire tractor with a bucket loader attached to the front and the backhoe attached to the rear.

Backhoes have several other attachments that can be used for special jobs. These include a ripper for hard or frozen material and roots; an auger for digging holes; a compactor; and a hydraulic breaker for breaking up pavement and other hard material.

The backhoe requires daily preoperation inspection checks just like any other piece of heavy equipment. Make sure the machine is operating properly before starting a job. Operating a broken or damaged backhoe could cause injury to yourself or others.

When using the backhoe for digging or trenching, the operator must be able to move the boom up and down while at the same time curling the bucket in order to pick up the material and retain it in the bucket. Dumping the material is the reverse operation. The operator must raise the boom while keeping the bucket curled enough to keep the material in the bucket.

Setting up the backhoe requires the use of stabilizers. When properly set, the machine will be level. The right and left stabilizers rest on their pads and the front bucket rests on the ground. The weight of the machine is off the wheels and on the stabilizer pads and the front bucket. This creates a stable platform for operating the machine.

For trenching, the backhoe should be lined up with the limits of the excavation. The backhoe works from a point of furthest reach of the stick to a point closest to and directly below the machine. The machine is then driven forward, away from the excavated area. Once repositioned, reset the backhoe to continue trenching.

The backhoe can be transported to a new job site either by roading or by loading it onto a trailer.

1. Because of its rubber tires, a typical backhoe _____.

    a. must be transported on a trailer no matter where it is moved
    b. can be operated on slopes that are greater than 33 degrees
    c. is able to be stabilized by reducing the tire air pressure
    d. can be driven short distances on public roads

2. One disadvantage of a backhoe is that it _____.

    a. is less stable than other equipment
    b. is heavy compared to its digging power
    c. cannot be used in small spaces
    d. can only be operated for a limited time

3. One of the most common uses of a backhoe is _____.

    a. heavy-duty hoisting of materials
    b. warehousing operations
    c. digging and cleaning narrow ditches
    d. replacing dozers for excavating large areas

4. Some backhoes have an option that provides operators with up to an additional 4 feet (1.2 meters) of boom reach. This option is called a(n) _____.

    a. boom stretcher
    b. extendable stick feature
    c. hydraulic breaker
    d. stabilizer outrigger

Identify the parts of the backhoe shown in *Review Question Figure 1* by writing the correct letter in each blank.

5. _____ Stabilizer

6. _____ Boom

7. _____ Bucket

8. _____ ROPS

Figure 1

22303-14_RQ01.EPS

9. The controls and instruments on the front control panel of a backhoe/loader are typically used to _____.
   a. move and position the machine
   b. control the backhoe operations
   c. disconnect the battery and fuel supply
   d. set up the machine for remote operation

10. On a backhoe/loader, the tachometer shows the _____.
    a. voltage in the electrical system
    b. degree of tilt in the stabilizers
    c. engine speed in revolutions per minute
    d. pressure in the hydraulic system

11. Some backhoe controls have a foot pedal to control the _____.
    a. transmission
    b. boom swing
    c. bucket
    d. hydraulic pressure

12. A large metal curved blade that can be used on frozen ground, asphalt, tree roots, and hard soils is a _____.
    a. dipper
    b. cold planer
    c. tamper
    d. ripper

13. An auger attachment is a tapered metal spiral blade that is _____.
    a. mounted on a rotating motor
    b. designed to grind asphalt or concrete
    c. used for backfilling trenches
    d. equipped with a large tooth

14. A backhoe attachment that is often used in demolition work is a _____.
    a. cold planer
    b. vibratory compactor
    c. ripper
    d. hydraulic breaker

15. When using a backhoe in an area where other equipment, traffic, or pedestrians are present, it is important to _____.
    a. allow a co-worker to ride on the backhoe as a lookout
    b. display flags inside the backhoe cab for added safety
    c. establish a safe work zone using cones or barriers
    d. position a co-worker directly beside the excavation area

16. Maintenance intervals for most machines like backhoes are established by the _____.
    a. International Society of Automation (ISA)
    b. Society of Automotive Engineers (SAE)
    c. American Society of Mechanical Engineers (ASME)
    d. American National Standards Institute (ANSI)

17. Continuous and hard operation of a backhoe's hydraulic system can cause _____.
    a. condensation to form in the hydraulic fluid
    b. transmission fluid problems, as well
    c. the backhoe control levers to shake uncontrollably
    d. the hydraulic fluid to boil and break down

18. To prevent spilling when loading a truck with a backhoe's bucket, it is best to _____.
    a. curl the bucket close to the stick
    b. fill the bucket only half full
    c. avoid swinging the boom
    d. have the truck back under each bucket load

19. When a backhoe is warming up, an operator should stop the engine and investigate the problem if the oil pressure light does not turn off _____.
    a. immediately after the engine fires
    b. within 10 seconds
    c. within 30 seconds
    d. within 1 minute

20. Steering is accomplished on a typical backhoe by using _____.
    a. control levers
    b. foot pedals
    c. a steering wheel
    d. a joystick

21. When positioning a backhoe on level ground, lower the front bucket and apply downward pressure until _____.
    a. most of the pressure is off the front wheels
    b. most of the pressure is off the rear wheels
    c. the bucket is 15 inches (38 centimeters) from the ground
    d. the bucket is empty

22. For safety, do not operate the backhoe on hills with a slope of more than ____.

    a. 10 degrees
    b. 15 degrees
    c. 25 degrees
    d. 30 degrees

23. What is the first step when loading a truck with a backhoe?

    a. Drag the bucket through the material.
    b. Position the backhoe so the boom reaches the spoils.
    c. Position the backhoe by the truck for easy loading.
    d. Move the bucket over the truck bed.

24. If a backhoe is being used to excavate a trench and load a haul truck at the same time, an operator should position the backhoe so that ____.

    a. the truck can be loaded from the rear
    b. it is in a straight line with trench markers that have been set
    c. it faces the truck being loaded
    d. the front loader bucket can do the initial trench preparation

25. A hydraulic breaker attachment that is used on a backhoe for demolition work is typically controlled using ____.

    a. loader controls
    b. a joystick
    c. foot- or button-operated controls
    d. a lever on the front instrument panel

26. If fill material has been placed correctly along the side of a trench, the front loader bucket of a backhoe can be used to backfill the trench by using a(n) ____.

    a. breaker attachment
    b. ripping procedure
    c. auger blade on the bucket
    d. dozing motion

27. When working on unstable soils, mats are placed under the ____.

    a. stabilizers
    b. wheels
    c. loader bucket
    d. backhoe bucket

28. When roading a four-wheel drive backhoe, it is best to ____.

    a. keep the backhoe in its standard four-wheel drive mode
    b. engage the transmission neutral lock to synchronize the wheel speeds
    c. disengage the four-wheel drive and road in two-wheel drive
    d. use the reverse gears and drive with the seat facing the backhoe attachment

29. Before roading a backhoe to another site, be sure to ____.

    a. lock the boom
    b. set the differential lock
    c. raise the bucket as high as possible
    d. secure the machine with proper tie-down equipment

30. How should the loader and backhoe buckets be positioned when a backhoe is being transported on a trailer?

    a. Both buckets raised to the travel position.
    b. Loader bucket raised; backhoe bucket lowered.
    c. Loader bucket lowered; backhoe bucket raised.
    d. Both buckets lowered onto the trailer floor.

Trade Terms Introduced in This Module

Bucket: A U-shaped closed-end scoop that is attached to the backhoe.

Crowd: The process of forcing the stick into digging, or moving the stick closer to the machine.

Curl: Rotate the bucket.

Draw: Move the stick back toward the operator.

Extend: Move the extendable stick outward.

Reach: Extend the stick away from the cab.

Retract: Move the extendable stick in.

Rubble: Fragments of stone, brick, or rock that have broken apart from larger pieces.

Spoils: Excavated material from a digging operation.

Trenching: Using a backhoe to dig a long, straight excavation with vertical walls.

Appendix

MAINTENANCE INTERVAL SCHEDULE

Maintenance Interval Schedule

Note: All safety information, warnings, and instructions must be read and understood before you perform any operation or maintenance procedure.

Before each consecutive interval is performed, all of the maintenance requirements from the previous interval must also be performed.

The normal oil change interval is every 500 service hours. If you operate the engine under severe conditions or if the oil is not Caterpillar oil, the oil must be changed at shorter intervals. Refer to the Operation and Maintenance Manual, "Engine Oil and Filter – Change" for further information. Severe conditions include the following factors: high temperatures, continuous high loads, and extremely dusty conditions.

Refer to the Operation and Maintenance Manual, "S.O.S. Oil Analysis" in order to determine if the oil change interval should be decreased. Refer to your Caterpillar dealer for detailed information regarding the optimum oil change interval.

The normal interval for inspecting and adjusting the clearance between the wear pads and the boom is 500 service hours. If the machine is working with excessively abrasive material then the clearance may need to be adjusted at shorter intervals. Refer to the Operation and Maintenance Manual, "Boom Wear Pad Clearance – Inspect/Adjust" for further information.

When Required

Battery – Recycle
Battery or Battery Cable – Inspect/Replace
Boom Telescoping Cylinder Air – Purge
Boom and Frame – Inspect
Engine Air Filter Primary Element – Clean/Replace
Engine Air Filter Secondary Element – Replace
Engine Air Filter Service Indicator – Inspect
Engine Air Precleaner – Clean
Fuel System – Prime
Fuel System Primary Filter – Replace
Fuel System Secondary Filter – Replace
Fuel Tank Cap and Strainer – Clean
Fuses and Relays – Replace
Oil Filter – Inspect
Radiator Core – Clean
Radiator Screen – Clean
Transmission Neutralizer Pressure Switch – Adjust
Window Washer Reservoir – Fill
Window Wiper – Inspect/Replace

Every 10 Service Hours or Daily

Backup Alarm – Test
Boom Retracting and Boom Lowering with Electric Power – Check
Braking System – Test
Cooling System Coolant Level – Check
Cooling System Pressure Cap – Clean/Replace
Engine Oil Level – Check
Fuel System Water Separator – Drain
Fuel Tank Water and Sediment – Drain
Indicators and Gauges – Test
Seat Belt – Inspect
Tire Inflation – Check
Transmission Oil Level – Check
Wheel Nut Torque – Check
Windows – Clean

Every 50 Service Hours or 2 Weeks

Axle Support – Lubricate
Bearing (Pivot) for Axle Drive Shaft – Lubricate
Boom Cylinder Pin – Lubricate
Boom Pivot Shaft – Lubricate
Brake Control Linkage – Lubricate
Carriage Cylinder Bearing – Lubricate
Carriage Pivot Pin - Lubricate
Compensating Cylinder Bearing – Lubricate
Cylinder Pin (Grapple Bucket) – Lubricate
Cylinder Pin and Pivot Pin (Bale Handler) – Lubricate
Cylinder Pin and Pivot Pin (Multipurpose Bucket) – Lubricate
Cylinder Pin and Pivot Pin (Utility Fork) – Lubricate
Fork Leveling Cylinder Pin – Lubricate
Frame Leveling Cylinder Pin – Lubricate
Pulley for Boom Extension Chain – Lubricate
Pulley for Boom Retraction Chain – Lubricate
Quick Coupler – Lubricate
Stabilizer and Cylinder Bearings – Lubricate

Initial 250 Service Hours (or after rebuild)

Boom Wear Pad Clearance – Inspect/Adjust
Service Brake – Adjust

Initial 250 Service Hours (or at first oil change)

Hydraulic System Oil Filter – Replace

Every 250 Service Hours or 3 Months

Axle Breathers – Clean/Replace
Belts – Inspect/Adjust/Replace
Boom Chain Tension – Check/Adjust
Differential Oil Level – Check
Drive Shaft Spline – Lubricate
Drive Shaft Universal Joint – Lubricate

22303-14_A01A.EPS

Engine Oil Sample – Obtain
Final Drive Oil Level – Check
Hydraulic System Oil Level – Check
Longitudinal Stability Indicator – Test
Transfer Gear Oil Level – Check
Transmission Breather – Clean

Every 500 Service Hours

Differential and Final Drive Oil Sample – Obtain

Every 500 Service Hours or 6 Months

Boom Wear Pad Clearance – Inspect/Adjust
Engine Oil and Filter – Change
Fuel System Primary Filter – Replace
Fuel System Secondary Filter – Replace
Fuel Tank Cap and Strainer – Clean
Hydraulic System Oil Filter – Replace
Hydraulic System Oil Sample – Obtain
Hydraulic Tank Breather – Clean
Service Brake – Adjust
Transmission Oil Sample – Obtain

Every 1,000 Service Hours or 1 Year

Differential Oil – Change
Engine Valve Lash – Check
Final Drive Oil – Change
Rollover Protective Structure (ROPS) and Falling
  Object Protective Structure (FOPS) – Inspect
Transfer Gear Oil – Change
Transmission Oil – Change
Transmission Oil Filter – Replace

Every 2,000 Service Hours or 2 Years

Fuel Injection Timing – Check
Hydraulic System Oil – Change

**Every 3 Years After Date of Installation or
Every 5 Years After Date of Manufacture**

Seal Belt – Replace

Every 3,000 Service Hours or 3 Years

Boom Chain – Inspect/Lubricate
Cooling System Coolant Extender (ELC) – Add
Cooling System Water Temperature Regulator –
  Replace
Engine Mounts – Inspect

Every 6,000 Service Hours or 6 Years

Cooling System Coolant (ELC) – Change

22303-14_A01B.EPS

Figure Credits

Deere & Company, Module opener, Figures 1, 4, 7, 14, 15, 30, 31, 37–39, 47

Reprinted Courtesy of Caterpillar Inc., Figures 2, 3, 5, 6, 8–13, 16–20, 22–29, 32–36, 40, 42, 44–46, 56, 57, Appendix

Tim Davis/NCCER, Figure 21

Associated General Contractors, Figure 41

Answer	Section Reference	Objective
Section One		
1.c	1.1.0	1a
2.d	1.2.0	1b
Section Two		
1.a	2.1.0	2a
2.c	2.2.0	2b
3.b	2.3.3	2c
4.d	2.4.6	2d
Section Three		
1.b	3.1.1	3a
2.d	3.2.1	3b
3.a	3.3.0	3c
Section Four		
1.c	4.1.0	4a
2.b	4.2.1	4b
3.a	4.3.1	4c
Section Five		
1.c	5.1.0	5a
2.b	5.2.0	5b
3.a	5.3.0	5c
4.c	5.4.0	5d
5.d	5.5.0	5e
6.c	5.6.4	5f
7.b	5.7.1	5g

NCCER CURRICULA — USER UPDATE

NCCER makes every effort to keep its textbooks up-to-date and free of technical errors. We appreciate your help in this process. If you find an error, a typographical mistake, or an inaccuracy in NCCER's curricula, please fill out this form (or a photocopy), or complete the online form at **www.nccer.org/olf**. Be sure to include the exact module ID number, page number, a detailed description, and your recommended correction. Your input will be brought to the attention of the Authoring Team. Thank you for your assistance.

Instructors – If you have an idea for improving this textbook, or have found that additional materials were necessary to teach this module effectively, please let us know so that we may present your suggestions to the Authoring Team.

NCCER Product Development and Revision

13614 Progress Blvd., Alachua, FL 32615

Email: curriculum@nccer.org
Online: www.nccer.org/olf

❏ Trainee Guide   ❏ Lesson Plans   ❏ Exam   ❏ PowerPoints   Other _____

Craft / Level: _____   Copyright Date: _____

Module ID Number / Title: _____

Section Number(s): _____

Description: _____

_____

_____

_____

_____

Recommended Correction: _____

_____

_____

_____

_____

Your Name: _____

Address: _____

_____

Email: _____   Phone: _____

22310-14

Off-Road Dump Trucks

OVERVIEW

Heavy construction, mining, and quarrying jobs require huge amounts of material to be moved from one area to another on site. One of the most common pieces of equipment used for this purpose is an off-road dump truck. A skilled operator can recognize different types of off-road dump trucks and understand which type of truck is best suited for the task at hand. Safe operators have a working knowledge of all of the instruments and controls, perform daily inspections and maintenance, and know the rules for operating on haul roads and excavated ground.

Module Four

Objectives

When you have completed this module, you will be able to do the following:

1. Identify and describe basic types, uses, and components of off-road dump trucks.
   a. Identify and describe rigid dump trucks.
   b. Identify and describe articulated dump trucks.
   c. Identify and describe off-road truck instrumentation.
   d. Identify and describe off-road truck control systems.
2. Identify and describe safety, inspection, and service guidelines associated with off-road dump trucks.
   a. Describe guidelines associated with off-road truck safety.
   b. Describe prestart inspection procedures.
   c. Describe preventive maintenance requirements.
3. Describe basic startup and operating procedures for off-road dump trucks.
   a. Describe startup, warm-up, and shutdown procedures.
   b. Describe safe driving maneuvers and loading and dumping procedures.

Performance Tasks

Under the supervision of your instructor, you should be able to do the following:

1. Complete a proper prestart inspection and maintenance for an off-road dump truck.
2. Perform the proper startup, warm-up, and shutdown procedures.
3. Carry out basic operations with an off-road dump truck:
   - Properly position a truck for loading.
   - Safely drive the truck to a designated dumping site.
   - Dump the load in the designated spot.
   - Set the retarder system to reduce wear on the service brakes.

Trade Terms

Articulated-frame dump truck
Automatic retarder control system
Canopy
Engine retarder
Exhaust body heating system
Haul truck

Hoist
Rigid-frame dump truck
Quarries
Traction control system
Underground mining dump truck

Industry Recognized Credentials

If you are training through an NCCER-accredited sponsor, you may be eligible for credentials from NCCER's Registry. The ID number for this module is 22310-14. Note that this module may have been used in other NCCER curricula and may apply to other level completions. Contact NCCER's Registry at 888.622.3720 or go to **www.nccer.org** for more information.

Contents

Topics to be presented in this module include:

1.0.0 Types, Uses, and Components of Off-Road Dump Trucks 1
  1.1.0 Rigid-Frame Dump Trucks 1
    1.1.1 Rigid-Frame Dump Truck Uses 2
    1.1.2 Rigid-Frame Dump Truck Components 3
    1.1.3 Rigid-Frame Dump Truck Bodies 5
  1.2.0 Articulated-Frame Dump Trucks 5
    1.2.1 Articulated Truck Uses 6
    1.2.2 Articulated Truck Components 6
    1.2.3 Underground Trucks 7
  1.3.0 Instrumentation 7
  1.4.0 Control Systems 10
    1.4.1 Engine Control 10
    1.4.2 Transmission Control 10
    1.4.3 Traction Control 11
    1.4.4 Braking Control 12
    1.4.5 Hoist Control 13
    1.4.6 Object Detection System 14
    1.4.7 Ground Level Controls 15
2.0.0 Safety, Inspection, and Maintenance 17
  2.1.0 Safety Guidelines 17
    2.1.1 Operator Safety 17
    2.1.2 Safety of Co-Workers and the Public 18
    2.1.3 Equipment Safety 18
  2.2.0 Prestart Inspection Procedures 19
  2.3.0 Preventive Maintenance 20
3.0.0 Startup and Operation 22
  3.1.0 Startup, Warm-Up, and Shutdown 22
    3.1.1 Startup 22
    3.1.2 Warm-Up 22
    3.1.3 Shutdown 23
  3.2.0 Safe Truck Operation 23
    3.2.1 Basic Maneuvering 24
    3.2.2 Backing Safely 25
    3.2.3 Climbing and Descending Hills 25
    3.2.4 Taking Curves 25
    3.2.5 Safe Loading Practices 26
    3.2.6 Safe Dumping Practices 26
    3.2.7 Emergency Situations 28

Figures

Figure 1   Off-road dump truck being loaded by a shovel excavator.............. 2
Figure 2   Rigid-frame dump truck... 2
Figure 3   Rigid-frame dump truck with diesel/electric drive system.............. 3
Figure 4   Rigid-frame dump trucks traveling over haul road......................... 4
Figure 5   Typical rigid-frame dump truck components 4
Figure 6   Common rigid-frame dump truck bodies.. 5
Figure 7   Articulated-frame dump truck 6
Figure 8   Articulated truck at highway construction site............................... 6
Figure 9   Articulated-frame dump truck components 7
Figure 10 Articulated truck with tailgate.. 7
Figure 11 Underground mining dump truck 8
Figure 12 Rigid-frame dump truck cab layout 9
Figure 13 Instrument panel 10
Figure 14 Articulated truck instruments and warning lights 11
Figure 15 Display panel 11
Figure 16 Throttle lock 12
Figure 17 Transmission shifter and display panel indication 12
Figure 18 The need for traction control... 12
Figure 19 Wet disc brake 13
Figure 20 Secondary brake pedal in cab 13
Figure 21 Automatic retarder control system 14
Figure 22 Hoist cylinders 14
Figure 23 Hoist control lever 14
Figure 24 Object detection system 15
Figure 25 Emergency shutoff switch 15
Figure 26 Electrical center 15
Figure 27 Fluid fill center 19
Figure 28 Grease fitting cluster.. 20
Figure 29 Restricted visibility around a typical off-road truck 24
Figure 30 Truck with balanced load 26
Figure 31 Articulated truck dumping from safe distance............................. 27

1.0.0 TYPES, USES, AND COMPONENTS OF OFF-ROAD DUMP TRUCKS

Objective

Identify and describe basic types, uses, and components of off-road dump trucks.

a. Identify and describe rigid dump trucks.
b. Identify and describe articulated dump trucks.
c. Identify and describe off-road truck instrumentation.
d. Identify and describe off-road truck control systems.

Trade Terms

Articulated-frame dump truck: A type of off-road dump truck that has a permanent pivoting point in the frame that allows the front of the truck to turn so that all of the wheels follow the same path.

Automatic retarder control system: A system that works with the engine brake and the traction control system to electronically slow the vehicle during downhill travel.

Canopy: A section of an off-road truck body that extends above the operator's cab to help protect the operator from falling material.

Engine retarder: An alternate braking system activated from the cab that slows the vehicle by reducing engine power.

Exhaust body heating system: A system that diverts some of the exhaust gas from a truck's exhaust pipe(s) to conduits in the body to help prevent material in the body from freezing.

Haul truck: A name that is sometimes used to describe a rigid-frame dump truck or a mining truck.

Hoist: The mechanism used to raise and lower the dump body, typically consisting of two hydraulic cylinders.

Quarries: Excavations or pits used to mine gravel, stone, and other material.

Rigid-frame dump truck: A type of off-road truck in which the cab and the body are mounted on a common, non-pivoting frame. Rigid-frame dump trucks are commonly referred to as haul trucks or mining trucks.

Traction control system: A computerized system that diverts torque from a spinning wheel to one or more of the other wheels to improve traction.

Underground mining dump truck: A type of articulated truck in which the vehicle height is reduced by moving the cab forward of the front axle, thus enabling the truck to be used in underground mining applications.

Dump trucks are widely used in the construction and mining industries. While some types of dump trucks are designed for on-road use, many are intended for use only on site. These off-road, or off-highway, trucks barely resemble on-road dump trucks. In fact, they are seldom referred to as dump trucks at all, even though they have bodies that carry large amounts of material and are lifted for dumping.

Off-road trucks are built for the demanding conditions that exist at most large construction sites, road building sites, quarries, and mines. They are usually much larger than on-road vehicles and are prohibited by law from routine operation on public roads because of their great size and weight. While on-road dump trucks require the operator to have a commercial driver's license (CDL), off-road trucks do not. Operators do, however, need to complete training to be certified to operate these vehicles. In addition, the Occupational Safety and Health Administration (OSHA) regulates certain aspects of these trucks in *Standard 29 CFR 1926.601*. The Mine Safety and Health Administration (MSHA) regulates other aspects.

Regardless of their size, off-road trucks all serve the same basic purpose: to safely and efficiently move large volumes of material from one area of a work site to another.

Off-road trucks are designed to be loaded on site with loaders, excavators, or other equipment (*Figure 1*). The loaded truck is then driven to a dump site, where hoist cylinders raise the body and dump the load. The efficiency of the loading, driving, and dumping process depends, in large part, on an operator's ability to safely maneuver the truck.

Off-road trucks vary a great deal in size and capacity. They can be divided into two basic categories: rigid-frame dump trucks and articulated-frame dump trucks.

1.1.0 Rigid-Frame Dump Trucks

A rigid-frame dump truck has a rigid, or non-pivoting, frame that is similar to the one is used on

22310-14_F01.EPS

Figure 1 Off-road dump truck being loaded by a shovel excavator.

a standard on-road dump truck. The cab and the body are mounted on a common frame and operate together. The rigid frame is incredibly strong so that it can handle the heavy loads the truck is designed to carry. These types of trucks are commonly called rigid-frame dump trucks or **haul trucks**. Some manufacturers also categorize their larger rigid-frame dump trucks as mining trucks. *Figure 2* shows a typical rigid-frame dump truck.

Rigid-frame dump trucks usually have two axles. The front axle has some type of conventional steering mechanism for steering the vehicle. The rear axle has the drivetrain gears needed to drive the vehicle. Most rigid-frame dump trucks

have dual wheels on each side of the rear axle for strength, traction, and stability.

All rigid-frame dump trucks have a large diesel engine, but many of them also use electric motors in combination with the diesel engine. In this diesel/electric powertrain configuration (*Figure 3*), the diesel engine drives an AC alternator or DC generator, which provides power to electric motors located inside the axle at the rear wheels. This arrangement increases efficiency by providing greater power to each drive wheel and better braking, but with less weight.

Rigid-frame dump trucks have huge capacities. Even smaller rigid-frame dump trucks can haul 30 tons (27.22 tonnes) of material. Larger trucks used in mining applications can haul up to 400 tons (362.87 tonnes). Any rigid-frame dump truck that is capable of hauling 272 tons (246.75 tonnes) or more is categorized as an ultra-class truck.

1.1.1 Rigid-Frame Dump Truck Uses

Rigid-frame dump trucks can be used in most heavy construction work, as long as the work site terrain is suitable. Because of their massive size and weight, these trucks are best suited for mining, quarrying, or heavy construction work where the terrain is relatively stable and/or haul roads have been established (*Figure 4*). In addition, these trucks require lots of space to operate. It is difficult for operators to see personnel, obstacles, or other vehicles unless they are a good distance from the truck itself.

22310-14_F02.EPS

Figure 2 Rigid-frame dump truck.

ELECTRIC GENERATING EQUIPMENT
FOR AC DRIVETRAIN MOTORS

AC DRIVETRAIN MOTOR
INSIDE REAR AXLE

22310-14_F03.EPS

Figure 3 Rigid-frame dump truck with diesel/electric drive system.

1.1.2 Rigid-Frame Dump Truck Components

Some of the major components of a typical rigid-frame dump truck are shown in *Figure 5*. The operator's cab is at the front of the vehicle. It has a built-in rollover protective structure (ROPS) and falling object protective structure (FOPS) and contains the controls for driving the truck and dumping the body.

The operator's cab is accessed by using the ladder or stairs on the front of the vehicle. A platform directly in front of the operator's cab provides access to the electrical generating equipment for the drivetrain motors. Operators can also use this platform to wash the cab windows and perform other routine tasks. A protective railing surrounds the platform outside of the cab to prevent falls.

The body of the truck is located behind the cab. The body has an extended section above the operator's cab to further protect the operator from falling material. This part of the body is called the canopy. Unlike the dump body on a conventional on-road truck, the body on an off-road truck does not have a tailgate. Hydraulic hoist cylinders are used to raise and lower the body. These are controlled from inside the cab.

Most rigid-frame dump trucks have two axles. The front axle is used for steering and the rear axle drives the vehicle.

The engine is in the front of the vehicle, much like a conventional on-road dump truck. Engine sizes vary, but the engine for this truck is rated at 2,500 horsepower.

World's Largest Haul Truck

The title for the world's largest haul truck changes periodically. But it is hard to call any of these behemoths small. Some of the largest trucks stand over 25 feet tall (7.62 meters), weigh over 250 tons (226.80 tonnes) empty, can haul 400 tons (362.87 tonnes), and have V-24 diesel engines capable of producing over 3,700 horsepower. The truck shown here is more than 15 feet (4.57 meters) tall at the top of the cab and is more than 35 feet (10.69 meters) long. Its front tires are nearly 14 feet (4.27 meters) wide.

REPRINTED COURTESY OF CATERPILLAR INC.

22310-14_SA01.EPS

Figure 4 Rigid-frame dump trucks traveling over haul road.

22310-14_F04.EPS

PROTECTIVE CANOPY

OPERATOR'S CAB

ELECTRICAL GENERATING EQUIPMENT FOR AC DRIVETRAIN MOTORS

DUMP BODY

PROTECTIVE RAILINGS

ENGINE COMPARTMENT

STAIRS FOR CAB ACCESS

FRONT AXLE (STEERING)

HYDRAULIC HOIST CYLINDER FOR DUMP BODY

REAR AXLE (DRIVE)

22310-14_F05.EPS

Figure 5 Typical rigid-frame dump truck components.

1.1.3 Rigid-Frame Dump Truck Bodies

Different types of bodies are available for rigid-frame dump trucks. The type that is used can depend on the need. Some manufacturers make different bodies that can be selected when the truck is purchased. These bodies are typically offered in different sizes and with different liners. Custom-built bodies and options such as tail extensions and sideboards can also be obtained. *Figure 6* shows a few examples of bodies that are available for a rigid-frame dump truck.

- *Mine specific design (MSD II)* – A lighter-weight body designed for mining work.
- *X-body* – A heavy-duty body designed for harsh applications.
- *Gateless coal* – A high-volume body designed for coal haulage.
- *Dual slope* – One type of standard body that works well in most tough applications.
- *Combination* – Similar to the dual-slope body. It is a type of standard, or multipurpose, body with a high volume that works well in light-density applications.

OSHA Standard 29 CFR 1926.601(b)(10) states that "trucks with dump bodies shall be equipped with positive means of support, permanently attached, and capable of being locked in position to prevent accidental lowering of the body while maintenance or inspection work is being done." This precaution must be taken into account anytime a truck body is changed or modified.

1.2.0 Articulated-Frame Dump Trucks

An articulated-frame dump truck (*Figure 7*) has a pivoting point in the frame between the cab and the dump body. This permanent hinge enables the front of the truck to turn in a way that all the wheels follow the same path. However, an articulated-frame dump truck should not be confused with a tractor trailer arrangement in which a separate tractor pulls a trailer that is connected to it. An articulated-frame dump truck is permanently connected in much the same way as an articulated loader or an articulated motor grader.

Articulated-frame dump trucks, which are sometimes called articulated haulers, are typically all-wheel drive vehicles. Most models have a front axle and two rear axles. Hydraulic cylinders are used to provide the steering by turning the entire front part of the vehicle. As *Figure 7* shows, articulated trucks are well adapted to rough terrain.

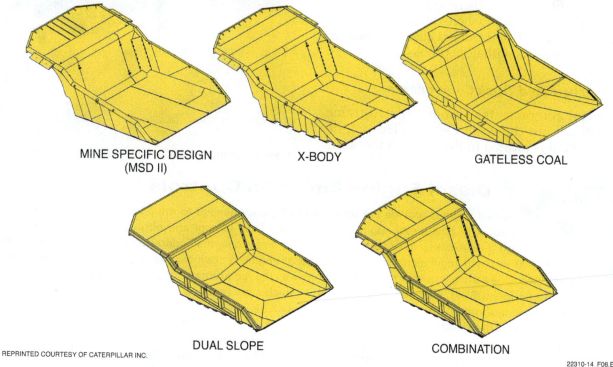

MINE SPECIFIC DESIGN (MSD II)   X-BODY   GATELESS COAL

DUAL SLOPE   COMBINATION

REPRINTED COURTESY OF CATERPILLAR INC.

22310-14_F06.EPS

Figure 6 Common rigid-frame dump truck bodies.

Figure 7 Articulated-frame dump truck.

1.2.1 Articulated Truck Uses

Articulated trucks are capable of handling loads from about 25 tons (22.68 tonnes) up to about 50 tons (45.34 tonnes). They can be used in basically any heavy hauling application. However, because of their all-wheel drive and maneuverability, they are most likely to be used in rugged applications that may require a lot of turning. Generally, they do not haul as much material as rigid-frame dump trucks, but they do a better job of moving material across muddy terrain and sites that may not have well-maintained haul roads.

One common use of articulated trucks is in highway construction (*Figure 8*). On many highway construction sites, there are numerous side roads and sloping areas where dump trucks need to travel. An articulated truck is able to maneuver around these areas with little problem.

1.2.2 Articulated Truck Components

The major components of an articulated-frame dump truck are shown in *Figure 9*. The engine compartment is located in the front, in the same basic area as an on-road dump truck engine. The operator's cab is directly behind the engine compartment. The cab houses the vehicle instruments and controls. It has an integrated ROPS and FOPS. The dump body and hydraulic hoist cylinders are behind the operator's cab. The main difference between this style of truck and a rigid-frame dump truck is the articulating joint located between the cab and the body. This pivot point means that the truck has a front frame and a rear frame, and it enables the front section of the truck to turn. The front axle is used for steering, but on an articulated truck it also serves as a drive axle. The two rear axles drive the vehicle as well. This all-wheel drive feature enables the truck to maneuver through rough terrain.

Articulated trucks typically have a standard body, but there are different options that can be added to the body to enhance it for certain work. One option is an exhaust body heating system, as shown in *Figure 9*. This system diverts some of the exhaust gas from the exhaust pipe(s) to conduits that line the sides and bottom of the body. By heating the body, material in the body is less likely to freeze and stick to the sides of the body.

Figure 8 Articulated truck at highway construction site.

Diesel Engine Emission Controls

Diesel engines provide the power and efficiency needed to drive heavy equipment. A major drawback of past diesel engines has been the level of soot and nitrous oxide (NOx) they emit through their exhaust. In order to reduce these emissions, the Federal Government's Clean Air Act of 1990 included regulations for reducing diesel engine emissions. The regulations were phased-in over time. One early element of these regulations was the requirement for diesel engines to use ultra-low sulfur diesel fuel (USDF). Between 2008 and 2015, vehicles are required to meet Tier IV exhaust emission standards by using a system for treating the engine exhaust. Most engine manufacturers adopted the selective catalytic reduction (SCR) exhaust treatment system. This system injects a solution known as diesel emission fluid (DEF) into the engine exhaust system. DEF is an ammonia-based solution known as urea. The urea is stored in a tank and then mixed with air in a mixing valve before being injected into the exhaust flow.

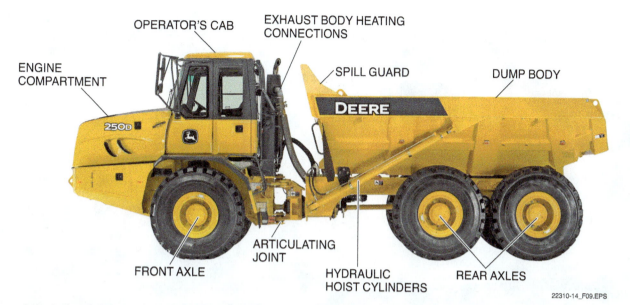

Figure 9 Articulated-frame dump truck components.

Exhaust body heating is particularly helpful in colder climates.

Another option that is often used with articulated truck bodies is a body liner. Some body liners are made of steel or rubber to protect against rocks; others are made of a polymer material to protect the body and reduce friction during dumping.

A front spill guard, also shown in *Figure 9*, is another option that is often used on articulated truck bodies. This bolt-on device helps prevent loose material from dropping onto the articulating pivot components.

Some trucks have an automatic tailgate at the rear of the body. The tailgate can be closed to prevent material from falling out of the body during travel over rough terrain and opened to allow material to be dumped. *Figure 10* shows an articulated truck that has a tailgate.

Side extensions can also be added to articulated truck bodies to increase the capacity of the body. These extensions basically raise the level of the body sides. They are typically used to increase the volume of the body when lighter weight material is being hauled. They should never be used to increase the payload of the truck beyond the manufacturer's ratings.

1.2.3 Underground Trucks

One type of specialized articulated truck is an **underground mining dump truck** (*Figure 11*). These trucks resemble above-ground articulated trucks, but are designed to work below ground in lower clearances. Like other articulated trucks, underground mining dump trucks vary in size and capacity. Typical payload capacities range from about 30 tons (27.22 tonnes) to 60 tons (54.43

tonnes) or more. Notice that the height of the vehicle is less than that of an above-ground truck. The cab height is reduced by moving the cab forward, so that the front axle is positioned at the rear of the cab.

The major components of an underground mining dump truck are basically the same as those of an aboveground truck. While some underground mining dump trucks have three axles (one front axle and two rear axles) with all-wheel drive, other models have a single rear axle.

1.3.0 Instrumentation

Off-road dump trucks, like on-road trucks, are designed so the driver can perform most duties from the cab. The layout and operation of the ve-

Figure 10 Articulated truck with tailgate.

22310-14_F11.EPS

Figure 11 Underground mining dump truck.

hicle's instruments and controls will vary among manufacturers, models, and uses of the truck. Some instruments and controls are optional, so they will not be found on all trucks. Each truck is slightly different, so the operator must study the manual for the particular vehicle. This should be the first step when assigned to a new truck. The operator's manual should be considered part of the equipment and of no less importance than other components such as the seat belt.

One example of a rigid-frame dump truck cab is shown in *Figure 12*. Locations for some of the major instruments and controls are identified. Depending on the manufacturer and the size of the vehicle, there may be a passenger or trainee/trainer seat in the cab in addition to the operator's seat. The extra seat might be beside or behind the operator. In either case, it allows a trainee or a trainer to observe the operation of the vehicle during training.

The basic trend in off-road truck design and use involves an increase in technology. Modern off-road trucks have complex monitoring systems that provide information to operators and mechanics about the condition of the engine, drivetrain, and other vital components. Those systems also monitor and help control the efficient operation of the vehicle. For example, a payload weighing system uses sensors to measure the weight of the payload and then displays that information to the operator. Some systems also illuminate exterior lights so that the person loading the truck can tell when the proper capacity has been reached. By tracking load weights and haul cycles over a period of time, a payload weighing system can provide valuable data about operating efficiency. Future development could include automated trucks that use guidance systems to maximize performance and efficiency. However, the use of integrated technology should not be a substitute for first-hand knowledge of the machine. Always study the operator's manual for the equipment and follow the recommended guidelines.

The instrument panel of an off-road truck (*Figure 13*) contains many of the same instruments found in an on-road dump truck. Among the instruments in this rigid-frame dump are the following:

- *Brake oil temperature gauge* – Indicates the temperature of the fluid used in the brake system and can alert the operator when the fluid is overheating.
- *Coolant temperature gauge* – Indicates the temperature of the liquid in the cooling system.
- *Engine overspeed indicator* – Alerts the operator to a condition in which the engine is revving beyond an established setpoint.
- *Fuel level* – Indicates the amount of fuel in the tank.
- *Hour meter* – Indicates the total number of hours that the vehicle has been operated.
- *Speedometer/odometer* – The speedometer shows the travel speed of the vehicle in miles per hour (mph) and kilometers per hour (km/h). The odometer indicates the total number of miles and kilometers that the vehicle has traveled.

INSTRUMENT PANEL

PRIMARY BRAKE PEDAL

SECONDARY BRAKE PEDAL

DISPLAY PANEL

TRANSMISSION SHIFTER

DUMP BODY CONTROL

THROTTLE LOCK

ACCELERATOR

FRONT VIEW

OPERATOR'S SEAT

TRAINEE/ TRAINER SEAT

REPRINTED COURTESY OF CATERPILLAR INC.

REAR VIEW

22310-14_F12.EPS

Figure 12 Rigid-frame dump truck cab layout.

- *Tachometer* – Indicates engine speed in revolutions per minute (rpm).
- *Powertrain temperature* – Indicates the temperature of the fluid used to lubricate the components in the powertrain.

The instrument panel also contains a number of warning lights. *Figure 14* shows the instrument panel of an articulated truck with various indicators and warning lights identified.

The cab of a rigid-frame dump truck has a separate display panel to the right of the instrument panel (*Figure 15*) that provides information about the condition of the vehicle, the transmission gear selection, and data about efficient operation. Most modern trucks have some sort of monitoring system that provides such information. Each manufacturer typically has its own system and display criteria.

1.4.0 Control Systems

Every off-road dump truck has specialized control systems that are needed for safe and efficient operation of the vehicle. Critical systems like traction control and braking control are needed to help start and stop these trucks. Parameters as varied as tire pressures, payload weight, and engine temperatures are monitored and displayed to the operator. All of these modern systems help ensure safety and performance, and the controls and displays are located in the cab of the truck. The specific aspects of these control systems can vary, depending on the manufacturer and the type and model of truck. The control systems for the truck are described in the operator's manual. This section covers typical control systems that are likely to be found on a modern off-road truck.

1.4.1 Engine Control

The engine in an off-road truck has to be powerful enough to drive the massive vehicle and its payload, yet efficient enough for a company to afford to operate it. It also has to meet stringent environmental regulations. To maximize performance and efficiency, the vehicle's control system monitors and regulates various engine parameters, and coordinates the engine performance with other systems. For example, some trucks have idle control systems that will automatically shift the transmission into and out of neutral during prolonged periods of idling in a forward gear. A similar system will automatically shut down the engine if the truck is idling in the park position beyond a preset amount of time. This feature saves fuel and can be set according to the needs of the job. Another engine control feature called speed limiting coordinates engine speed with the transmission gear selection. This feature enables the truck to achieve its optimal fuel efficiency during travel.

Some trucks have an electronic throttle lock (*Figure 16*) that enables the operator to lock the throttle pedal at full throttle. This feature is designed to reduce operator fatigue during prolonged uphill travel. If the vehicle brakes are applied, or if the throttle pedal is depressed, the throttle lock will release.

1.4.2 Transmission Control

Most off-road trucks today have some form of automatic transmission. A typical truck might have a seven-speed automatic powershift transmission. This provides seven forward speeds, as well as reverse, neutral, and park. The top speed for a large off-road truck is typically about 40 to 45 miles per hour (about 64 km/h to 72 km/h). *Figure 17* shows the automatic shifter for an off-road truck, along with the display panel indicator for the transmission setting.

Some of the automatic controls that are associated with the transmission include coasting and shifting inhibitors, a neutral start switch, and a shift management control. A reverse neutralizer

COOLANT TEMPERATURE

BRAKE OIL TEMPERATURE

TACHOMETER

POWERTRAIN TEMPERATURE

FUEL LEVEL

LED DISPLAY FOR OTHER INDICATIONS

22310-14_F13.EPS

Figure 13 Instrument panel.

TACHOMETER

LEFT TURN · COOLANT TEMP · ACTION LAMP · PARK BRAKE · FUEL LEVEL · RIGHT TURN

TRACTION CONTROL
HYDRAULIC SYSTEM
ENGINE OIL PRESSURE
TRANSMISSION HOLD
MAIN BEAM
CHARGING SYSTEM

STEERING SYSTEM
POWERTRAIN TEMP
POWERTRAIN FAULT
BRAKE WARNING
HIGH EXHAUST SYSTEM TEMPERATURE (HEST)

RETARDER UPSHIFT · FILTER WARNING · LIQUID CRYSTAL DISPLAY (LCD) · BODY TIP · CAT REGENERATION SYSTEM (CRS)

22310-14_F14.EPS

Figure 14 Articulated truck instruments and warning lights.

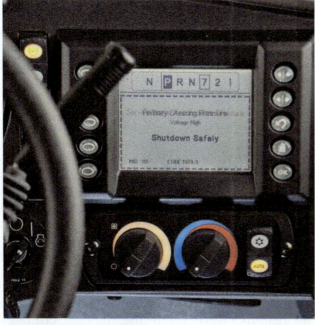

22310-14_F15.EPS

Figure 15 Display panel.

function automatically shifts the transmission out of reverse and into neutral during dumping. These controls are designed to prevent unsafe conditions from occurring during operation and to improve performance and efficiency.

1.4.3 Traction Control

Off-road trucks operate in harsh conditions that can include steep slopes, rutted and muddy terrain, and rocky soil (*Figure 18*). Ensuring that the vehicle maintains traction in these conditions is critical to safety and efficiency. For those reasons, most off-road trucks have some type of traction control system to limit any slipping or spinning of the vehicle's wheels.

Unlike an on-road dump truck that uses a locking differential to improve traction, an off-road truck uses a computerized system to basically divert torque from a spinning wheel to one or more of the other wheels. This type of system usually involves applying a braking action to the spinning wheel and increasing the spinning action of the other wheel or wheels. Some of the advantages of a traction control system are:

- *Improved steering* – The system is able to distinguish between wheel slip caused by spinning and wheel slip caused by turning the vehicle at high speed, and adjust accordingly.

REPRINTED COURTESY OF CATERPILLAR INC.

22310-14_F16.EPS

Figure 16 Throttle lock.

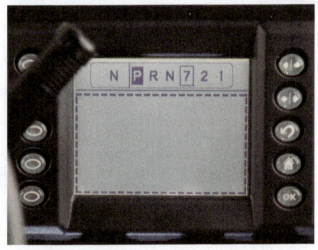

REPRINTED COURTESY OF CATERPILLAR INC.

22310-14_F17.EPS

Figure 17 Transmission shifter and display panel indication.

REPRINTED COURTESY OF CATERPILLAR INC.

22310-14_F18.EPS

Figure 18 The need for traction control.

- *Shorter loss of traction time* – Modern traction control systems respond faster and at slower vehicle speeds to restore traction sooner.
- *Reduced tire wear* – Faster and more efficient traction control system response helps reduce tire wear.
- *Improved efficiency* – Less time is lost during vehicle operation, which results in lower fuel use and faster haul cycles.

1.4.4 Braking Control

Stopping an off-road truck can be a challenge, because the truck and its payload can easily weigh over one million pounds (453,590 kilograms). The enormous momentum that these trucks develop makes it difficult for any kind of conventional braking system to slow or stop one of these vehicles. To ensure that a truck of this size can be controlled, off-road truck manufacturers often use heavy-duty hydraulic braking systems in combination with engine braking and automatic retarding control. *OSHA Standard 1926.601(b)(1)* states that "all vehicles shall have a service brake system, an emergency brake system, and a parking brake system. These systems may use common components, and shall be maintained in operable condition."

In the past, air brake systems were common on off-road trucks. However, air brakes can be slower to respond during braking. In cold climates, moisture in the air can freeze and cause serious problems. Hydraulic brakes are more dependable. Hydraulics are used extensively on off-road trucks for hoisting and steering mechanisms, so manufacturers are finding it less expensive and more practical to use hydraulic braking systems.

A typical hydraulic braking system on an off-road truck uses either dry or wet disc brakes at all four corners of the truck. Wet disc brakes use a cooling oil to control overheating. Some trucks also have an anti-lock braking system (ABS) to further reduce the likelihood of skidding or sliding. These brakes are designed to resist fading and overheating during prolonged braking. *Figure 19* shows a cut-away view of a wet disc brake with its major parts labeled.

As *Figure 19* shows, there are also components for a secondary and parking brake system. A secondary brake system is used as a backup for the primary system. The secondary brake can be manually activated using a pedal in the operator's cab (*Figure 20*). Also, if hydraulic pressure to the primary braking system drops below a specific point, the secondary brakes will be applied automatically.

No matter how good the service brakes are on an off-road truck, using them alone to stop the vehicle can cause a great amount of wear on the braking system. In addition, when a truck is driving downhill, constantly applying the brakes can cause them to overheat, fade, or fail. For this reason, off-road trucks are often equipped with an **engine retarder** (engine brake), which works by using the power coming from the engine to slow the truck.

Another system that is used to enhance off-road truck braking is an **automatic retarder control system**. As *Figure 21* shows, this system works in conjunction with the engine brake and the traction control system to electronically control retarding during downhill travel. An automatic retarder control system helps prevent engine overspeeding and enables an operator to maintain better control of the truck.

1.4.5 Hoist Control

The hoist mechanism is used to raise and lower the dump body. Off-road trucks typically use two double-acting hydraulic cylinders to accomplish this—one on each side of the body—as shown in *Figure 22*.

The control lever for the hoist mechanism is located in the cab, typically to the side of the operator's seat (*Figure 23*). As with many dump trucks, the hoist control for an off-road truck typically

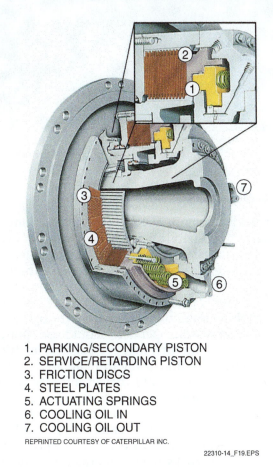

1. PARKING/SECONDARY PISTON
2. SERVICE/RETARDING PISTON
3. FRICTION DISCS
4. STEEL PLATES
5. ACTUATING SPRINGS
6. COOLING OIL IN
7. COOLING OIL OUT

REPRINTED COURTESY OF CATERPILLAR INC.

22310-14_F19.EPS

Figure 19 Wet disc brake.

SECONDARY
BRAKE PEDAL

REPRINTED COURTESY OF CATERPILLAR INC.

22310-14_F20.EPS

Figure 20 Secondary brake pedal in cab.

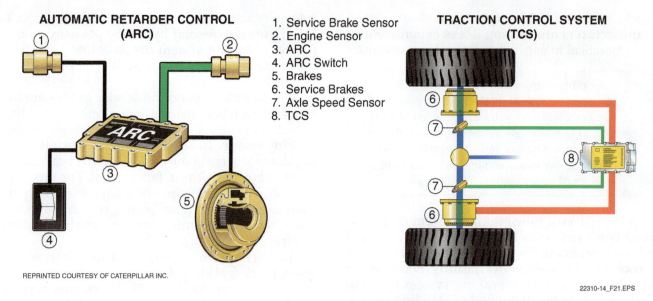

AUTOMATIC RETARDER CONTROL (ARC)

1. Service Brake Sensor
2. Engine Sensor
3. ARC
4. ARC Switch
5. Brakes
6. Service Brakes
7. Axle Speed Sensor
8. TCS

TRACTION CONTROL SYSTEM (TCS)

REPRINTED COURTESY OF CATERPILLAR INC.

22310-14_F21.EPS

Figure 21 Automatic retarder control system.

22310-14_F22.EPS

Figure 22 Hoist cylinders.

has a four-position control. The four positions are raise, hold, float, and lower. As a safeguard, an automatic control feature will shift the transmission from reverse to neutral if the dump body is raised. The control positions are described as follows:

- *Raise* – Pull the lever up to raise the dump body and empty the load.
- *Hold* – Move the lever down to the hold position and the dump body will not move.
- *Float* – Push the lever down to the float position and the dump body will seek its own level. This is the primary and default position. The truck should ordinarily be operated with the hoist lever in this position.
- *Lower* – Push the lever all the way down to lower the dump body. When the lever is released, it will return to the float position.

REPRINTED COURTESY OF CATERPILLAR INC.

22310-14_F23.EPS

Figure 23 Hoist control lever.

1.4.6 Object Detection System

Off-road truck manufacturers go to great lengths to maximize visibility from the operator's cab. Most trucks today have a lot of glass, lights, and mirrors to enhance visibility and eliminate as many blind spots as possible. However, on many trucks it simply is not possible to have a clear view of everything around the truck from the operator's cab. For that reason, operators should use spotters on the ground anytime they are backing up, turning around, or pulling away from an area. To further minimize the risk of runovers, off-road trucks are equipped with object detection systems.

Figure 24 shows two essential components of an object detection system: a camera mounted under the body at the rear axle and a monitor inside the cab. Most trucks have cameras positioned at

CAMERA

REPRINTED COURTESY OF CATERPILLAR INC.

22310-14_F24.EPS

Figure 24 Object detection system.

different points around the truck. Some systems also use radar to detect objects and provide visual or audible warnings when people or objects are close to the truck. The thing to remember is to use all available systems on the truck, watch the mirrors, check the area before starting and moving the truck, and always be alert to traffic or workers in the area.

> **WARNING!**
>
> The risk of running over objects and people is the greatest hazard associated with off-road dump trucks. Operators of off-road dump trucks have limited visibility from the operator's cab. Operators must always use the truck's object detection system and, when needed, spotters on the ground to minimize the risk of injuries and damage.

1.4.7 Ground Level Controls

Nearly all the controls for an off-road truck are located inside the operator's cab. However, there are a few controls located at ground level. These are basically emergency shutoff controls, exterior lighting, and maintenance-related controls.

Figure 25 shows an emergency shutoff switch that is located at ground level. Beside the shutoff switch is a separate switch that can be used to turn on exterior lights near the steps that lead to the cab.

Another set of ground level controls that are common on off-road trucks is an electrical center that includes a battery disconnect switch and ports for connecting diagnostic equipment (*Figure 26*). Having these devices at ground level makes routine switching and maintenance easier.

REPRINTED COURTESY OF CATERPILLAR INC.

22310-14_F25.EPS

Figure 25 Emergency shutoff switch.

REPRINTED COURTESY OF CATERPILLAR INC.

22310-14_F26.EPS

Figure 26 Electrical center.

1.0.0 Section Review

1. Which of the following is true of most rigid-frame dump trucks?

   a. Most of them use only a large electric motor for power.
   b. They typically have two axles—one for steering and one to drive the vehicle.
   c. Most of them have three or more axles for maneuverability.
   d. They are usually all-wheel drive vehicles.

2. The pivoting point in an articulated-frame dump truck enables the front of the truck to turn so that _____.

   a. the load weight shifts to counterbalance the vehicle
   b. the dump body can be tilted without cab interference
   c. backing up to dump the load is much easier
   d. all of the wheels follow the same path

3. Which of these instruments would most likely indicate a problem during a long downhill run that involved prolonged use of the primary hydraulic brakes?

   a. Engine overspeed indicator
   b. Coolant temperature gauge
   c. Brake oil temperature gauge
   d. Secondary air pressure indicator

4. If an off-road truck idles beyond a set time while it is in a forward gear, an engine idle control system is most likely to _____.

   a. shift the transmission to neutral
   b. shut down the engine
   c. shift the transmission to a higher gear
   d. apply the parking brake

2.0.0 Safety, Inspection, and Maintenance

Objectives

Identify and describe safety, inspection, and service guidelines associated with off-road dump trucks.

a. Describe guidelines associated with off-road truck safety.
b. Describe prestart inspection procedures.
c. Describe preventive maintenance requirements.

Performance Task 1

Complete a proper prestart inspection and maintenance for an off-road dump truck.

It should go without saying that an off-road dump truck can present a significant safety concern to operators, co-workers, and anyone else in the vicinity. In addition, the truck and other equipment at the job site are susceptible to accidents and damage. Operators are responsible for their own safety as well as that of their co-workers, the public, and the equipment itself. They must have a thorough understanding of their company's safety policies and all safety guidelines provided by the equipment manufacturer. A safety-conscious operator has greater value than any safety device that might be found on a piece of equipment.

It is also the operator's responsibility to make sure the equipment is working properly. Off-road dump trucks require frequent inspections and periodic maintenance to ensure that they function safely and efficiently. Operators may not perform maintenance on the equipment, but they are responsible for inspecting the truck before and after each shift to ensure that it is safe to operate. They must also know the maintenance schedule and make sure that the maintenance is performed when required. Most companies have policies and guidelines about safety, inspections, and machine maintenance. It is critical for operators to be thoroughly familiar with those policies and always follow them.

2.1.0 Safety Guidelines

There are a number of things that can be done to protect yourself and others from getting hurt on the job. Always be alert and work to avoid accidents.

Safety can be divided into three basic areas: safety of the operator, safety of other personnel, and safety of the equipment. The operator is responsible for working safely, protecting the public and co-workers from harm, and protecting the equipment from damage.

2.1.1 Operator Safety

Be sure to know and follow the employer's safety rules. Your foreman, supervisor, or the company safety officer should provide the requirements for proper dress and safety equipment. The following are recommended general safety procedures for off-road dump trucks:

- Keep all steps, grab irons, platforms, protective railings, and the operator's cab of the truck clean and free of debris and personal items. Clean mud off boots as much as possible.
- Mount and dismount the truck only where steps and handholds are provided. Use three-point contact when appropriate to avoid slips and falls. Do not jump from high vehicle steps or platforms.
- When necessary, use the exterior truck lighting to enhance visibility when entering or exiting the truck.
- Recognize and understand the use of the primary exit and alternative exit on the vehicle.
- Keep the windshield, windows, and mirrors on the operator's cab clean at all times.
- Make sure any object detection system components, such as cameras and monitors, are clean and functioning properly.
- Always wear the seat belt.
- Always wear and use the appropriate personal protective equipment (PPE), such as a hardhat, safety glasses, hearing protection, and gloves.
- Avoid wearing loose clothing or jewelry, especially rings, that could catch on railings, controls, or moving parts.
- Be aware of pinch points, especially between the truck and the dump body and the pivot point on articulated trucks.
- Never operate equipment under the influence of alcohol or drugs.
- Never smoke while refueling or while checking batteries or fluids.
- Never remove protective railings, guards, or panels from the equipment.

- Always follow the manufacturer's guidelines about parking and securing the truck before performing any service or before leaving the equipment unattended.
- Use extreme caution when operating around low overhead obstructions, such as power lines.
- Do not refuel the truck with the engine running.
- Avoid any conditions that could cause the truck to tip. Avoid dumping the truck if the truck is on an upward slope. Avoid ditches, ridges, and other obstacles. Avoid operating the truck across a slope. Instead, operate up and down slopes whenever possible.
- Know the maximum dimensions of the vehicle.
- Never exceed the truck's capacity. Pay close attention to the vehicle's monitoring system displays to ensure that the equipment is operating within its design guidelines.

All off-road dump trucks have warning signs and labels attached at specific locations. Take time to read and become familiar with these safety signs. Make sure all of them can be easily read. Clean or replace any that cannot be read or understood.

Every off-road dump truck has an operator's manual that clearly states safety hazards associated with the truck. Read and follow the operating guidelines in the manual. If you have questions or need clarification about a specific safety issue, always refer to the manual or ask the supervisor.

2.1.2 Safety of Co-Workers and the Public

Operators are responsible for their own personal safety, as well as the safety of other individuals who may be working nearby. Be especially careful when working in areas that are close to pedestrians or motor vehicles. Remember these basic safety points when working around other people:

- Make sure there is a clear view in all directions. Off-road dump trucks are notorious for having blind spots. Never assume that the area is clear; always verify that it is safe to move the truck. Use a signaler or spotter whenever possible.
- Understand and use any equipment on the truck that is designed to improve visibility, including mirrors and object detection cameras.
- Be sure everyone is clear of the equipment before starting and moving.
- In general, do not allow workers to be located at a pinch point area, either on the machine or around the machine, while it is in operation.
- Do not stand or allow others to stand or work under a raised dump body. If work must be performed on the truck with the body raised,

the body should be empty and suitable locking pins should be in place to hold the body up according to the manufacturer's instructions.
- Know the limitations of the truck. Never exceed the truck's capacity.
- Always follow site guidelines for direction of travel and speed limits when driving over excavated ground or haul roads.
- Never allow riders on the equipment. If a trainer or trainee is inside the cab, make sure they are correctly seated in the passenger seat and properly belted.
- Be sure to clearly understand all warning devices being used at the work site. Know the rules and the meaning of all flags, hand signals, signs, and markers.

2.1.3 Equipment Safety

In terms of equipment safety, the basic rule of thumb is to know the equipment and its limitations. All off-road dump trucks are designed with certain safety features to protect the operator as well as the equipment. For example, these trucks have safety devices such as guards, protective railings, protective canopies, cabs designed for high visibility, shields, warning lights and horns, on-board monitoring equipment, and seat belts, to name a few. Be sure to know the truck's safety devices and be sure they are in working order.

- Do a walk-around inspection at the beginning and end of each shift. Check for loose or worn components, leaks, or visible damage.
- Make sure all shields, protective railings, hinge pins, and related items are secure.
- Never park the truck on an incline. Park on a level surface and install wheel chocks before leaving the vehicle.
- Never exceed the manufacturer's limits for operating on slopes and inclines.
- Always turn the engine off and secure the controls before leaving the vehicle.
- Never exceed the manufacturer's recommendations for capacity.
- Do not allow material to be loaded onto the protective canopy of the truck.

WARNING!

Serious injury could result from tipping of the machine. Never exceed the rated load capacity of the machine. If the machine is not on level ground, load capacities will be less than specified.

Never operate an off-road dump truck if it is not in good working order. Avoid obstacles such as rocks, curbs, and ditches. These massive trucks are capable of running over and crushing other vehicles and pieces of equipment. Always be aware of other equipment, people, and obstacles in the area and avoid them.

2.2.0 Prestart Inspection Procedures

Routine inspections and maintenance can significantly reduce breakdowns and improve safety and performance. The first thing that should be done before starting any work is to conduct a daily walk-around inspection of the truck. This should be done before mounting the truck and starting the engine. The inspection will identify any potential problems that could cause a breakdown and will indicate whether the truck can be operated. It is also a good idea to inspect the truck during and after operation.

The operator's manual for the truck must be consulted for the specific checks and the procedures used to accomplish them. The following list describes typical prestart inspections and actions that should be performed on off-road dump trucks:

- Perform a walk-around inspection to check for any fluid leaks or loose, worn, or missing parts.
- Check the axles, the body, and the hoist cylinders for cracks or leaks.
- Check the tires for wear, cuts, and proper inflation. Modern off-road dump trucks typically have an on-board tire pressure monitor that displays information in the operator's cab.
- Check all lights, horns, cameras, and mirrors that are accessed from the exterior of the truck.
- Check the engine compartment and radiator for trash or dirt buildup.
- Check all engine belts and hoses for wear or leaks.
- Check the coolant level in the cooling system.
- Check the engine oil level.
- Check the hydraulic oil level.
- Check the fuel level.
- Check the differential oil level.
- Top off any fluid levels that are below normal.

As a convenience, many modern off-road dump trucks have centralized fluid filling centers, or clusters, at ground level. *Figure 27* shows a filling cluster where all the major fluids can be added and extracted.

Some trucks also group together grease fittings (*Figure 28*) that need to be accessed routinely during daily maintenance. Automatic grease systems are also used on some trucks. Both the fluid fill

REPRINTED COURTESY OF CATERPILLAR INC.

22310-14_F27.EPS

Figure 27 Fluid fill center.

center and the clustered grease fittings enable an operator or mechanic to easily perform routine fluid checks and top-offs without having to go to numerous areas in and around the truck.

> **NOTE**
>
> Trucks that have diesel engines designed to meet Tier IV emission standards use ultra-low sulfur diesel fuel. This includes truck engines manufactured since 2010. Most of these trucks have a diesel exhaust fluid (DEF) tank containing urea, an ammonia-based chemical. The DEF is used to treat engine exhaust in order to reduce particulate matter and nitrous oxide emissions. The DEF tank must be checked regularly and refilled as needed. The fluid must also be periodically replaced in accordance with the manufacturer's instructions. The vehicle control system should be equipped with a means to alert the operator when the DEF is low or when is has become contaminated.

Once inside the cab of the truck, inspect the gauges, lights, switches, and alarms to make sure they are functioning properly. Make sure the interior of the cab, the windows, and the mirrors are clean. Check the windshield wipers and washers, as well. Ensure that all safety devices, such as a first aid kit and a fire extinguisher, are available and operational.

The vehicle monitoring system provides important information about the condition of the various systems on the truck. Before using the truck, make sure the system is functioning properly and displaying the appropriate information. If any of the information being displayed seems wrong, check it before proceeding.

REPRINTED COURTESY OF CATERPILLAR INC.

22310-14_F28.EPS

Figure 28 Grease fitting cluster.

2.3.0 Preventive Maintenance

Preventive maintenance is an organized effort to regularly perform periodic lubrication and other service work on the vehicle to avoid poor performance and breakdowns at critical times. An operator is expected to perform daily inspections of the vehicle. These walk-around inspections help identify potential problems. The operator may or may not be involved in performing the actual maintenance work, but ensuring that preventive maintenance is done on a scheduled basis will keep the vehicle operating efficiently and safely, and reduce the possibility of costly failures in the future.

The leading cause of premature equipment failure is putting things off. This is true whether the equipment is a small loader or a massive off-road truck. Preventive maintenance should become a habit, performed on a regular basis.

Maintenance time intervals for most heavy equipment are established by the Society of Automotive Engineers (SAE) and adopted by most equipment manufacturers. Examples of time intervals are 10 hours (daily), 50 hours (weekly), 100 hours, 250 hours (monthly), 500 hours, 1,000 hours, and 2,000 hours. The vehicle monitoring systems on modern off-road trucks will provide information about the condition of fluids and when they need to be changed. You can also use the hour meter on the equipment to determine when inspection and service are required. Instructions for inspections and maintenance, including any special servicing requirements, are provided by the equipment manufacturer. Some information may also be provided on labels and notices attached to the truck.

The fluid capacities for a large off-road truck are much greater than those for a smaller piece of heavy equipment. It is not uncommon for a large engine crankcase to hold over 25 gallons (over 94 liters) of oil, or for a cooling system to hold over 60 gallons (over 227 liters) of coolant. The expense involved in changing the basic fluids in a large off-road truck can be enormous. So while it is important to change these fluids and filters according to the manufacturer's recommended schedule, it is also important to not change them too frequently.

When an off-road truck is ready for maintenance, follow these basic procedures to prepare the vehicle:

Step 1 Park the vehicle on a level surface to ensure that fluid levels are indicated correctly. Make sure there is adequate room around the vehicle for maintenance personnel and that the vehicle is away from other vehicles, personnel, and traffic.

Step 2 Move the transmission shifter to the neutral position (some manufacturers might specify placing the shifter in the park position).

Step 3 Apply the parking brake and perform any additional steps that the manufacturer specifies to ensure that the vehicle is stable and ready for service.

Step 4 If the manufacturer suggests shutting off any monitoring or control systems before vehicle shutdown, make sure it is done.

Step 5 Shut off the engine following normal shutdown procedures. If required by the company, place a lockout/tagout notice on the switch.

Step 6 Secure the cab and move to ground level.

Step 7 Turn the battery disconnect switch key to the Off position.

Step 8 Chock the wheels, if necessary or required.

Preventive Maintenance Records

Accurate, up-to-date maintenance records are essential for knowing the history of a piece of equipment, including like an off-road trucks. Each truck should have a record that describes any inspection or service that is to be performed and the corresponding time intervals. An operator's manual and some sort of inspection sheet should be kept with the vehicle at all times. Operators should update maintenance records as part of their daily routine. Following these few simple guidelines can prevent problems and breakdowns, and it can help predict when a recurring problem might reappear.

REPRINTED COURTESY OF CATERPILLAR INC.

22310-14_SA02.EPS

2.0.0 Section Review

1. To help maximize visibility and to protect co-workers and personnel, operators of large off-road trucks should use the vehicle's mirrors and _____.

   a. curb finding probes
   b. telescopic sighting
   c. laser leveling systems
   d. object detection systems

2. Who is normally responsible for performing daily prestart inspections on an off-road truck?

   a. Only a qualified mechanic
   b. The operator, or driver
   c. The work site supervisor
   d. An OSHA representative

3. A driver can typically identify when preventive maintenance is needed on a truck by _____.

   a. listening closely to how the vehicle sounds during operation
   b. monitoring fluctuations in fluid system pressures
   c. using the vehicle monitoring information and the hour meter display
   d. getting opinions from drivers of similar equipment

SECTION THREE

3.0.0 STARTUP AND OPERATION

Objective

Describe basic startup and operating procedures for off-road dump trucks.

a. Describe startup, warm-up, and shutdown procedures.
b. Describe safe driving maneuvers and loading and dumping procedures.

Performance Tasks 2 and 3

Perform the proper startup, warm-up, and shutdown procedures.

Carry out basic operations with an off-road dump truck:

- Properly position a truck for loading.
- Safely drive the truck to a designated dumping site.
- Dump the load in the designated spot.
- Set the retarder system to reduce wear on the service brakes.

Off-road truck operators are responsible for making sure that their vehicles perform safely and efficiently. This responsibility begins prior to startup, with daily walk-around inspections. It continues with following proper startup, warm-up, operation, and shutdown procedures.

3.1.0 Startup, Warm-Up, and Shutdown

As with other pieces of heavy equipment, there are specific steps involved in starting up, warming up, and shutting down off-road trucks. Operator's manuals typically include these steps. It is important to become familiar with these steps and always follow the manufacturer's guidelines.

3.1.1 Startup

After a thorough walk-around inspection has been completed, follow these basic guidelines to start up an off-road truck:

Step 1 Set the battery disconnect switch to the ON position. Large trucks typically have an electrical panel at ground level that includes this switch.

Step 2 Mount the truck using the appropriate ladders and steps. Maintain three-point contact during the climb to the operator's cab.

Step 3 Adjust the seat to a comfortable and safe operating position. Make sure you can reach all controls and pedals with your back firmly against the seat back.

Step 4 Inspect and fasten the seat belt. Adjust the belt as needed.

Step 5 Check and adjust all mirrors.

Step 6 Place the transmission shifter in the neutral position.

Step 7 Set the parking brake.

Step 8 Make sure that the body is grounded to the frame.

Step 9 Place the hoist control lever in the float position.

Step 10 Place the starting aid switch in the automatic position.

Step 11 Ensure that all personnel are clear of the vehicle.

Step 12 Test the backup alarm to verify that it is working properly.

Step 13 Use the key switch to start the vehicle. When the engine starts, release the switch.

Step 14 Perform the appropriate startup test on the vehicle monitoring system.

3.1.2 Warm-Up

Once a truck has been started up, it is important to allow a few minutes for the engine and operating systems to warm up before they are put to use. In cold climates, the warm-up period should be longer. Follow these basic guidelines for warming up an off-road truck:

Step 1 Allow the vehicle to warm up for five minutes at low idle. This enables the critical fluids and components to reach their normal operating temperature and pressure ranges.

> **NOTE**
> Failing to allow a vehicle to adequately warm up before operation can cause equipment damage and reduce the life expectancy of the fluids and parts.

Step 2   Perform a secondary brake test.

Step 3   Monitor all gauges, indicators, and warning lights closely. If any of the monitored variables is out of its ordinary range or value, shut down the truck until the problem can be investigated.

Step 4   Verify fluid levels using the instrumentation, especially the hydraulic oil tank and transmission oil levels.

3.1.3 *Shutdown*

Shutting down the engine and operating systems of an off-road truck too quickly after it has been working under load can result in overheating and accelerated wear of components. A gradual shutdown is necessary to allow the engine and systems to cool properly. Follow these basic steps, as well as any special procedures in the manufacturer's instructions, every time the vehicle is shut down:

Step 1   Park the truck on a smooth, level area that is well out of the flow of traffic.

Step 2   Place the gearshift in the neutral position.

Step 3   Set the parking brake.

Step 4   Move the throttle control to the low idle position and allow the engine to run for five minutes.

Step 5   Turn the key switch to the Off position and remove the key.

Step 6   Secure the operator's cab and dismount the vehicle using three-point contact during the climb down.

Step 7   Set the battery disconnect switch to the OFF position and remove the key.

Step 8   Chock the wheels if necessary.

Step 9   If the vehicle has air brakes, purge the air reservoirs of fluid by following the manufacturer's recommendation and your employer's policy.

> **NOTE**
>
> Most air brake systems have air dryers to remove moisture from the compressed air that feeds the brake system. However, some moisture can get through and build up in the air tanks. This is a serious concern in cold weather, when the moisture can freeze and potentially disable the brakes. So, even if air dryers are present, it is still necessary to bleed off the moisture as part of the daily routine.

Step 10 Perform a walk-around inspection and note any leaks and broken or missing parts.

3.2.0 Safe Truck Operation

After the proper startup procedures have been performed, the truck is ready for use. Because of the size of off-road trucks, they pose a significant hazard to operators, co-workers, and the public if they are used improperly. The top speed for an off-road truck is less than that of an on-road truck, but an operator needs to know how to maintain safe control of the vehicle at all times. This includes not only driving the truck, but also safe loading and unloading.

While working on a job, it will be necessary to coordinate the vehicle's movements with those of other vehicles. Usually, one person at a site, known as a signalperson or spotter, is appointed to direct the movement of vehicles in congested areas. Find out who this is and then watch for and obey his or her signals. When the signalperson is not in view, exercise caution when moving the truck. Look and think ahead to avoid situations that could disrupt the smooth flow of traffic. When operating the vehicle near workers on foot, keep them in sight.

A driver behind the wheel of an off-road truck is responsible for guiding many tons of material around a work site or on a haul road. Avoid actions that can weaken your skills or judgment on the job. Never use drugs or alcohol on the job. They can impair alertness, judgment, and coordination and may be grounds for termination. When required to take prescription or over-the-counter medications, seek medical advice about whether you can safely operate machinery. Only you can judge your physical and mental condition; don't take chances with your life or someone else's.

Never perform any unsafe maneuver or operations, even if told to do so by another person. To be qualified, the operator must understand the manufacturer's written instructions; have training, including actual operation of the vehicle; and know the safety rules for the job site. Current federal regulations state that at the time of your initial assignment and then at least annually, your employer must instruct you in the safe operation and servicing of equipment that you will use. Most employers have rules about operation and maintenance of equipment; make sure you know them. Finally, your employer must authorize you to operate a vehicle.

The key to staying safe is to develop good habits while learning a new skill. That way safety becomes second nature. The following are some good safety habits to practice:

- Read the operator's manual thoroughly. Know the machine.
- Always follow the job site's safety precautions.
- Always use a seat belt. Use other personal protective equipment (PPE) as required by the job site.
- Do not haul people in the dump body or on the exterior of the vehicle.
- Never get under the dump body when it is raised unless it is securely blocked using an approved method.
- Make sure the cab is free of objects such as chains, lunch boxes, etc., that could act as projectiles.
- Keep the windshield, windows, and mirrors clean at all times.
- Use a proper three-point mounting and dismounting technique when entering or leaving the vehicle.
- Never use cell phones or other devices that can be a distraction while driving.
- Never get under the truck unless the wheels are securely blocked.
- Clean mud and other slippery materials off your shoes. Clean shoes will help prevent slipping on steps or having your feet slip from the pedals.
- Set the parking brake and chock the wheels when parking trucks.

3.2.1 Basic Maneuvering

When preparing to drive the truck, the operator needs to complete a few basic steps. The following guidelines apply to typical off-road trucks:

Step 1 Make sure that the area around the truck is clear. Check the mirrors and use any object detection cameras that are available. If necessary, have a co-worker check around the vehicle.

Step 2 Ensure that the retarder control system is activated.

Step 3 Depress the service brake pedal.

Step 4 Release the parking brake.

Step 5 Move the gearshift to the appropriate direction and gear.

Step 6 Release the service brake and press the accelerator pedal.

The basic mechanisms for driving an off-road truck are similar to those used for on-road vehicles. There is a steering wheel for turning, an accelerator for speeding up, and a brake pedal for stopping. Keep in mind, though, that the braking system on an off-road truck is much more sophisticated than the one used on an on-road truck. More information about using the braking systems on an off-road truck is covered later in this section.

When driving the truck, observe all gauges and indicators. Look, listen, and feel for defects, strange noises, and changes in engine or braking power. Drive slowly at first to make sure that all the truck's systems are functioning properly. Keep the hoist mechanism in the float position while driving. Since visibility from the operator's cab is limited, always watch for other vehicles and pedestrians. *Figure 29* shows some of the blind spots that off-road truck operators have to deal with.

Even with all the safety features that have been built into the design of off-road dump trucks, there are still many dangers. Drive defensively; watch for unsafe conditions in the terrain or unsafe situations created by other equipment operators. Anticipate turns and stops ahead of time to allow the vehicle's control systems to maximize traction and braking. Obey all regulations and policies for off-road truck operation.

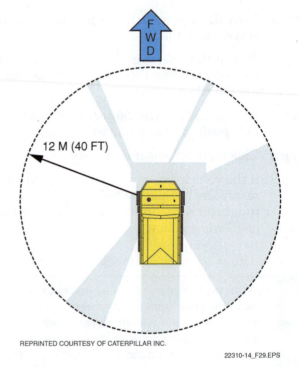

12 M (40 FT)

22310-14_F29.EPS

Figure 29 Restricted visibility around a typical off-road truck.

To be a safe truck operator, one must know and practice safe operating principles and procedures on a continual basis. Several areas of operation require particularly close attention, including backing, climbing and descending hills, and negotiating curves.

3.2.2 Backing Safely

Backing is a hazardous maneuver under any circumstance. The operator is always responsible for knowing what is behind the truck. Avoid backing if you can drive forward instead. Even when using the truck's mirrors and object detection systems, there will always be blind areas around an off-road truck. So whenever possible, use a spotter or signal person to help guide the truck and reduce the chance of injury or property damage.

The following tips should help in safe backing:

- Back slowly. Make sure there is adequate clearance behind and on each side of the truck.
- Remain properly seated when backing the vehicle. Use the mirrors and any available object detection systems to maximize visibility.
- Avoid long backing runs. Drive forward and position the truck as closely as possible to the dumping point to minimize the backing distance.
- Avoid backing downhill and never dump a load when the truck is positioned on a downhill grade.

3.2.3 Climbing and Descending Hills

Off-road trucks typically use heavy-duty automatic transmissions, traction control systems, and other mechanisms to prevent unsafe conditions from occurring during operation and to improve performance and efficiency. In most cases, the vehicle's monitoring system will use measured variables to automatically adjust engine speed, shifting, and traction control. However, sloping terrain can affect the operation of these systems. Operators must be completely aware of the terrain and anticipate hills, curves, and obstacles.

Haul roads and excavated terrain can be rutted, muddy, and rocky. Climbing hills in these conditions can be challenging. When climbing, make sure that the truck is in the appropriate transmission setting. Usually, the drive position is the preferred setting. The truck's shift management control will coordinate the engine speed and the proper transmission gear. If wheel spinning occurs, the traction control system will divert power away from the spinning wheel to other wheels. During prolonged climbs, use the throttle-lock feature to minimize fatigue.

When descending a hill, do not depend solely on the conventional braking system. These brakes can overheat and fail. The best approach is to allow plenty of time for the truck's engine brake and/or automatic retarder control system to slow the vehicle. In fact, it should only be necessary to use the brake pedal when coming to a complete stop.

The following are safety tips for descending a steep downhill grade:

- Reduce speed ahead of time and set the retarder for minimal use of the service brakes.
- Check the monitoring display frequently to remain aware of speed, engine rpm, and other variables.
- Take extra care when weather and road conditions are unfavorable. Never exceed the advised speed for any downgrade.
- Keep the truck under complete control at all times.

3.2.4 Taking Curves

Taking a curve in an off-road truck with a full load is dangerous. At higher speeds, centrifugal force will push the truck toward the outer edge of the curve and make the vehicle unstable. The only resistance is the friction of the tires on the road. An increase in load will raise the center of gravity, increase the force that causes the truck to slide sideways, and also increase the risk of tipping over. If the outward force is greater than the vehicle can handle, the truck will tip over or go into a skid. If the terrain is rough or rocky, the risk of skidding is greater. The safest action is to reduce speed before going into the curve, and accelerate gently as the vehicle is coming out of the curve.

The traction control system is designed to help maintain equal traction at all four corners of the truck. This improves the truck's stability, but operators must use caution when taking curves to avoid exceeding the truck's capabilities.

3.2.5 Safe Loading Practices

Off-road trucks are usually loaded by large excavators or other equipment. An operator's job involves correctly positioning the truck for loading. Loading sites typically have their own loading procedures that must be followed. If possible, position the truck for loading so that you can drive away without having to back up. If it is necessary to back the truck, use the route that reduces the time moving in reverse and use a spotter when possible.

Material must be evenly distributed in the dump body so the load will be balanced (*Figure 30*). If the truck is being loaded with a chute, there will most likely be a spotter who will signal when to pull forward to evenly distribute the load. Material should never be loaded onto the protective canopy of the truck body. When the truck is being loaded, the truck should be positioned so that the load does not pass over the operator's cab.

Observe the truck's display panel to ensure that the vehicle's capacity has not been exceeded. Overloading a truck can cause serious problems for hauling and dumping. In addition, the force of excess weight against the sides of the dump body can damage the equipment and create an unsafe situation. Watch the loading operation and make sure the recommended capacity of the truck is not exceeded.

Once loading has been completed, pull well away from the loading area if you need to do anything before beginning the haul. During the drive to the dump site, make sure the hoist mechanism is set to the float position.

> **WARNING!**
>
> Never drive with an unbalanced load. It can cause loss of control and may cause an accident. Ask the workers at the loading site for help to balance the load or contact a supervisor.

3.2.6 Safe Dumping Practices

Dump trucks are at the greatest risk for tipping over while they are being dumped. This is especially true if the load is unbalanced, or if dumping is being done on a slope. Always stay alert for problems during the dumping procedure and be ready to stop the procedure at any time the conditions warrant.

Many dump areas require the operator to turn the truck and back to the edge of the fill. If possible, the fill should be arranged so that any turn is made near the dump spot. The turn spot should be wide enough so that reverse gear is used only once. Turns in reverse should be toward the driver's left to give the driver maximum visibility. The truck should be level or facing uphill for dumping. Never try to dump on a downhill grade or across a slope. Use extra caution if the ground is soft or rough, or when complicated steering is required. Use a spotter to help guide the truck to the edge of the fill.

When dumping off the edge of a fill, back so that both rear wheels will be the same distance from the edge, rather than at an angle. Check the stability of the ground at the dump area before driving on it. If one wheel sinks in the ground deeper than the other, it may not be possible to dump the truck safely or to pull out with a load.

22310-14_F30.EPS

Figure 30 Truck with balanced load.

The distance from the edge of a fill that can be considered safe is determined by site conditions and the judgment of the operator. If the truck has all-wheel drive, or if the fill is shallow, a close approach can be made (*Figure 31*). If the fill is soft, slippery, sandy, or otherwise unstable, keep the rear wheels six feet or more (about 2 meters) away from the edge. This can be a difficult judgment for an inexperienced driver, so it may be necessary to ask a supervisor or a more experienced driver for advice. A berm should be placed close to the edge of the hill to alert the driver.

When the dump body is raised, the center of gravity for the truck changes to a much higher position, placing the truck at risk of tipping over, especially if the truck is unbalanced. Off-road trucks are often equipped with an automatic control feature that will shift the transmission from reverse to neutral if the dump body is raised. This helps prevent tipovers when the truck is backing to the dump site. However, tipovers can still occur—either to the back or to the side of the dump bed. The following are the leading reasons for backward tipovers:

- *Top-heavy load* – The rear portion of the load is dumped, leaving the rest of the load stuck at the front of the bed.
- *Material stuck together* – Either the material is frozen together or to the bed of the truck.
- *Uneven loading* – This can be either front-to-back or side-to-side.
- *Uneven dumping* – This is similar to uneven loading, but is caused by trying to dump part of a load in a particular spot.

When part of a load remains stuck in the body of a truck, it places a great strain on the hydraulics and causes an unstable condition for the whole truck. This is a hazardous condition that could cause tipping or damage to the dump body. This situation is sometimes referred to as a split load. A split load can be caused by frozen or sticky material that attaches to the bottom and one side of

22310-14_F31.EPS

Figure 31 Articulated truck dumping from safe distance.

the body. Compacting the material in the body during loading may also cause the material to stick on a side and in the corners.

Off-road trucks that operate in cold climates are often equipped with an engine body heating system. This system diverts some of the exhaust gas from a truck's exhaust to conduits in the body to help prevent material in the body from freezing. Some trucks use specially designed bodies or body liners to maximize the amount of material they can haul and help ensure that the material can be dumped properly.

If there is a problem with dumping the material from the body, the best thing to do is lower the body and pull away from the dump area. Then notify your supervisor and ask for instructions. You may be asked to drive to an area where another piece of equipment, such as an excavator, can be used to dislodge the material in the body.

Runaway Truck Ramps

Haul roads with long, steep downgrades often have ramps designed to bring a runaway truck to a safe stop. The ramp is usually made from sand or gravel, which will rapidly retard the speed of the truck. The ramps have a steep upward grade, which further helps in retarding speed. If a ramp can't be built, the soft material is sometimes piled along the shoulder of the haul road. The truck can be steered into the material and brought under control.

The following actions will help prevent tipover due to materials problems:

- Make sure the body is empty and clean before loading.
- Before using the truck each day, test the hoist mechanism with the body empty to ensure proper operation.
- If the truck is equipped with a tailgate, check its operation before starting a shift.
- Do not dump with the truck on a slope.
- When dumping, raise the body slowly.
- Make sure that the material is loaded into the body evenly.

It is the operator's responsibility to ensure that the equipment is working properly and that the material is being handled safely. Never perform any unsafe maneuver or operations, even if told to do so by another person.

3.2.7 Emergency Situations

No operator wants to be in an emergency situation in an off-road truck. Most accidents can be avoided by following required safety procedures and using common sense. Prevention of accidents involving off-road trucks depends primarily on the person operating the equipment. The four most common causes of off-road truck accidents are steep grades in the terrain, defective brakes, operator error, and overloading the vehicle.

Large off-road trucks use several systems to help control traction and speed and to stop the truck. This reduces the likelihood of a catastrophic braking failure in one of these vehicles, as long as the systems are functioning properly. Properly inspecting and testing the retarding and braking systems before and during truck use is critical. Staying alert and anticipating the need to slow down and brake well ahead to time is just as important. Wearing the seat belt is another absolute necessity. The operator stands a much greater chance of surviving and avoiding injury if properly belted.

If the truck is not slowing down adequately with the retarding systems, use the service brake pedal. These are the conventional brakes for the vehicle and should be capable of stopping the vehicle. If these brakes fail, use the secondary brake pedal. There is also a parking brake that can be activated in an emergency. The transmission can also be downshifted to slow down the truck in an emergency situation.

If the vehicle starts to skid, stay calm. Avoid hard braking, since slamming on the brakes will lock the wheels, cause further loss of traction, and increase the skid. The secret is to make small steering corrections. Steer in the direction of the skid; as the vehicle begins to correct, straighten the front wheels carefully and be ready to correct in the other direction. Some skids need more than one correction to regain control. Avoid oversteering; turning the steering wheel too far will whip the rear end of the truck into a skid in the opposite direction. Keep the transmission in drive. Holding the vehicle in gear helps reduce speed and provides the most control. Avoid lifting your foot from the accelerator suddenly, as this action can worsen the skid.

It is better to try to prevent a skid than to have to recover from one. Adjusting speed to the conditions in the haul road or work site will reduce the chance of a skid.

If the vehicle starts to run out of control down a hill, the last resort is to ditch the vehicle. Running it off the road against a bank at a gradual angle will slow and stop the vehicle. This must be done promptly before the runaway vehicle has gained too much speed. Proper action in such an emergency may prevent a much more serious accident. Remaining belted in the cab is the safest position.

3.0.0 Section Review

1. It is necessary to warm up an off-road truck for about five minutes before using it to allow _____.

   a. time for the operator to become familiar with the instruments and controls
   b. tire pressures to equalize
   c. time for personnel to get into position for the day's work
   d. fluids and components to reach normal operating temperatures and pressures

2. One key factor in safe truck operation is _____.

   a. overcoming the habit of monitoring instruments and displays
   b. anticipating curves and slopes well ahead of time
   c. remembering to always dump the load on a downhill grade
   d. trying to back up as much as possible to stabilize the load

Summary

Off-road dump trucks are common pieces of heavy equipment on construction sites, road building sites, mines, and quarries. Operating an off-road truck requires a high level of skill and knowledge, much like that required to operate other heavy equipment. The operator must be able to operate the truck on job sites and perform difficult maneuvers such as backing up and dumping the load. The operator must be able to do these tasks safely and efficiently.

An off-road dump truck consists primarily of a dump body and a cab mounted on either a rigid frame or an articulated frame. A powerful diesel engine and, in many cases, AC drive motors are used to power these vehicles. The transmission is typically automatic, with up to seven forward speeds. Rigid-frame dump trucks are usually two-axle vehicles, with the front axle used for steering and the rear axle used for driving the truck. Articulated trucks often have a single front axle, two rear axles, and all-wheel drive. The dump body of the truck is operated by a pair of hydraulic hoist cylinders mounted on the frame. Older off-road trucks used an air brake system, but newer trucks tend to use hydraulic brakes. In either case, the operator must know how to use and maintain them in proper working order.

An off-road truck operator is responsible for its preventive maintenance and safety. This includes performing daily walk-around inspections to make sure the truck is operating properly and safely. Proper preventive maintenance will prolong the life of the truck and must be performed per the manufacturer's schedule. Do not drive a truck that is not operating properly or is unsafe. Follow the procedures in the operator's manual for proper inspection and operation activities.

Operating an off-road truck requires specialized knowledge and the skill to properly use the engine brake, retarding systems, and service brakes in order to negotiate hills and curves. It is especially important to know how to manage the vehicle in an emergency such as a skid or loss of brakes.

Dump trucks may tip over if loaded improperly or when dumping material that is frozen or stuck together. Tipping usually occurs when the dump body is being hoisted and the material is being unloaded. Tipping can also occur if a large load causes the center of gravity to be too high and the truck is moving too fast around a curve.

1. Many rigid-frame dump trucks are powered by a large diesel engine that is used in combination with _____.

   a. a gasoline booster engine
   b. pneumatic motors
   c. electric motors
   d. LPG engines

2. A rigid-frame truck is best suited for situations where the _____.

   a. overhead clearance is low
   b. terrain is stable and/or haul roads have been established
   c. ground is steeply sloped and no roads exist
   d. site is heavily rutted, rocky, and/or muddy

3. The body of a rigid-frame dump truck has an extended section above the operator's cab called a protective _____.

   a. ROPE
   b. railing
   c. tailgate
   d. canopy

4. What device enables the front of an articulated-frame dump truck to turn in a way that all of the wheels follow the same path?

   a. A pivoting point in the frame
   b. A traction control cylinder
   c. A pair of AC motors at the front wheels
   d. An independent suspension at the rear axle

5. Articulated trucks are typically _____.

   a. front wheel drive vehicles
   b. rear wheel drive vehicles
   c. all-wheel drive vehicles
   d. towed vehicles

6. Articulated trucks are most likely to be used in _____.

   a. applications where a rigid-frame dump truck is too small
   b. rugged applications that may require a lot of turning
   c. situations where larger equipment can tow them
   d. places where diesel engines cannot be used

7. To prevent material from freezing in the dump body, some articulated trucks use _____.

   a. an exhaust body heating system
   b. an engine steam diverter
   c. a pressurized coolant system
   d. a wooden bed liner

8. In addition to a basic instrument cluster, most off-road trucks have a separate display that provides information from _____.

   a. other work site vehicles
   b. regulatory agencies
   c. loading site personnel
   d. the vehicle's monitoring system

9. A computerized traction control system works by diverting _____.

   a. air pressure from the drivetrain to atmosphere
   b. torque from a spinning wheel to other wheels
   c. hydraulic oil to change the truck's center of gravity
   d. power from the wheels back to the engine

10. To help verify that no one is around the vehicle before moving it, use _____.

    a. the throttle control to rev the engine
    b. a signalman or spotter
    c. the vehicle's retarder system
    d. a warning flag in the cab

11. The only person who should be accompanying the operator inside the cab of an off-road truck is a(n) _____.

    a. OSHA representative
    b. job site coordinator
    c. trainer or trainee
    d. FOPS specialist

12. Many modern off-road trucks allow fluids to be added and extracted at _____.

    a. a ground-level fluid filling center
    b. various points all over the truck
    c. a single cluster inside the cab
    d. the vehicle's electrical panel

13. When an off-road truck is parked for maintenance, the operator should _____.

    a. lock the primary, secondary, and parking brakes
    b. place the transmission in neutral and apply the parking brake
    c. set the battery disconnect switch to On and chock the wheels
    d. leave the truck in low idle until the service crew arrives

14. An off-road truck should be started up only after _____.

    a. a thorough walk-around inspection has been completed
    b. it has been allowed to cool down for at least 15 minutes
    c. the vital fluids have reached their normal temperatures
    d. the loading site has been notified of the loading plan

15. When is it permissible to exceed the payload capacity of an off-road truck?

    a. If the loader indicates that it is advisable to do so.
    b. When a supervisor gives a written order.
    c. Only when you feel certain that you can control the extra weight.
    d. The truck's capacity should never be exceeded.

Trade Terms Introduced in This Module

Articulated-frame dump truck: A type of off-road dump truck that has a permanent pivoting point in the frame that allows the front of the truck to turn so that all of the wheels follow the same path.

Automatic retarder control system: A system that works with the engine brake and the traction control system to electronically slow the vehicle during downhill travel.

Canopy: A section of an off-road truck body that extends above the operator's cab to help protect the operator from falling material.

Engine retarder: An alternate braking system activated from the cab that slows down the vehicle by reducing engine power.

Exhaust body heating system: A system that diverts some of the exhaust gas from a truck's exhaust pipe(s) to conduits in the body to help prevent material in the body from freezing.

Haul truck: A name that is sometimes used to describe a rigid-frame dump truck or a mining truck.

Hoist: The mechanism used to raise and lower the dump body, typically consisting of two hydraulic cylinders.

Quarries: Excavations or pits in which gravel, stone, and other material are mined.

Rigid-frame dump truck: A type of off-road truck in which the cab and the body are mounted on a common, non-pivoting, frame. Rigid-frame dump trucks are commonly referred to as haul trucks or mining trucks.

Traction control system: A computerized system that diverts torque from a spinning wheel to one or more of the other wheels to improve traction.

Underground mining dump truck: A type of articulated truck in which the vehicle height is reduced by moving the cab forward of the front axle, thus enabling the truck to be used in underground mining applications.

Figure Credits

Komatsu America Corp., Module opener, Figures 1, 2, 4

Deere & Company, Figures 3, 7–10, 22, 31

Reprinted Courtesy of Caterpillar Inc., Figures 5, 6, 11–21, 23–30, SA01, SA02

Answer	Section Reference	Objective
Section One		
1. b	1.1.0	1a
2. d	1.2.0	1b
3. c	1.3.0	1c
4. a	1.4.1	1d
Section Two		
1. d	2.1.2	2a
2. b	2.2.0	2b
3. c	2.3.0	2c
Section Three		
1. d	3.1.2	3a
2. b	3.2.1	3b

NCCER CURRICULA — USER UPDATE

NCCER makes every effort to keep its textbooks up-to-date and free of technical errors. We appreciate your help in this process. If you find an error, a typographical mistake, or an inaccuracy in NCCER's curricula, please fill out this form (or a photocopy), or complete the online form at **www.nccer.org/olf**. Be sure to include the exact module ID number, page number, a detailed description, and your recommended correction. Your input will be brought to the attention of the Authoring Team. Thank you for your assistance.

Instructors – If you have an idea for improving this textbook, or have found that additional materials were necessary to teach this module effectively, please let us know so that we may present your suggestions to the Authoring Team.

NCCER Product Development and Revision

13614 Progress Blvd., Alachua, FL 32615

Email: curriculum@nccer.org
Online: www.nccer.org/olf

❏ Trainee Guide   ❏ Lesson Plans   ❏ Exam   ❏ PowerPoints   Other _____

Craft / Level: _____ Copyright Date: _____

Module ID Number / Title: _____

Section Number(s): _____

Description: _____

_____

_____

_____

_____

Recommended Correction: _____

_____

_____

_____

_____

_____

Your Name: _____

Address: _____

_____

Email: _____ Phone: _____

22302-14

Dozers

OVERVIEW

A dozer is one of the most useful pieces of heavy equipment in the excavation and grading industry. Heavy equipment operators need to understand how a dozer functions and how to set the blade and other accessories so that the equipment works effectively. Operators must also be aware of potential dangers that exist with dozers, know how to conduct daily preventive maintenance on the machine, and practice safety at all times. Finally, operators should understand the general properties of materials such as clay, sand, mud, different types of rock, and different types of manufactured materials like asphalt and concrete that can affect dozer operation.

Module Five

Trainees with successful module completions may be eligible for credentialing through NCCER's National Registry. To learn more, go to **www.nccer.org** or contact us at **1.888.622.3720**. Our website has information on the latest product releases and training, as well as online versions of our *Cornerstone* magazine and Pearson's product catalog.

Your feedback is welcome. You may email your comments to **curriculum@nccer.org**, send general comments and inquiries to **info@nccer.org**, or fill in the User Update form at the back of this module.

This information is general in nature and intended for training purposes only. Actual performance of activities described in this manual requires compliance with all applicable operating, service, maintenance, and safety procedures under the direction of qualified personnel. References in this manual to patented or proprietary devices do not constitute a recommendation of their use.

Objectives

When you have completed this module, you will be able to do the following:

1. Identify and describe the uses and components of a dozer.
   a. Identify and describe common uses and types of dozers.
   b. Identify and describe major parts of a dozer.
   c. Identify and describe dozer instrumentation.
   d. Identify and describe dozer controls.
   e. Identify and describe common dozer blades.
   f. Identify and describe common dozer attachments.
2. Identify and describe safety, inspection, and service guidelines associated with a dozer.
   a. Describe guidelines associated with dozer safety.
   b. Describe prestart inspection procedures.
   c. Describe preventive maintenance requirements.
3. Describe the basic startup and operating procedures for a dozer.
   a. Describe startup, warm-up, and shutdown procedures.
   b. Describe basic maneuvers and operations.
   c. Describe common work activities.

Performance Tasks

Under the supervision of your instructor, you should be able to do the following:

1. Demonstrate a proper prestart inspection and preventive maintenance on a dozer.
2. Perform proper startup, warm-up, and shutdown procedures.
3. Perform basic maneuvers on a dozer, including moving forward, moving backward, turning with the blade up, and straight dozing.
4. Demonstrate basic dozer operation by:
   - Creating a level pad that measures approximately 20 feet by 20 feet ±1/10 foot (6 meters by 6 meters ±3 centimeters).
   - Pushing a stockpile while maintaining proper windrows and berms.

Trade Terms

Bank state	Charging hoppers	Pawl
Blade	Dozing	Ripping
Blade float	Grouser	Slot dozing
Blade pitch	Grubbing	Stockpile
Blade tilt	KG blade	Winch

Industry Recognized Credentials

If you are training through an NCCER-accredited sponsor, you may be eligible for credentials from NCCER's Registry. The ID number for this module is 22302-14. Note that this module may have been used in other NCCER curricula and may apply to other level completions. Contact NCCER's Registry at 888.622.3720 or go to **www.nccer.org** for more information.

Contents

Topics to be presented in this module include:

1.0.0 Uses and Components of Dozers 1
  1.1.0 Common Uses and Types of Dozers 1
    1.1.1 Dozer Uses 1
    1.1.2 Dozer Types and Configurations 1
  1.2.0 Major Parts of a Dozer 1
    1.2.1 Operator's Cab 2
  1.3.0 Instrumentation 4
    1.3.1 Engine Coolant Temperature Gauge 5
    1.3.2 Torque Converter Oil Temperature Gauge 6
    1.3.3 Fuel Level Gauge 6
    1.3.4 Hydraulic Oil Temperature Gauge 6
  1.4.0 Controls 7
    1.4.1 Vehicle Movement Controls 7
    1.4.2 Blade Controls 9
    1.4.3 Ripper Control 9
  1.5.0 Blades 9
    1.5.1 General-Purpose Blades 10
    1.5.2 Production Blades 11
    1.5.3 Special Application Blades 11
  1.6.0 Attachments 13
    1.6.1 Rippers 13
    1.6.2 Winches 14
    1.6.3 Other Attachments 14
2.0.0 Safety, Inspection, and Maintenance 17
  2.1.0 Safety Guidelines 17
    2.1.1 Operator Safety 17
    2.1.2 Safety of Co-Workers and the Public 18
    2.1.3 Equipment Safety 18
  2.2.0 Prestart Inspection Procedures 19
  2.3.0 Preventive Maintenance 20
3.0.0 Basic Startup and Operations 23
  3.1.0 Startup, Warm-Up, and Shutdown 23
    3.1.1 Startup 23
    3.1.2 Warm-Up 24
    3.1.3 Shutdown 24
  3.2.0 Basic Maneuvers and Operations 25
    3.2.1 Moving Forward 26
    3.2.2 Changing Direction and Speed 26
    3.2.3 Steering and Turning 27
    3.2.4 Operating the Blade 27
  3.3.0 Work Activities 28
    3.3.1 Dozing 28
    3.3.2 Cutting and Building Slopes or Stockpiles 29
    3.3.3 Clearing Land 30

3.3.4 Backfilling 30
3.3.5 Sidehill Cutting 31
3.3.6 Moving Large Objects 31
3.3.7 Finishing 32
3.3.8 Excavating Work in Confined Areas 32
3.3.9 Working in Unstable Soils 33
3.3.10 Pushing Other Equipment 33
3.3.11 Winching 34
3.3.12 Ripping 35
3.3.13 Using Towed Attachments 36
3.3.14 Using a Side Boom 39
3.3.15 Moving the Dozer 39
Appendix Typical Daily Inspection Check Sheet 45

Figures and Tables

Figure 1   Various styles of dozers 3
Figure 2   Basic parts of a typical dozer 3
Figure 3   Operator's cab for a wheel dozer with a steering wheel 4
Figure 4   Operator's cab for a wheel dozer with joystick controls 4
Figure 5   Operator's cab for a track dozer 5
Figure 6   Instrument panel 6
Figure 7   Joystick control for vehicle movement 8
Figure 8   Joystick with rocker switch for vehicle speed 8
Figure 9   Decelerator and service brake pedals 9
Figure 10 Dozer blade control joystick 9
Figure 11 Ripper control 9
Figure 12 PAT blade 10
Figure 13 Straight blade 10
Figure 14 Variable-radius blade 10
Figure 15 SU blade 11
Figure 16 Heavy-duty U blade 12
Figure 17 Push plate blade being used to push a scraper 12
Figure 18 Landfill blade 12
Figure 19 Dozer using a rake blade 13
Figure 20 KG blade 13
Figure 21 V-tree cutter blade 13
Figure 22 Coal blade 13
Figure 23 Ripper 14
Figure 24 Winch 15
Figure 25 Towed sheepsfoot rollers 15
Figure 26 Dozer pulling a disc attachment 15
Figure 27 Dozers with side boom attachments 16
Figure 28 Check all fluids daily 18
Figure 29 Work up a hill 19
Figure 30 Battery disconnect switch 24
Figure 31 Discuss concerns with the foreman or engineer 25

Figures and Tables (continued)

Figure 32 Blade movement 27

Figure 33 Soil with good moisture content...................................... 28

Figure 34 Straight dozing 28

Figure 35 Slot dozing 29

Figure 36 Building a stockpile 29

Figure 37 Ditching using blade tilt 29

Figure 38 Clearing brush and small trees 30

Figure 39 Backfilling... 31

Figure 40 Sidehill cutting.. 31

Figure 41 Finish grading.. 32

Figure 42 Filling unstable areas 33

Figure 43 Pushing a scraper 34

Figure 44 Secure the winch cable after use 35

Figure 45 Ripping compacted soil with a three-shank ripper........... 35

Figure 46 Ripping soil with a single-shank ripper 36

Figure 47 Dozer towing a sheepsfoot roller 36

Figure 48 Dozer with side boom laying pipe 38

Table 1 Characteristics of Small and Large Dozers........................ 2

1.0.0 USES AND COMPONENTS OF DOZERS

Objective

Identify and describe the uses and components of a dozer.

   a. Identify and describe common uses and types of dozers.
   b. Identify and describe major parts of a dozer.
   c. Identify and describe dozer instrumentation.
   d. Identify and describe dozer controls.
   e. Identify and describe common dozer blades.
   f. Identify and describe common dozer attachments.

Trade Terms

Blade: The main attachment on the front end of a dozer that is used for moving material.

Charging hoppers: The process of filling a hopper or temporary storage container with a material, such as coal, that will be dispersed later.

Dozing: Using a blade to scrape or excavate material and move it to another place.

Grubbing: Digging out roots and other buried material.

KG blade: A type of dozer-mounted blade used in forestry and land clearing operations. On some blades, a single spike called a stinger splits and shears stumps at the base.

Stockpile: Material that is dug and piled for future use.

Winch: An attachment commonly mounted on the rear of a dozer that uses an electric motor and wire rope to pull equipment, trees, and other objects.

Dozers are powerful pieces of equipment. They are designed to provide maximum power and stability at speeds of up to 10 miles per hour (16 kilometers per hour), although a few dozers are designed for faster speeds. While most dozers are not very fast, their good traction, stability, and low center of gravity make them especially suited to work on irregular terrain and steep slopes. Because dozers are available in many different models and can accept numerous

attachments, they can be used for a wide variety of grading and excavation jobs.

1.1.0 Common Uses and Types of Dozers

It is a rare sight to pass by a construction project and not see some type of dozer at work. Since the 1920s, dozers have become increasingly prevalent in construction, excavation, grading, agricultural, and even military tasks. The following sections describe some of the common uses and types of dozers that heavy equipment operators are likely to encounter.

1.1.1 Dozer Uses

Some of the primary uses of a dozer include the following:

- Excavating, cutting, filling, backfilling, and spreading material
- Moving soil, rock, and other material
- Performing rough and finish grading
- Maintaining haul roads and temporary roads through rocky or rough terrain
- Clearing construction sites of debris, clearing land, and pulling out tree stumps and other buried items
- Towing accessory equipment
- Pushing other heavy equipment

Some of the **blades**, attachments, and control techniques that enable a dozer to perform these tasks are covered in detail later in this module.

1.1.2 Dozer Types and Configurations

Not only can dozers be configured in different ways to serve different purposes, they are often called by different names. For instance, they are commonly referred to as crawler tractors, track-type tractors, crawlers, dozers, or crawler dozers. But no matter the name, dozers come in a wide range of sizes.

Generally, small dozers are used in light or finish grading and other tasks that require work in tight spaces. Larger dozers are typically used for clearing and rough grading on large excavation or construction projects for roads, airports, dams, and buildings. The largest dozers are used in mining operations. See *Table 1* for a side-by-side comparison of small and large dozers.

Figure 1 shows several different styles of dozers. Two basic types of crawler dozers are low-track dozers and high-track dozers. The low-track machine is used primarily for grading. The high-track type is used primarily for pushing. Some manufacturers also make heavy-duty four-wheel drive dozers. Wheel dozers have excellent traction, and

Table 1 Characteristics of Small and Large Dozers

Approximate Specifications	Smallest Dozer	Largest Dozer
Engine power	43 horsepower	1,150 horsepower
Track shoe width	12 inches (30.48 centimeters)	34 inches (86.36 centimeters)
Overall length	11 feet (3.35 meters)	38 feet (11.58 meters)
Blade width	7 feet (2.13 meters)	24 feet (7.32 meters)
Blade capacity	1 cubic yard (0.76 cubic meters)	90 cubic yards (68.81 cubic meters)
Machine weight	8,690 pounds (3,941.72 kilograms)	336,420 pounds (152,5967.55 kilograms)

a steering wheel makes them easy to operate and maneuver. High-speed dozers are capable of operating at up to 16 miles per hour (25.7 kilometers per hour). The one shown in *Figure 1D* has an articulated frame to enhance maneuverability and rubber tracks that enable it to move across concrete or paved material without damaging it.

1.2.0 Major Parts of a Dozer

Before operating a dozer, it is important to understand its basic parts and controls. Proper knowledge of the features, instruments, controls, and attachments helps prevent component damage and operator injuries. There are many different makes and models of dozers, so the cab setup, instruments, and controls can vary. This section focuses on features likely to be found on a typical dozer, but operators should always refer to the manufacturer's recommendations for operating a particular machine.

The basic parts of a typical dozer are shown in *Figure 2*. The operator's cab is contained within the rollover protective structure (ROPS) and is mounted behind the engine. Wide tracks are mounted on both sides of the machine. The blade is mounted in the front of the machine. Attachments such as rippers and winches can be mounted on the rear of the machine.

The blade is the primary operating component of the dozer. It is mounted on two push arms on either side of the machine. The blade is controlled with hydraulic cylinders and can be moved in several directions. The hoist cylinder lifts and lowers the blade. The blade can be tilted from side to side. Some blades can be angled from side to side to control the direction of the push.

1.2.1 Operator's Cab

Dozer operations are performed from the operator's cab. The cab contains the instruments and controls for vehicle movement, blade operation, and auxiliary functions. An operator must understand the instruments and controls before operating a dozer. Study the operator's manual to become familiar with the instruments and controls and their functions.

The configuration of the operator's cab varies among different makes and models of dozers. Typically, vehicle movement lever controls are operated with the right hand and blade controls with the left. The operator's cab in a wheel dozer differs slightly from that of a track dozer. Always read the operator's manual and understand the cab configuration for the dozer being operated.

Figure 3 shows one configuration for a wheel dozer that uses a steering wheel for vehicle movement and levers for blade controls. This setup is similar to what is used for some types of wheel loaders.

Some wheel dozers have joystick steering controls, with a joystick on the left that controls vehicle movement. The joystick on the right controls the blade. The instrument panels and foot pedals are in front of the operator's seat. The cab shown in *Figure 4* has this configuration.

Track dozers typically have joystick controls. Older models may have lever or foot pedal controls. A typical cab for a track dozer is shown in *Figure 5*. The joystick on the left controls vehicle movement. The instrument panel is directly in front of the operator's seat. Engine speed is controlled with the governor, decelerator, and service brakes. The blade is controlled with the joystick on the right. Much of the information in this module is based

Something for Everyone

When it comes to dozers, one size does not fit all. As the *Table 1* below indicates, there are huge differences in size and capacity between the smallest and largest dozers. This wide range of choices means that there is a size and type of dozer for practically every need.

(A) LOW-TRACK DOZER

CAT D11T TRACK-TYPE TRACTOR. REPRINTED COURTESY OF CATERPILLAR INC.

(B) HIGH-TRACK DOZER

CAT 854K WHEEL DOZER. REPRINTED COURTESY OF CATERPILLAR INC.

(C) WHEEL DOZER

(D) HIGH-SPEED DOZER

22302-14_F01.EPS

Figure 1 Various styles of dozers.

22302-14_F02.EPS

Figure 2 Basic parts of a typical dozer.

REPRINTED COURTESY OF CATERPILLAR INC.

22302-14_F03.EPS

Figure 3 Operator's cab for a wheel dozer with a steering wheel.

on this control configuration. However, operators should always review the operator's manual for the machine they are using to be familiar with the control configuration for that machine.

Dozer cabs are designed for maximum visibility. Most cabs have an open design or have large windshields or windows that extend around the front of the cab. This allows the operator to see the area to the front and sides of the machine. Windshield wipers help keep the windshield clean. Lights aid vision at night, in excavations, or in poor lighting conditions.

The seat in a dozer cab is adjustable. An operator should adjust the seat so that all of the controls can be reached easily. An operator's legs must be able to fully depress the foot pedals when his or her back is flat against the back of the seat. Armrests should be adjusted so that the operator can operate the controls easily.

1.3.0 Instrumentation

An operator must pay attention to the instrument panel, which includes the gauges that indicate

REPRINTED COURTESY OF CATERPILLAR INC.

22302-14_F04.EPS

Figure 4 Operator's cab for a wheel dozer with joystick controls.

engine speed and operating temperatures. There are also numerous warning lights and gauges that must be monitored for safe operations. If the operator ignores alarms and indications on the instrument panel, the machine can be seriously damaged.

The instrument panel varies on different makes and models of dozers. A typical instrument panel is shown in *Figure 6*. It contains switches, alert indicators, cab comfort controls, and a quad-gauge cluster. Some machines may include additional instruments and displays, so it is important to read the operator's manual for the machine being used and be familiar with all of the instruments.

A dozer has sensors in various parts of the engine and related systems. These sensors are connected to a monitoring system. The monitoring system display has alert indicators that light when the machine is not functioning properly. If the situation needs immediate attention, the indicators may flash or remain lit, or an audible alarm will sound. Machines may have different indicator lights, or they may be arranged differently. Always read the operator's manual to fully understand all warning lights before operating the equipment.

Below the indicator lights is a display window. This display is a digital readout. It can be set to show a digital reading of any of the gauges or set to show other readings. These other readings in-

VEHICLE MOVEMENT CONTROL

INSTRUMENT PANEL

BLADE CONTROL

SERVICE BRAKES

DECELERATOR

ANGLED SEAT (FOR GREATER COMFORT AND VISABILITY)

REPRINTED COURTESY OF CATERPILLAR INC.

22302-14_F05.EPS

Figure 5 Operator's cab for a track dozer.

clude total operating hours, current engine speed, or diagnostic codes. The default display shows the current gear.

The engine key switch is also located on the instrument panel. In some models, the key switch is on the steering control joystick. In that case, the key switch will also lock the parking brake. Check the operator's manual for the specific functions of the key switch and any other starting aids.

All dozers also have indicator lights and an alert system similar to those of tractors and loaders. Review the operator's manual to be familiar with all alert and indicator lights before operating the machine.

> **WARNING!**
>
> Operators must know what each indicator light and alert indicator means before operating the equipment. When these action lights are lit, the operator must take immediate action to prevent machine damage or personal injury. Familiarize yourself with all controls and instruments, including action and indicator lights, before operating any equipment.

The gauges on the instrument panel include the engine coolant temperature gauge, torque converter oil temperature gauge, fuel level gauge, and hydraulic oil temperature gauge. These gauges are described in the following sections.

1.3.1 Engine Coolant Temperature Gauge

The engine coolant temperature gauge indicates the temperature of the coolant flowing through the cooling system. Refer to the operator's manual to determine the correct operating range for normal operations. Temperature gauges normally read left to right, with cold on the left and hot on the right. Most gauges have a red segment. If the needle is in the red zone, the coolant temperature is excessive. Some machines may also activate warning lights if the engine overheats.

> **CAUTION**
>
> Operating equipment when temperature gauges are reading in the red zone may severely damage it. Stop operations, determine the cause of the problem, and resolve it before continuing operations.

GAUGES

ALERT INDICATIONS

DIGITAL DISPLAY

ENGINE KEY SWITCH

COMFORT CONTROLS

REPRINTED COURTESY OF CATERPILLAR INC.

22302-14_F06.EPS

Figure 6 Instrument panel.

If the engine temperature gets too high, operations should be stopped immediately. Get out of the machine and investigate the problem. There are several checks that can be performed. First, check the engine coolant level. Add more fluid if it is too low. Check if the fan belt is loose or broken. Replace it if necessary. Check that the radiator fins are not fouled and clean them if necessary. These are the three primary causes of the engine overheating. If initial troubleshooting fails to resolve the problem, stop operations and take the machine out of service.

WARNING!

Engine coolant is extremely hot and under pressure. Check the operator's manual and follow the procedure to safely check and fill engine coolant.

1.3.2 Torque Converter Oil Temperature Gauge

The torque converter oil temperature gauge indicates the temperature of the oil flowing through the transmission. This gauge also reads left to right in increasing temperature. It has a red zone

that indicates excessive temperatures. If the temperature is in the red zone, shut down the machine and have it checked by a mechanic.

NOTE

The torque converter temperature will rise if the wrong transmission speed is used.

1.3.3 Fuel Level Gauge

The fuel level gauge indicates the amount of fuel in the fuel tank. On diesel engines, the gauge may contain a low fuel warning zone. Some models have a low fuel warning light. Avoid running out of fuel on diesel engines because the fuel lines and injectors must be bled of air and the injector pump may need to be primed before the engine can be restarted. Always check the operator's manual for the correct restarting procedures if the dozer has run out of fuel, and before running down the batteries.

1.3.4 Hydraulic Oil Temperature Gauge

The hydraulic oil temperature gauge indicates the temperature of the oil flowing through the hydraulic system. This gauge also reads left to right in increasing temperature. It has a red zone that indicates excessive temperatures. If the temperature is in the red zone, shut down the machine and have it checked by a mechanic.

1.4.0 Controls

A dozer uses various levers, pedals, and joysticks to control vehicle movement as well as the blade and any attachments that might be connected to the machine. It is critical that operators be able to identify these controls and understand how they work before operating a dozer. Operating the controls incorrectly can cause serious injuries. Once operators develop skills with the controls, they will be able to perform different tasks with ease and efficiency. Always refer to the manufacturer's manuals for the particular piece of equipment being operated.

1.4.1 Vehicle Movement Controls

There are several controls used to move and maneuver the dozer. On some dozers, the transmission and steering are controlled with a single joystick. Machine ground speed is controlled with a governor lever (throttle control), a service brake pedal, a decelerator pedal, and a speed selector.

The joystick shown in *Figure 7* is one that controls the primary vehicle movement. The parking brake knob (1) is depressed to release the parking brake. The knob is pulled upward to engage the parking brake. This also locks the steering controls and the transmission in neutral. On some models, the engine key switch (2) is used to lock and unlock the parking brake.

Modern Cabs, Modern Features

As technology improves, the instruments, controls, and other features inside the operator's cab become more sophisticated. Pushbuttons and switches become touch screen features. On-board computers monitor equipment performance and display information about system conditions, efficiency, and safety. Ergonomic improvements help increase operator visibility and comfort. These improvements are especially true for larger dozers. After mastering basic dozer operations, operators can advance to more powerful and complex machines.

REPRINTED COURTESY OF CATERPILLAR INC.

22302-14_SA01.EPS

The transmission controls include direction selection and gear selection. Once the parking brake is unlocked, the joystick can be rotated to select forward, neutral, or reverse (FNR). Rotate the joystick forward to engage in forward motion. Rotate the joystick back to move in reverse. The central position is neutral. The direction (FNR) selected is indicated on the cuff of the joystick (3).

The transmission gears are changed using buttons (4). Push the top button to shift the transmission to the next highest gear. Push the lower button to downshift. The selected gear and direction will be shown on the digital display on the instrument panel.

Steering is controlled by moving the joystick (5). When the machine is moving forward, push the joystick forward to make a left turn. To make a right turn, pull the joystick toward you. When the joystick is released, it will return to the central position and the machine will continue to move in straight line in the direction selected.

The joystick shown in *Figure 8* is a slightly different version. In this case, vehicle speed is controlled using the rocker switch that is shown with the arrow. Moving the switch forward increases the speed. Moving the switch backward decreases the speed.

Additional movement controls that are not contained on a joystick are shown in *Figure 9*. Depressing the decelerator pedal will override the throttle control and reduce engine speed. The decelerator is used when changing direction or maneuvering in small areas.

The service brakes are used to slow and stop the machine. Depress the pedal to apply the service brakes and release the pedal to allow the machine to move. Use the service brakes on a downhill slope to prevent the engine from overspeeding.

REPRINTED COURTESY OF CATERPILLAR INC.

22302-14_F08.EPS

Figure 8 Joystick with rocker switch for vehicle speed.

Symbols

Indicators, warning lights, and controls typically use symbols to help operators identify them and understand their function. Not all manufacturers uses identical symbols for their equipment, but many of the more common symbols are the same. Below are some examples of symbols commonly used on dozer instruments and controls. Always read the operator's manual to make sure that you are thoroughly familiar with the symbols used on the equipment that you are operating.

BRAKE SYSTEM

PARKING BRAKE LIGHT

ENGINE RUN

COOLANT

FUEL

HYDRAULIC OIL

TRANSMISSION

WINCH

22302-14_SA02.EPS

REPRINTED COURTESY OF CATERPILLAR INC.

22302-14_F07.EPS

Figure 7 Joystick control for vehicle movement.

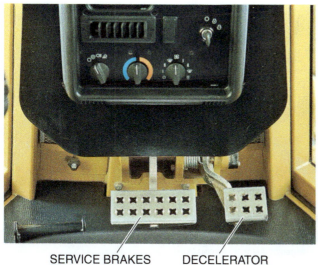

SERVICE BRAKES   DECELERATOR

REPRINTED COURTESY OF CATERPILLAR INC.

22302-14_F09.EPS

Figure 9 Decelerator and service brake pedals.

1.4.2 Blade Controls

A dozer blade is used to perform three tasks: cut, carry, and deposit material. It can be raised and lowered to control the depth of the cut. It can be tilted so one side digs deeper than the other and can be angled to direct the placement of material to either side. On some machines, the pitch of the blade—the angle it forms with the ground—can also be adjusted. There are several different configurations for the lever that controls the dozer blade. Always read the operator's manual to become familiar with the controls of the machine you are operating.

A typical dozer blade control joystick is shown in *Figure 10*. Move the joystick forward (2) to lower the blade. When the joystick is moved all the way forward (1) the blade will float and follow the contour of the ground. Move the joystick toward the back (4) to raise the blade. Moving the joystick to the right (5) will tilt the blade to the right; that is, the right side of the blade will be lower than the left side. Move the joystick to the left (6) and the blade will tilt to the left. When the joystick is released, it will return to the central position (3) and the blade will stop moving.

On some machines, blade control is accomplished using a lever with a trigger switch. When the trigger is depressed, the lever controls the blade angle. When the trigger is depressed and the lever is moved to the left, the blade will angle to the left. When the lever is moved to the right with the trigger depressed, the blade will angle to the right. Releasing the lever will stop blade movement.

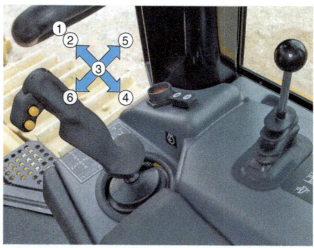

REPRINTED COURTESY OF CATERPILLAR INC.

22302-14_F10.EPS

Figure 10 Dozer blade control joystick.

1.4.3 Ripper Control

On dozers equipped with a ripper, the ripper control is typically located next to the blade control (*Figure 11*). Moving the control to the left lowers the ripper. Moving it to the right raises the ripper. When released, the lever will return to the central hold position and the ripper will stop moving.

1.5.0 Blades

Blades come in many different shapes and sizes. Blade selection is determined by the type of material to be moved and the dozer's limitations. There are specialized blades designed for land clearing, coal mining, managing wood chips, and landfills.

REPRINTED COURTESY OF CATERPILLAR INC.

22302-14_F11.EPS

Figure 11 Ripper control.

Properly matching the blade with the machine is a basic requirement for maximizing production.

There are three basic types of blades made for dozers: general purpose, production, and special application blades. Within each category there are many different types and sizes of blades, each designed for a specific purpose. You should know which type of blade to use for the work assigned. Using a dozer with the wrong blade results in poor equipment performance and may even cause damage to the machine or the work area.

Before using a dozer for a specific task, check the blade to make sure it can effectively handle the work. Also, check the dozer controls to determine if they perform the intended functions for the attached blade.

REPRINTED COURTESY OF CATERPILLAR INC.

22302-14_F12.EPS

Figure 12 PAT blade.

> **NOTE**
>
> Always consult the operator's manual and attachment literature to ensure that any blades and/or attachments being used are suitable for the machine.

1.5.1 General-Purpose Blades

General-purpose blades are used for general construction, land improvement, soil conservation, stripping, and site development. General-purpose blades include power angle and tilt, variable radius, and straight blades.

The angling blade can be hydraulically angled so that one side is more forward than the other. The power angle and tilt blade (*Figure 12*) is often called a PAT blade or six-way blade. The blade can be angled and tilted using hydraulic power. This gives the operator greater control of the placement of the material. These blades are designed for side casting, pioneering roads, backfilling, cutting ditches, and performing other smaller tasks. They can reduce the amount of maneuvering required to do these jobs. They are not recommended for rock or severe applications.

The straight blade (*Figure 13*) is known as an S blade. S blades tend to be shorter and have less lateral curve than PAT blades. They have small side wings to prevent too much soil from spilling off the blade ends. The design of the straight blade makes it useful for fine grading.

On the variable-radius blade (*Figure 14*), the moldboard is thicker at the edges. Dirt and other materials follow the path of least resistance. The variable-radius moldboard causes material to move to the center of the blade and fold over itself. This action increases blade capacity and requires less tractor effort to move the load.

REPRINTED COURTESY OF CATERPILLAR INC.

22302-14_F13.EPS

Figure 13 Straight blade.

REPRINTED COURTESY OF CATERPILLAR INC.

22302-14_F14.EPS

Figure 14 Variable-radius blade.

1.5.2 Production Blades

Production blades are used for large-scale production operations where big loads must be moved over long distances, such as in land reclamation, stockpile work, and charging hoppers. The wings provide increased load retention while maintaining the blade's ability to penetrate and quickly load tightly packed materials.

Production blades include the universal, semi-universal, and heavy-duty universal blades. Universal blades are often called U blades and semi-universal blades are known as SU blades or four-way blades. The two sides of a U blade are angled toward the center. This creates a rolling action within the load and increases blade capacity. Semi-universal or SU blades are shorter and flatter than U blades, as shown in *Figure 15*.

Heavy-duty U blades (*Figure 16*) are designed for use in tougher environments like mining work and work in rocky soils. They are built with high-tensile steel and feature bolt-on moldboard covers, heavy-duty rock guards, and tilt cylinder guards.

1.5.3 Special Application Blades

Many manufacturers make dozers designed for specific applications. Some manufacturers make dozers with special blades for specific applications such as forestry work, coal mining, tree cut-

REPRINTED COURTESY OF CATERPILLAR INC.

22302-14_F15.EPS

Figure 15 SU blade.

ting, and handling trash in landfills. Each blade is designed to do specific jobs based on the type of material and the type of terrain. However, specialization may reduce the versatility of the overall machine.

The push plate (*Figure 17*) and cushion blades are used for push-loading scrapers or pushing other equipment. The center of the blade is reinforced. Cushion blades have rubber cushions to absorb the impact of contacting a scraper push block. When not push-loading, the dozer can be

GPS Technology

An important addition to dozer control involves the use of Global Positioning System (GPS) technology. A GPS system on a dozer is similar to that found on automobile navigation systems. Instead of providing travel directions, however, a dozer system uses satellite guidance to relate the blade position to a computerized model of the site. The onboard computer then signals the operator or the dozer's hydraulic system to adjust the blade position to achieve the design requirements. This system increases grading accuracy and speed, and it removes some of the burden on the operator. The result is higher productivity and lower operating costs.

GPS RECEIVERS

22302-14_SA03.EPS

Figure 16 Heavy-duty U blade.

Figure 18 Landfill blade.

used for cut maintenance and other general *dozing* jobs.

The landfill blade (*Figure 18*) is specifically designed for landfill operations where the blade channels trash to the compactor tracks. A unique hydraulic tipping system enhances its channeling ability. Tipping the blade forward increases its ability to funnel trash to the tracks. The blade's integrated trash rack allows good visibility.

The rake blade (*Figure 19*) is used in land clearing and *grubbing* applications and some types of demolition. A rake blade can be attached to the dozer blade or replace the dozer blade. It can handle vegetation up to various sizes of trees and offers good soil penetration for the removal of small

stumps, rocks, and roots. To use this blade effectively, dig the end of the blade into the top layer of the earth and move forward. The tines of the rake will pass through the soil, but will catch any large plants, bushes, and trees.

Two blades used primarily for specific applications in the forest products industry are the *KG blade* and the V-tree cutter blade. The KG blade (*Figure 20*) is used for land clearing operations. In addition to cutting trees, this versatile blade can pile vegetation, cut V-shaped drainage ditches, and build roads and firebreaks through heavily wooded areas.

The V-tree cutter blade (*Figure 21*) is used for shearing trees, stumps, and brush at ground level. A sharp angle or V, formed by two cutting blades, directs the tractor's weight and horsepower through the center line of the cutter. The use of this force allows most growth to be cut at a steady

REINFORCEMENTS

Figure 17 Push plate blade being used to push a scraper.

REPRINTED COURTESY OF CATERPILLAR INC.

22302-14_F19.EPS

Figure 19 Dozer using a rake blade.

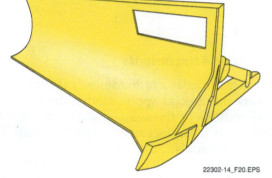

22302-14_F20.EPS

Figure 20 KG blade.

pace and then cast to the sides. The moldboard on the blade has teeth for clearing scrub, while the sharp edge in the center is used for splitting larger trees.

Coal or wood chip blades (*Figure 22*) are designed for maximum productivity when handling these materials. The moldboard is higher on a coal blade to help roll the load. The side wings are angled to reduce side spillage and retain the load.

1.6.0 Attachments

Dozers are versatile pieces of equipment. Besides the main operation of dozing with a blade, many functions can be performed using attachments. Some common attachments that are used with dozers include rippers, winches, drilling tools, pusher plates, and rollers.

> **WARNING!**
>
> Using an attachment that is not compatible with the machine can be extremely dangerous. Incompatible attachments may fail and cause severe injury or property damage.

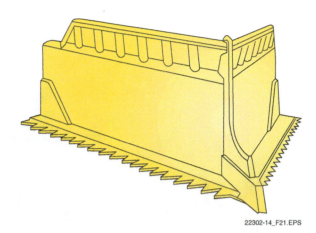

22302-14_F21.EPS

Figure 21 V-tree cutter blade.

REPRINTED COURTESY OF CATERPILLAR INC.

22302-14_F22.EPS

Figure 22 Coal blade.

1.6.1 Rippers

Rippers are attachments used for ripping up hard or compacted surfaces, pavements, or rocky ground, and for breaking roots. The dozer blade alone would not be able to penetrate these materials. Once the material is broken up, it can be moved with the dozer blade. Rubble can be removed and the surface can be graded.

Rippers are mounted on a frame and can be raised or lowered hydraulically (*Figure 23*). There are several different styles and configurations used in various applications.

The ripper has one or more pointed teeth that vary in length and width. The teeth are made up of a shank and a point, tip, or cap. The points are replaceable and should be replaced whenever they are worn. Dull teeth are inefficient. If neglected, continued use could damage the shank. Ripper points should be checked as part of the daily inspection.

Typically, the ripper is used when the dozer is moving forward. However, backup ripper at-

Figure 23 Ripper.

tachments can be mounted onto certain types of blades. These rippers are used when the dozer is moving in reverse.

1.6.2 Winches

Winches (*Figure 24*) are standard attachments on many dozers. They are usually rear-mounted and use large-diameter wire rope. A powerful motor mechanically winds the wire rope around the drum. The end of the rope can be attached to equipment, trees, or other objects to be moved. The winch is enclosed in a steel cage mounted on the back of the machine. This is a safety function to prevent damage or injury in case the rope breaks. Dozers are often used to rescue equipment that has become stuck in mud.

Never use damaged or frayed wire rope. Always inspect the wire rope as part of the daily maintenance check. Replace damaged rope. Wire rope that has been damaged from wear, corrosion, kinking, or other factors will no longer function at its original strength. If the wire rope breaks, it can backlash violently and injure the operator or anyone else working in the area. In addition, whatever is being winched with the wire rope will be released suddenly to fall, slide, or roll away.

> **WARNING!**
>
> Always wear gloves when working with wire rope. Never stand next to the cable when it is taut. If it breaks, it can cause severe injury. It is recommended that the rear of the cab be caged when using a cable.

1.6.3 Other Attachments

Other attachments are made by various manufacturers for special applications. The functions and operations of these attachments are usually described in manufacturers' sales literature and operating manuals.

A tree shear can be used on dozers to harvest trees in a logging operation. This attachment allows the operator to hold the trunk of a tree and then shear off the trunk close to the ground. After cutting through the trunk, the operator can rotate the tree horizontally and gently lay down the log.

Rollers and discs are attachments that are usually towed by the dozer instead of attached directly to the frame. These attachments are used for compacting or digging up surface material. *Figure 25* shows a pair of sheepsfoot compacting rollers. *Figure 26* shows a dozer pulling a disc attachment.

Side booms, which are sometimes called pipe booms, are special attachments that can be used for heavy hoisting and carrying, particularly in pipe or cable-laying projects (*Figure 27*). The boom is attached to one side of the dozer with the hoisting mechanism. Counterweights are attached on the opposite side. A power takeoff drives two drums that are controlled through separate clutches and brakes. One drum controls the boom height, while the other one controls the load line that is attached to a hoist block. The counterweight is hinged so that it can be

REPRINTED COURTESY OF CATERPILLAR INC.

22302-14_F24.EPS

Figure 24 Winch.

22302-14_F26.EPS

Figure 26 Dozer pulling a disc attachment.

brought in close to the dozer for traveling and handling light loads. It can be extended away from the dozer by hydraulic pistons to counterbalance heavy loads on the boom. Because of their weight, dozers provide a stable platform for hoisting pipe and cable that have been spliced together and moving it the short distance from the side of the trench into the trench.

22302-14_F25.EPS

Figure 25 Towed sheepsfoot rollers.

22302-14_F27.EPS

Figure 27 Dozers with side boom attachments.

1.0.0 Section Review

1. One of the many uses of a dozer is _____.

   a. hoisting equipment and material from low areas
   b. pushing other heavy equipment
   c. loading trucks with excavated material
   d. deep trenching with a backhoe attachment

2. The primary operating component of a dozer, which is mounted on two push arms on either side of the machine, is the _____.

   a. track mechanism
   b. operator's cab
   c. underbelly
   d. blade

3. The torque converter oil temperature gauge indicates the temperature of the oil flowing through the _____.

   a. transmission
   b. engine
   c. attachment hoses
   d. hydraulic system

4. Raising and lowering a dozer blade will control _____.

   a. which side digs deeper
   b. the placement of material to either side
   c. the depth of the cut
   d. the angle the blade forms with the ground

5. A general purpose six-way dozer blade is known as a(n) _____.

   a. heavy-duty U-blade
   b. power angle and tilt (PAT) blade
   c. variable-radius blade
   d. SU blade

6. Ripper attachments are typically used on a dozer to _____.

   a. dig trenches as the dozer grades
   b. prevent damage to the main blade
   c. break up compacted surfaces
   d. enhance pipe or cable laying projects

SECTION TWO

2.0.0 SAFETY, INSPECTION, AND MAINTENANCE

Objective

Identify and describe safety, inspection, and service guidelines associated with a dozer.

a. Describe guidelines associated with dozer safety.
b. Describe prestart inspection procedures.
c. Describe preventive maintenance requirements.

Performance Task 1

Demonstrate a proper prestart inspection and maintenance on a dozer.

Trade Terms

Grouser: A ridge or cleat across a track shoe that improves the track's grip on the ground.

Dozers are among the most common and the most powerful pieces of heavy equipment found at a job site. Because of their size and power, dozers that are operated safely and efficiently can perform an enormous amount of work. To ensure that a dozer functions safely and efficiently, operators must be familiar with all safety guidelines that apply to the dozer, and they must be able to inspect and maintain the dozer in good working order.

> **WARNING!**
> Dozers are complex and powerful machines that require complete attention for proper control during operation.

2.1.0 Safety Guidelines

Companies have safety policies and meetings, and many have safety committees and safety officers. Ultimately, however, safety is an individual responsibility. An operator is responsible for three areas of safety: personal safety, safety of others, and safety of the equipment. With planning and attention to detail, operators can keep themselves safe and protect those around them.

2.1.1 Operator Safety

Many accidents happen when people are not paying attention to what is happening around them. Safety can be as simple as remaining alert and avoiding accidents. This means thinking about potential problems and planning how to avoid of them.

Employers will tell operators about company safety policies. Many companies have a written safety manual. However, the company relies on individual responsibility to put the policies into action. Safety begins with an operator taking personal responsibility to stay safe. It begins before the machine is started, and continues until an operator leaves the job site. The following are recommended safety procedures for all situations:

- Mount and dismount the equipment carefully using three-point contact and facing the machine.
- Clean steps, grab handles, and the operator's compartment. Mud on steps or other surfaces can cause an injury.
- Wear the personal protective equipment required for the job when operating the equipment.
- Do not wear loose clothing or jewelry that could catch on controls or moving parts.
- Keep the windshield, windows, and mirrors clean at all times.
- Never operate equipment under the influence of alcohol or drugs.
- Never smoke while refueling or checking batteries or fluid.
- Never remove protective guards or panels. Replace guards that break or become damaged.
- Never use your hands to check for leaks. Instead, hold a piece of wood, cardboard or other solid material underneath any areas where leaks might occur. Hydraulic and cooling systems operate at high pressure. Fluids under high pressure can cause serious injury.
- Observe extreme caution working along the top of banks or slopes. Stay away from the edge. The edges could collapse and cause a rollover or severe injury.

> **WARNING!**
> Getting in and out of equipment can be dangerous. Always face the machine and maintain three points of contact when you are mounting and dismounting. That means you should have three out of four of your hands and feet on the equipment. That can be two hands and one foot or one hand and two feet.

2.1.2 Safety of Co-Workers and the Public

Equipment operators rarely work alone. A typical job site includes other operators, grade checkers, flaggers, or other support personnel. It is often difficult to see ground personnel. The operator must know where all ground personnel are at all times. Additional safeguards are needed when working in public areas to protect pedestrians.

Dozers are some of the largest and most powerful pieces of equipment around. Operators must allow themselves time to recognize any potential dangers and try to avoid them. The main safety points when working around other people include the following:

- Walk around the equipment to make sure that everyone is clear of the equipment before starting and moving it.
- Before working in a new area, take a walk to locate any cliffs, steep banks, holes, power or gas lines, or other obstacles that could cause a hazard to safe operation.
- Maintain a clear view in all directions. Do not carry any equipment or material that obstructs your view.
- Always look before changing directions.
- Never allow riders on the dozer.

- When working in traffic, find out what warning devices are required. Always know the rules and the meaning of all flags, hand signals, signs, and markers being used.

2.1.3 Equipment Safety

In addition to being responsible for the safety of themselves and others, operators are also responsible for the safety of the equipment. This starts with performing regular inspections and maintenance. However, the operator must continue to look and listen to the machine throughout the day to make sure it is functioning properly.

It also includes not taking risks with the equipment. Do not push the equipment to the design limits. Always leave a margin for safety or error. An accident or equipment failure will cost more time than the time saved by operating beyond the machine's capacity. Use the following guidelines to keep equipment in good working order:

- Perform prestart inspection and lubrication daily (*Figure 28*).
- Look and listen to make sure the equipment is functioning normally. Stop if it is malfunctioning. Correct or report trouble immediately.

REPRINTED COURTESY OF CATERPILLAR INC.

22302-14_F28.EPS

Figure 28 Check all fluids daily.

- Always travel with the blade low, just high enough to clear obstructions.
- Never exceed the manufacturer's limits for speed or operating on inclines.
- Never park on an incline.
- Always lower the blade and other attachments, engage the parking brake, turn off the engine, and secure the controls before leaving the equipment.
- When loading, unloading, and transporting equipment, know and follow the manufacturer's recommendations for loading, unloading, and tie down.
- Make sure clearance flags, lights, and other required warnings are on the equipment when transporting it.

The basic rule is to know the equipment. Learn the purpose and use of all gauges and controls, as well as the equipment's limitations. Never operate a dozer if it is not in good working order. Some basic safety rules of operation include the following:

- Do not operate the dozer from any position other than the operator's seat.
- Do not coast. Neutral is for standing still only.
- Maintain control when going downhill. Do not shift gears.
- Whenever possible, avoid obstacles such as rocks, fallen trees, curbs, or ditches.
- If it is necessary to cross over an obstacle, reduce speed and approach at an angle to reduce the impact on the equipment and yourself.
- Use caution when undercutting high banks, backfilling trenches, and removing trees.
- Avoid any condition that could cause the machine to tip over, including hills, banks, and slopes. Tipping can also occur when crossing ditches, ridges, or other unexpected obstacles.
- When pulling a load with the winch, never stand near the cable because it may break and injure you.
- Whenever possible, work up or down a hill rather than across (*Figure 29*).

2.2.0 Prestart Inspection Procedures

Routine inspections and maintenance can significantly reduce equipment breakdowns and improve performance. The first thing that should be done before starting any work is to conduct a daily inspection. This should be done before starting the engine. The inspection will identify any potential problems that could cause a breakdown and will indicate whether the machine can be operated. The equipment should be inspected before, during, and after operation.

22302-14_F29.EPS

Figure 29 Work up a hill.

The daily inspection is often called a walk-around. The operator should walk completely around the machine checking various items. This general inspection of the equipment should include checking for the following:

- Leaks (oil, fuel, hydraulic, and coolant)
- Worn or cut hoses
- Loose or missing bolts
- Trash or dirt buildup
- Broken or missing parts
- Damage to gauges or indicators
- Damaged hydraulic lines
- Dull, worn, or damaged blades
- General wear and tear

> **WARNING!**
> Do not check for leaks with your bare hands. Use wood, cardboard or other solid material. Pressurized fluids can cause severe injuries to unprotected skin. Long-term exposure can cause cancer or other chronic diseases.

> **NOTE**
> If a leak is observed, locate the source of the leak and fix it. If leaks are suspected, check fluid levels more frequently.

Some manufacturers require that daily maintenance be performed on specific parts. These parts are usually those that are the most exposed to dirt or dust and may malfunction if not cleaned or serviced. For example, a typical service manual may recommend lubricating specific bearings every 10 hours of operation, or always cleaning the air filter before starting the engine.

Before beginning operation, check the fluid levels and top off any that are low. Check all of the machine's major functions including the following components:

- *Air cleaner* – If the machine is equipped with an air cleaner service indicator, observe the indicator. If the indicator shows red, the air filter and intake chamber need to be cleaned. If the machine does not have a service indicator attachment, remove the air cleaner cover and inspect the filter. Clean out any dirt at the bottom of the bowl. Check the operator's manual for the specific procedures.
- *Battery* – Check the battery cable connections.
- *Blade and attachments* – Check hardware to make sure there is no damage that would create unsafe operating conditions or cause an equipment breakdown. Make sure the blade is not cracked or broken. Check the wire rope on the winch for frayed wire. If there is an implement such as a roller attached, make sure the hitch is properly set and that the safety pin is in place.
- *Cooling system* – Check the coolant level and make sure it is at the level specified in the operating manual.
- *Drive belts* – Check the condition and adjustments of drive belts on the engine.
- *Engine oil* – Check the engine oil level and make sure it is in the safe operating range.
- *Environmental controls* – The machine may be equipped with lights and windshield wipers. If operating under conditions where these accessories are needed, make sure they work properly.
- *Fuel level* – Check the level in the fuel tank(s). Do this manually with the aid of the fuel dipstick or marking vial. Do not rely on the fuel gauge at this point. Check the fuel pump sediment bowl if one is fitted on the machine.
- *Hydraulic fluid* – Check the hydraulic fluid level in the reservoir.
- *Idlers* – These components keep tension on the tracks so they do not come off the sprockets. Check for broken or cracked rollers or tension springs, or any other damage.
- *Pivot points* – Clean and lubricate all pivot points.

> **NOTE**
>
> In cold weather it is sometimes preferable to lubricate pivot points at the end of a work shift when the grease is warm. Warm the grease gun before using it for better grease penetration.

- *Sprockets* – The teeth on a worn sprocket will not be well-defined or may even be broken. This may cause the dozer to throw a track, causing damage to the track, the drive shaft, or the bearings.
- *Track* – Tracks have a habit of wearing out very fast because they are the one part of the dozer that is constantly in use whenever the equipment is being operated. Check for broken or missing shoes and bolts. Also, inspect the grousers to see if they will need replacing soon. If the grousers are worn down too far, it is difficult to get good traction. Also check the tracks for proper tension.
- *Transmission fluid* – Measure the level of the transmission fluid to make sure it is in the operating range.

The operator's manual usually has detailed instructions for performing periodic maintenance. If an operator finds problems with the machine and is not authorized to fix the problems, he or she should inform the foreman or field mechanic. Get the problem fixed before beginning operations. An example of a typical daily inspection check sheet is shown in the *Appendix*.

2.3.0 Preventive Maintenance

Preventive maintenance involves an organized effort to regularly perform periodic lubrication and other service work in order to avoid poor performance and breakdowns at critical times. By performing preventive maintenance on a dozer, operators keep it operating efficiently and safely and avoid the possibility of costly failures in the future.

Preventive maintenance of equipment is essential and it is not that difficult if the right tools and equipment are available. The leading cause of premature equipment failure is putting things off. Preventive maintenance should become a habit, performed on a regular basis.

Maintenance time intervals for most machines are established by the Society of Automotive Engineers (SAE) and have been adopted by most equipment manufacturers. Instructions for preventive maintenance and specified time intervals are usually found in the operator's manual for each piece of equipment. Common time intervals are: 10 hours (daily), 50 hours (weekly), 100 hours, 250 hours, 500 hours, and 1,000 hours. The operator's manual will also include lists of inspections and servicing activities required for each time interval.

When servicing a dozer, follow the manufacturer's recommendations and service chart. Any special servicing for a particular piece of equipment will be highlighted in the manual. Normally, the service chart recommends specific intervals, based on hours of run time, for such things as changing oil, filters, and coolant.

Hydraulic fluids should be changed whenever they become dirty or break down due to overheating. Continuous and hard operation of the hydraulic system can heat the hydraulic fluid to the boiling point and cause it to break down. Filters should also be replaced during regular servicing.

Before performing maintenance procedures, always complete the following steps:

Step 1   Park the machine on a level surface to ensure that fluid levels are indicated correctly.

Step 2   Lower all equipment to the ground. Operate the controls to relieve hydraulic pressure.

Step 3   Engage the parking brake.

Step 4   Lock the transmission in neutral.

Preventive Maintenance Records

Accurate, up-to-date maintenance records are essential for knowing the history of a piece of equipment. Each machine should have a record that describes any inspection or service that is to be performed and the corresponding time intervals. Typically, an operator's manual and some sort of inspection sheet are kept with the equipment at all times. Operators should update maintenance records as part of their daily routine.

REPRINTED COURTESY OF CATERPILLAR INC.

22302-14_SA04.EPS

2.0.0 Section Review

1. A key factor in operator safety is to mount and dismount the equipment ____.

   a. using three-point contact and facing the machine
   b. only with the help of a co-worker
   c. using the fastest method possible
   d. facing away from the machine toward the work area

2. A walk-around inspection of a dozer should occur ____.

   a. at the end of a job, before the equipment is moved to another location
   b. during off hours, when operators are not present
   c. only when the equipment is being shut down for scheduled maintenance
   d. first thing in the morning, before operations begin

3. Service work for things like changing oil, filters, and coolant is typically specified ____.

   a. in the job contract, based on the work being performed
   b. by the manufacturer, based on hours of service
   c. by the site manager, based on past experience
   d. by the operator, based on his or her judgment

3.0.0 BASIC STARTUP AND OPERATIONS

Objective

Describe basic startup and operating procedures for a dozer.

a. Describe startup, warm-up, and shutdown procedures.
b. Describe basic maneuvers and operations.
c. Describe common work activities.

Performance Tasks 2, 3 and 4

Perform proper startup, warm-up, and shutdown procedures.

Perform basic maneuvers on a dozer, including moving forward, moving backward, turning with the blade up, and straight dozing.

Demonstrate basic dozer operation by:

- Creating a level pad that measures approximately 20 feet by 20 feet ±1/10 foot (6 meters by 6 meters ± 3 centimeters).
- Pushing a stockpile while maintaining proper windrows and berms.

Trade Terms

Bank state: A term indicating that material, such as soil, is in its natural state. Also referred to as virgin ground.

Blade float: When the blade is allowed to float over the surface for a smooth finish.

Blade pitch: The angle of the blade from the vertical.

Blade tilt: The angle of the blade from the horizontal.

Grubbing: Digging out roots and other buried material.

Pawl: Pivoted lever which has a free end to engage with the teeth of a cogwheel or ratchet so that the wheel or ratchet can only move one way.

Ripping: Loosening hard soil, concrete, asphalt, or soft rock.

Slot dozing: A method that creates a trench for controlling spillage from the blade ends.

A well-maintained dozer that is operated properly can accomplish an amazing amount of work. With a wide array of attachments, a dozer has almost unlimited potential for grading and excavation jobs. But before operation begins, operators must understand the importance of proper startup, warm-up, and shutdown procedures.

3.1.0 Startup, Warm-Up, and Shutdown

The startup, warm-up, and shutdown of an engine are very important. Proper startup lengthens the life of the engine and other components.

3.1.1 Startup

There may be specific startup procedures for the piece of equipment being operated. Check the operator's manual for a specific procedure. In general, the startup procedure should follow this sequence:

Step 1 Be sure all controls are in neutral. Blade and accessory controls should be in the hold position.

Step 2 Engage the parking brake. This is done with a lever or knob, depending on the dozer make and model.

Step 3 Make sure there are no people close to the equipment.

> **NOTE**
>
> When the parking brake is engaged, an indicator light on the dash will light up or flash. If it does not, stop and correct the problem before operating the equipment.

Step 4 Turn the battery disconnect switch (*Figure 30*) to the On position (if equipped).

Step 5 Engage the transmission safety control lock (if equipped).

Step 6 Place the ignition switch in the On position. The engine oil pressure light should come on.

Step 7 Move the throttle control to low idle. Press the starter button or turn the engine key switch to the start position until the engine starts.

Step 8 Warm up the engine for at least 5 minutes.

Figure 30 Battery disconnect switch.

Step 9 Check all the gauges and instruments to make sure they are working properly. Many machines perform an automatic self-test when first started.

Step 10 Shift the gears to low range.

Step 11 Release the parking brake and depress the service brakes.

Step 12 Check all the controls for proper operation.

Step 13 Check the service brakes for proper operation.

Step 14 Check the steering for proper operation.

Step 15 Manipulate the controls to be sure all components are operating properly.

Step 16 Shift the gears to neutral and lock.

Step 17 Reset the parking brake.

Step 18 Make a final visual check for leaks, unusual noises, or vibrations.

Some machines have special procedures for starting the engine in cold temperatures. Some units are equipped with glow plugs or starting aids. Review the operator's manual to fully understand the procedures for using these aids.

3.1.2 Warm-Up

A slow warm-up is essential for proper operation of the machine under load. As soon as the engine starts, release the starter switch and keep the engine speed low until the oil pressure registers. The oil pressure light should come on briefly

and then go out. If the oil pressure light does not turn off within 10 seconds, stop the engine, investigate, and correct the problem. Many machines do a self-test upon first starting. Make sure that all systems are operating normally.

Let the engine warm up to operating temperature before moving the dozer. A longer warm-up is necessary in cold weather. If the temperature is at or above freezing, let the engine warm up for 15 minutes. If the temperature is between 32°F and 0°F (0°C and –18°C), warm the engine for 30 minutes. If the temperature is less than 0°F (–18°C) or hydraulic operations are sluggish, additional time is needed. Follow the manufacturer's procedures for cold starting.

Check the other gauges and indicators to see that the engine is operating normally. Check that the water temperature and oil pressure indicators are in the normal range. If there are any problems, shut down the machine and investigate or get a mechanic to look at the problem.

3.1.3 Shutdown

The shutdown of the machine is critical because of all the hot fluids circulating through the system. These fluids must cool so that they can cool the metal parts before the engine is switched off. Proper shutdown will reduce engine wear and possible damage to the machine.

Shutdown should follow a specific sequence, such as the following:

Step 1 Find a dry, level spot to park the dozer. Stop the dozer by decreasing the engine speed and placing the direction lever in neutral. Depress the service brakes and bring the machine to a full stop.

Step 2 Place the transmission in neutral and engage the parking brake.

Step 3 Lower the blade so that it rests on the ground. If there are any attachments, be sure the attachment is also lowered.

Step 4 Place the speed control in low idle and let the engine run for approximately five minutes.

Step 5 Turn the engine key switch to the Off position and remove the key.

> **CAUTION**
>
> Failure to allow the machine to cool down can cause excessive temperatures in the turbocharger, which could result in damage.

Step 6 Release the hydraulic pressure by moving the control levers until all movement stops.

Some machines have disconnect switches, guards, or locks for added security. Check the machine and secure any additional features. Close the battery disconnect switch and fuel shutoff valve. Close and lock any vandal guards. Lock the parking brake and transmission. These controls provide additional safety and deter unauthorized users. Always engage any additional security systems when leaving the dozer unattended.

3.2.0 Basic Maneuvers and Operations

Operation of a dozer requires constant attention to the controls, instruments, and the surrounding environment. Dozers have a tremendous amount of power and force. Once that force is put in motion, it is not an easy task to stop or redirect it quickly. Operators should plan their work and movements in advance and be alert to other operations going on around their machine.

Do not take risks. If there is doubt about the capability of the machine to do some work, stop the equipment and investigate the situation. Discuss it with the foreman or engineer in charge (*Figure 31*). Whether it is a slope that may be too steep, or an area that looks too unstable to work, know the limitations of the equipment. Decide how to do the job before starting. Once an operator gets in the middle of something, it may be too late, causing extra work for others or an unsafe condition.

The following are suggestions to improve operating efficiency when operating a dozer:

REPRINTED COURTESY OF CATERPILLAR INC.

22302-14_F31.EPS

Figure 31 Discuss concerns with the foreman or engineer.

- Observe all safety rules and regulations.
- Keep equipment clean. Make sure the cab is clean so that nothing affects the operation of the controls.
- Calculate and plan operations before starting.
- Set up the work cycle so that it will be as short as possible.
- When grading, stop occasionally and spread the material. Usually, it is easier to spread several small piles than to spread one big pile.

Mount the equipment using the grab rails and foot rests. Maintain three points of contact facing the machine. Preparing to work involves getting yourself organized in the cab, fastening the seat belt, and starting the machine.

> **WARNING!**
>
> Always maintain three points of contact when mounting equipment. Keep grab rails and foot rests clear of dirt, mud, grease, ice, and snow.

Next, adjust the seat to a comfortable operating position. Make sure you can see clearly and reach all the controls. The seat should be adjusted to allow full brake pedal travel with your back against the seat back. This will permit the application of maximum force on the brake pedals. Typical seat adjustments include the following:

- Back angle adjustment
- Seat height adjustment
- Fore/aft adjustment
- Angle adjustment for bottom cushion
- Back height adjustment
- Weight adjustment and weight setting indicator

Operator stations vary, depending on the manufacturer, size, and model of the equipment. However, all stations have gauges, indicators and switches, levers, and pedals.

Gauges show the status of critical items such as water temperature, oil pressure, and fuel level. Typically, there will be at least a tachometer, temperature gauge, oil pressure gauge, and hour meter. Read the operator's manual and learn the purpose and normal operating settings for the gauges on the machine.

Indicators alert the operator to low oil pressure, engine overheating, clogged air and oil filters, and electrical system malfunctions. Switches are for activating the glow plugs, starting the engine, and operating accessories such as lights. Typical instruments and controls were described previously.

Some dozers use joysticks for steering, some use levers or pedals, and some use a steering wheel. The particular arrangement and operation may differ, but the basic operation is similar for all equipment. Review the operator's manual to know the specifics of the machine being operated.

Dozer controls are generally grouped as follows:

- Blade and accessory controls
- Steering
- Engine controls
- Movement or speed changes

Basic maneuvers for a dozer include being able to start, go forward and backward, and turn. Because many dozers do not use a steering wheel, it will take some practice to coordinate the control of the hand controls and foot pedals to steer the machine while operating the blade or other attachments.

3.2.1 Moving Forward

The operation of the controls on a particular piece of equipment will depend on whether one of the models with direct drive or a model with some type of power shift is being used. The procedure given below is for a power shift or torque converter type machine.

Step 1 Raise the lowered blade and other attachments high enough to clear obstructions, but not high enough to block your line of sight.

Step 2 Push down on the service brake pedal to keep the machine from moving.

Step 3 Disengage the parking brake and unlock the transmission lever.

Step 4 Turn the directional control lever to the desired direction and turn the speed selector to the desired speed.

Step 5 Release the brake pedal.

Step 6 Move the governor control lever to the desired engine speed.

Step 7 Drive the machine forward for best visibility and control.

Tips for forward operation of the machine include the following:

- Reduce engine speed when maneuvering in tight quarters or when breaking over a rise.
- Select the speed necessary before starting down grade. Do not change speed while going downhill. Use the same speed going downhill as you would going uphill.
- Do not allow the engine to overspeed downhill. Use the service brake pedal or decelerator to reduce engine overspeed going downhill. Speed is automatically selected on newer models.

3.2.2 Changing Direction and Speed

Speed and directional changes at full engine speed are possible on newer machines with power shift or hydrostatic transmission. However, to ensure operator comfort and maximize the service life of transmission components, decelerating and/or braking when changing directions is recommended.

Follow these steps for changing direction:

Step 1 If applicable, decrease the engine speed by pushing the throttle control forward, or by pushing the decelerator pedal down.

Step 2 Push the brake pedal down to stop the machine.

Step 3 Turn the transmission control lever to neutral.

Step 4 Shift the transmission to the desired direction and speed.

Step 5 Release the brake pedal.

Step 6 Increase the engine speed by releasing the decelerator pedal or pulling back on the governor control lever.

3.2.3 Steering and Turning

Although the type of device used for turning differs from one machine to another, the turning principle is the same. To turn, disengage or brake the track on the side to which you wish to turn, while keeping power to the other side. This will cause the machine to turn in the direction of the track that does not have power to it.

Newer machines typically use joystick controls for steering. However, controls vary by manufacturer and the operator's manual should always be checked. This control can be used for steering only, or in combination with another function such as transmission speed and direction. For these machines, the movement of the joystick forward or backward will cause the machine to turn. For example, moving the steering joystick forward, or away from the operator, results in a left turn when moving forward and a right turn when moving in reverse.

Moving the joystick back toward the operator results in a right turn when moving forward and a left turn when moving in reverse.

For older machines that use levers and pedals, operators must learn to coordinate the action of applying power to one track and braking the other to accomplish the turn. In this case, the quickness of the turn is proportional to the amount of power and braking that is applied.

For a gradual turn, disengage power to one track while powering the other. For a sharp turn, disengage and brake one track while powering the other.

3.2.4 Operating the Blade

Before moving the dozer, you should understand how to operate the blade and its controls. The blade position can be changed with lift, angle, tilt, and pitch (see *Figure 32*).

- *Lift* – The term *lift* means to lower or raise the blade. This is done by moving the blade lever forward or back. Lowering the blade allows the operator to change the amount of bite, or depth to which the blade will dig into the material. Raising the blade permits the dozer to travel, shape slopes, or create stockpiles. Another feature controlled by the lift lever is float, which permits the blade to adjust freely to the contour of the ground. The **blade float** position is commonly used in reverse to smooth the surface along the contour of the land.
- *Angle* – The term *angle* means to adjust the blade in relation to the direction of travel. When moving a load, the blade should be perpendicular to the line of travel. For filling a ditch, the

blade should be angled to permit the load to be pushed off to the side. Angling is accomplished by turning the blade control lever to the side desired.

- *Tilt* – Tilt changes the angle of the blade relative to the ground. This permits the blade to cut deeper on one side than on the other. This process is very useful for performing sidehill work where the blade tends to hang lower on the downhill side. It is also useful for crowning roads and for grading slopes and curves. **Blade tilt** is accomplished by moving the blade control lever to the side desired.
- *Pitch* – Pitch adjusts the slope of the blade from top to bottom. The greater the slope, the more the blade tends to dig in. On most dozers, **blade pitch** must be changed manually. Because of the difficulty of controlling it, pitch is only changed for unusual conditions.

The blade must be properly matched to the job and the dozer in order to maximize production. First, consider the type of work the machine will be doing. Then, evaluate the material to be moved and the limitations of the equipment. Most materials can be moved with a dozer. However, dozer performance will vary with different materials.

- *Particle size* – The larger the individual particle size, the harder it is for a cutting edge to penetrate. Particles with sharp edges resist the natural rolling action of a dozer blade. These particles require more horsepower to move than a similar volume of material with rounded edges.

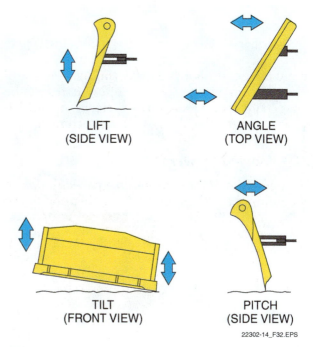

Figure 32 Blade movement.

- *Voids* – Few voids, or the absence of voids, means the individual particles have most or all of their surface area in contact with other particles. This forms a bond that must be broken. A well-graded material, which lacks voids, is generally heavy and will be hard to remove from the **bank state**.
- *Moisture content* – In most materials, the lack of moisture increases the bond between particles and makes the material difficult to remove from the bank state. High moisture content makes dozing difficult because the material is heavy and requires more force to move. Optimum moisture reduces dust and offers the best condition for dozing ease and operator comfort (*Figure 33*).

The weight and horsepower of the machine determine its ability to push. No dozer can exert more pounds of push than the machine itself weighs and its power train can develop. Various terrain and underfoot conditions on the job limit the dozer's ability to use its weight and horsepower.

There are three basic categories of blades that are manufactured for dozers. Within each category, there are many different types and sizes of blades, each designed for a specific purpose as described previously. Know which type of blade to use for the work assigned. Using a dozer with the wrong blade results in poor equipment performance and may even cause damage to the machine or the work area.

Before you begin using a dozer, check the blade to make sure it can effectively handle the specific task. Also, check the dozer controls to determine if they perform the intended functions for the attached blade.

3.3.0 Work Activities

There are many operations that can be performed with a dozer. In addition to dozing, these include moving large objects, doing finish work, using the dozer to push or pull other equipment, and using attachments. There are a number of very useful attachments that can be used to serve specific purposes. This section includes some of the more commonly used attachments, but other attachments are available. Before operating any attachment, read the operator's manual for the dozer. Also, read the manual for the attachment. Make sure that the attachment can be used safely with the machine. In addition, special care must be taken when working in confined areas or in unstable soils.

3.3.1 Dozing

The first operation operators should perform is called drifting or dozing. This is the process of pushing the material straight forward. *Figure 34* shows a dozer pushing material directly in front of it. This is straight dozing.

This process is most efficient when the blade carries the greatest amount of material possible. Maximize the amount of earth drifted by minimizing the amount of spillage around the outer edge of the blade. If the blade digs in and the rear

22302-14_F33.EPS

Figure 33 Soil with good moisture content.

22302-14_F34.EPS

Figure 34 Straight dozing.

of the machine rises, raise the blade to continue an even cut. When moving a heavy load causes travel speed to drop, shift to a lower gear and/or raise the blade slightly.

Excessive operation of the blade control will cause the blade to make uneven cuts over which the dozer will have to travel. The uneven surface will cause the front of the dozer to move up and down. This will cause the blade to cut still more uneven surfaces, thereby increasing the up and down action of the dozer.

After the blade is loaded, raise the blade to ground level and push the material to the point where it is to be stockpiled. To carry the load on the blade, anticipate and compensate for the up-and-down movement of the front of the dozer.

Slot dozing is used to move large amounts of material over a distance with a minimum amount of spillage. The dozer makes several light cuts that form a trench the width of the blade. The spoil material can now be moved without spillage off the side of the blade, as shown in *Figure 35*. This technique is used when large amounts of material must be moved from an excavation.

3.3.2 Cutting and Building Slopes or Stockpiles

Building a stockpile is a very common job for a dozer. First, place several piles of material in a row. Then, use the blade to create a ramp. The height of the ramp can then be increased by placing a second row on top, and leveling again to

REPRINTED COURTESY OF CATERPILLAR INC.

22302-14_F35.EPS

Figure 35 Slot dozing.

form a higher ramp. The size of the stockpile can be increased by repeating this process or by spilling the material off the end of the ramp. *Figure 36* shows a dozer building a stockpile using a wood-chip blade.

> **NOTE**
>
> There should always be windrows on the sides of the ramp and a berm around the top of the stockpile.

Another common job is called ditching. This can be done either by tilt dozer ditching or by straight dozer ditching. To cut a small ditch, first tilt the blade with the low side in the center of the ditch, as shown in *Figure 37*. This forces the blade to cut one side of the V-shaped ditch. Repeat the process for the other side. If the blade is angled, it

REPRINTED COURTESY OF CATERPILLAR INC.

22302-14_F36.EPS

Figure 36 Building a stockpile.

REPRINTED COURTESY OF CATERPILLAR INC.

22302-14_F37.EPS

Figure 37 Ditching using blade tilt.

will cast the material away from the center of the ditch when the cut is made.

For larger ditches, work at right angles to the center line of the ditch. When the desired depth of the ditch is reached, doze the length of the ditch in order to smooth the sides and bottom. Cut the slope from the ditch line by pushing the material up and out of the ditch.

3.3.3 Clearing Land

Land clearing consists of clearing the area of trees, brush, rubbish, and other materials that might interfere with the project. Dozers used for clearing and felling trees should have adequate engine protection, as well as a cage to protect the operator from injury.

To clear brush and small trees, place the blade below ground and push forward at a slow speed (*Figure 38*). Make a second pass to clear away the fallen trees and other trash.

When you encounter medium-sized trees of approximately 10 to 20 inches (25.4 to 50.8 centimeters) in diameter, first raise the blade to the highest possible level. Then, apply pressure slowly. This should provide enough leverage to push the tree over. As the tree falls, back up quickly to allow the root to break free and prevent the dozer from riding up on the root ball. Next, lower the blade and remove the root by pushing forward while lifting the blade. The felled tree can then be pushed to a disposal area.

22302-14_F38.EPS

Figure 38 Clearing brush and small trees.

For trees over 20 inches (50.8 centimeters) in diameter, exercise more care. The procedure is as follows:

Step 1 Inspect the tree to be sure that there are no large dead branches that might fall on you or the machine.

Step 2 Make a side cut deep enough to cut some of the main roots. The cut should be on the opposite side from where you want the tree to fall.

Step 3 Cut the roots on the adjacent sides in a similar manner.

Step 4 Build up an earth ramp on the side opposite the direction of the fall. Raise the blade as high as possible. This will allow the blade to contact the trunk as high as possible.

Step 5 Begin pushing slowly with the blade raised against the tree. Be alert so that when the tree begins to fall, you can quickly back away from the rising root to keep from riding up on the root ball.

3.3.4 Backfilling

A job often performed by dozers is the backfilling of material around culverts and large structures. Dozers are the best equipment for backfilling because the material is pushed directly ahead of the machine.

When performing backfilling, place the material against the structure or pipe in layers, taking care not to concentrate the weight of the fill in one place. Also, be careful not to push against the structure with the material or blade. To reduce pressure on the material and provide better visibility, be sure to raise the blade as you approach the structure.

When backfilling a trench that has pipe or conduit in it, care must be taken to protect the pipe from bending or breaking due to uneven or excessive pressure caused when the material is pushed into the trench. Material is usually placed in layers, depending on the amount of compaction required around the pipe. This could range from 6 inches to several feet (about 15 centimeters to a meter or more). Also, make sure there is sufficient cover on the top of the pipe or any buried structure before driving the dozer over it.

There are two basic methods of placing backfill with a dozer. In the first method, the dozer moves parallel to the trench. In the second method, the dozer moves perpendicular to the trench. For smaller excavations, parallel work is recommended. Perpendicular work is used for deeper excavations.

The first way involves a straightforward push, with the material being delivered at a 90-degree angle to the open area. Move the dozer parallel to the trench on the spoils side with the blade angled slightly in the direction of the excavation. This allows the material to roll off the end of the blade and into the excavation.

The second way involves backfilling perpendicular to the trench and in front of the open area. *Figure 39* shows a dozer backfilling around a foundation by approaching head-on and pushing the material directly into the excavated area.

3.3.5 Sidehill Cutting

One of the more important uses of dozers is cutting a roadway on the side of a hill or slope. This work is best done with an angle or tilt blade. This blade is designed to push at an angle in the direction of travel of the dozer. The most common angle is 25 to 30 degrees. To begin the cut, work from the uphill side of the cut. Push the earth downward and outward to create a level roadbed.

If the slope of the hill is too steep, operators will have to work sidehill. Use the edge of the blade to cut into the side of the hill while pushing along the side of the hill. Make as many passes as necessary to create a level roadbed. Be sure to keep the slope of the bank gradual so as to avoid collapse. Also, keep the slope of the cut toward the hill in order to increase stability.

Whenever possible, move the material downhill in order to increase productivity (*Figure 40*). To finish side slopes, work parallel to the roadway. Start from the top so that material from each pass will fall downhill, forming a windrow that will be picked up on the next pass. This will help keep the dozer from sliding downhill, making a smoother grading operation.

3.3.6 Moving Large Objects

Moving large objects such as boulders is best done by pushing one side of the object, then pushing the other side, and repeating the process until the object has been moved to the proper location. Use a tilted blade and follow these six steps:

Step 1 Approach the rock or boulder with the blade tilted to make contact with the ground at the edge of the boulder.

Step 2 Hook the right lower corner of the blade under the boulder. Then, gradually lift the blade while making a left turn with the dozer.

Step 3 If the rock is deeply imbedded, use a side cut to help get the corner of the blade under the boulder.

Step 4 If the rock is too large to push all at once, line the dozer up on one end, push forward, and lift the blade slowly.

22302-14_F39.EPS

Figure 39 Backfilling.

22302-14_F40.EPS

Figure 40 Sidehill cutting.

Step 5 Reverse direction and line up the dozer on the other end. Push from this side and raise the blade.

Step 6 Continue this procedure until the boulder is removed.

3.3.7 Finishing

Finishing with a dozer is one of the more difficult jobs. The first step is to start level with the finish grade. Before lowering the blade, place the dozer in motion. Then, lower the blade gradually and feed it into the ground. The dozer cutting edge must be parallel to the grade. Always keep the blade at least half full of dirt (*Figure 41*). This will help the blade cut easier and the extra dirt will fill the low spots.

The following is the step-by-step procedure:

Step 1 Start the forward motion of the dozer and slowly start to lower the blade.

Step 2 Raise and lower the blade very slightly in order to maintain a level footing for the machine.

Step 3 Tilt and steer the machine to skim material from high spots to low areas.

Step 4 Raise the blade gradually to loosen materials.

Step 5 Shift to reverse and travel back to the starting point.

Step 6 Move the dozer over one blade width.

22302-14_F41.EPS

Figure 41 Finish grading.

Step 7 Repeat this cycle until the area is cleaned up.

To distribute a thin layer of material, place the blade slightly above the level of the ground. This will allow the desired level of material to pass under the blade. To smooth a surface, place the blade in float and make a pass in reverse.

The float position may also be used for forward operation to push dirt or snow on pavement or other smooth surfaces that are hard enough to support the weight of the blade.

Reverse floating should only be used when operating forward is not feasible. The dragging of the blade in reverse will increase wear on the back of the blade.

One method of finishing side slopes is to work diagonally, starting the dozer at the bottom and working up the slope. This will form a windrow on the downhill side that can be picked up on the next pass. This is one of the few instances where uphill dozing is efficient.

3.3.8 Excavating Work in Confined Areas

Work must sometimes be done in small spaces or confined areas. This requires careful thought regarding what needs to be done and a plan for carrying it out. If you are excavating, remember to leave enough of a ramp to get out of the hole.

Follow this procedure when excavating in confined areas:

Step 1 Line the dozer blade up on the inside edge of the stringline.

Step 2 Lower the blade to the ground, keeping the dozer far enough back from the edge of excavation to allow for ramping down into an excavation.

Step 3 Select the first forward gear and begin cutting a shallow lift of material.

Step 4 When the blade becomes full, raise the blade to ground level and push the material outside the excavation.

Step 5 Raise the blade and shift to reverse.

Step 6 Return to the area where the blade was lifted.

Step 7 Lower the blade to the ground and shift to first forward gear.

Step 8 Repeat this cycle until the layer is cut.

Step 9 Raise the blade gradually to begin ramping out on the opposite end.

Step 10 To maintain a smooth slope, lower the blade as the tracks begin climbing out.

Step 11 Raise the blade gradually to lose material.

Step 12 Shift to reverse and return to the opposite side.

Step 13 Continue this cycle until the proper depth is reached.

3.3.9 Working in Unstable Soils

Working in mud or unstable soils that will not support equipment can be difficult. This is a problem even for experienced operators. Dozers are very difficult to pull out of mud if they get stuck, especially on small jobs where there is only one dozer and nothing else is heavy enough to do the job.

When you see that you are about to enter a wet area, go very slowly. If you feel the front of the dozer start to settle, stop and back out immediately. That settling is the first indication that the ground is too soft. The engine will lug slightly and the front end will start to settle.

After backing out, examine how deep the tracks sank into the ground. If they are deep enough that the mud hits the underside of the dozer, the ground is too soft to work in a normal way. The dozer will sink until it becomes high-centered. At that point, the tracks will be of no use because the force is pushing against the bottom of the dozer and not against the tracks.

Figure 42 shows a dozer pushing some good material out in front so that it will have a stable platform on which to work.

To work in soft material, follow this approach:

- Start from the edge and work forward slowly.
- Push the mud ahead of the dozer blade and be sure the ground underneath is firm.
- Do not try to move too much material on any one pass.
- Keep the tracks from slipping and digging in.

Partially stable material can also be a hazard because the operator may drive in and out over relatively firm ground many times, while it slowly gets softer from the weight of the dozer pumping more water to the surface. If this happens, the tracks will sink a little more each time until the dozer finally gets high-centered.

To keep this from happening, avoid running in the same track each time entering or leaving an area. Move over slightly in one direction or the other so the same tracks aren't pushed deeper into the unstable material each time.

3.3.10 Pushing Other Equipment

On large excavation jobs, dozers are frequently used to push scrapers to help them take bigger bites of material. Typically, dozers that are used for this work have a special blade that provides extra strength and a wearing surface for contact with the scraper push block.

REPRINTED COURTESY OF CATERPILLAR INC.

22302-14_F42.EPS

Figure 42 Filling unstable areas.

The procedure for pushing a scraper is as follows:

Step 1 Spot the scraper and allow it to get into position.

Step 2 Move forward toward the scraper at low speed. Allow rpms to decrease and select a lower gear.

Step 3 Line up directly behind the scraper and position the blade to push as shown in *Figure 43*. Make sure you are centered behind the scraper push block or you will push the scraper sideways.

Step 4 Downshift to first gear and decelerate to soften the impact.

Step 5 On contact with the scraper, accelerate to full power.

Step 6 Continue pushing and watching the scraper bowl to see when the bowl is full.

Step 7 Signal the scraper operator when the bowl is full.

Step 8 Decelerate as the scraper pulls away from the machine.

Step 9 Place the transmission in reverse as the dozer comes to a stop.

Step 10 Look behind the dozer and back up to the starting area in order to reposition the dozer for pushing the next scraper.

3.3.11 Winching

Because the dozer is usually the most powerful piece of equipment around the construction site, it

REPRINTED COURTESY OF CATERPILLAR INC.

22302-14_F43.EPS

Figure 43 Pushing a scraper.

is often used to pull out other pieces of machinery that may become stuck or bogged down in unstable material or mud. Before starting this operation, observe the stuck machine and decide on the best method to use for doing the job.

Follow these procedures when winching:

Step 1 Back the dozer into position for attaching the cable to the stuck machine.

Step 2 Place the dozer transmission in neutral and lock the shifter.

Step 3 Set the parking brake.

Step 4 Release the winch jaw clutch and set the brake with slight drag.

Step 5 Move to the winch hook and release it. Pull the cable to the disabled machine and attach it to the tow hook.

Step 6 Inform the machine operator of the procedures you are going to use to free the disabled machine. Some hand signals may be needed.

Step 7 Re-engage the winch jaws and take up any slack in the line.

> **WARNING!**
> When there is a load on the cable, make certain all personnel are clear of the cable. It can break and cause personal injury.

Step 8 Increase engine rpm and apply power to the winch. When power is applied, the other machine should be in gear with the wheels driving.

Step 9 Observe the winch to be sure the cable is feeding onto the drum correctly. If the counterforce of the stuck machine is greater than the pulling force of the dozer, the dozer will be pulled backward toward the machine. In order to get more leverage, place the dozer in gear and spin down until a bank is created behind the dozer. If this method does not work, it will probably be necessary to anchor the dozer by using blocks or a chain attached to a solid object.

Step 10 Release the brake, place the winch in the locked position, and release the transmission lock.

Step 11 Place the dozer in first forward gear and increase the engine rpm.

Step 12 Continue pulling until the disabled machine is completely clear of the mud.

Step 13 Select first reverse gear and back up toward the machine until there is enough slack to unhook the cable.

Step 14 Have the operator of the other machine disconnect the cable.

Step 15 Place the winch in gear, checking to see that the cable is feeding onto the drum correctly, until all the cable is on the drum.

Step 16 Disconnect the winch and secure the machine.

Step 17 Secure the cable by placing the cable hook in the eye or bracket as shown in *Figure 44*.

3.3.12 Ripping

A ripper is a rear-mounted dozer attachment that has one or more shanks to break up pavement and hard packed or rocky soils (*Figure 45*). It can be raised or lowered by a control lever in the cab.

The process of **ripping** is more art than science, and success will depend on operator skill and experience. Ripping for scraper loading may call for different techniques than if the same material is to be dozed away.

The general procedure for ripping is as follows:

Step 1 Begin at low speed.

Step 2 Raise the blade to clear obstructions.

Step 3 Lower the ripper into the material while moving forward.

Step 4 Move the ripper shank to the forward position when prying out lodged material.

Step 5 Keep the tracks on an unripped area.

Step 6 Raise the ripper out of the material whenever you stop for long periods of time or shut down the machine.

Always use the center shank when ripping with one shank. If material breaks up satisfactorily, more shanks can be used. Some rippers only have one shank, as shown in *Figure 46*.

If you are using an adjustable ripper, you can vary the angle of the shank to achieve maximum performance throughout each ripping pass. Some ripping attachments have a shank that can be adjusted manually by placing the retaining pin in different holes. On some rippers, the shank angle can be adjusted hydraulically.

REPRINTED COURTESY OF CATERPILLAR INC.

22302-14_F44.EPS

Figure 44 Secure the winch cable after use.

REPRINTED COURTESY OF CATERPILLAR INC.

22302-14_F45.EPS

Figure 45 Ripping compacted soil with a three-shank ripper.

When ripping boulders or rock layers with a dozer, it is usually best to work from north to south or south to north. Cracks in rocks usually run in that direction. In most cases, working north to south makes ripping easier. If this is not the case in your area, study the rock layers to determine the best direction to rip.

When operating a dozer in a rocky area, remember the following safety tips:

- Make a smooth area to work from.
- Don't try to overpower boulders.
- Be careful when working parallel on a rocky slope. If both tracks get on a rock cap at the same time, the dozer may slide and turn over.
- If you are in a precarious position on a steep slope, stop and then lower the blade to the ground. This will add some stability until a cable can be attached to keep the dozer from rolling over.

3.3.13 Using Towed Attachments

The two main attachments that are towed by a dozer are a roller and a disc. Both attachments are used for material preparation. A roller is used to compact material. A disc is used to mix or break up a top layer of material to a depth of approximately 20 inches (50 centimeters). *Figure 47* shows a high-speed dozer towing a sheepsfoot roller.

Procedures for using a roller are as follows:

Step 1   Back the dozer up so that the hitch on the back of the dozer is aligned with the drawbar of the roller.

Step 2   Align the drawbar with the hitch on the dozer and make the connection. Make sure the safety pin has been inserted.

Step 3   Depress the decelerator and brake.

REPRINTED COURTESY OF CATERPILLAR INC.

22302-14_F46.EPS

Figure 46 Ripping soil with a single-shank ripper.

Step 4   Place the transmission in neutral position.

Step 5   Select the first forward gear and travel to the work area.

Step 6   Observe the forward speed and align the right side of the roller with the right side of the pad.

22302-14_F47.EPS

Figure 47 Dozer towing a sheepsfoot roller.

Step 7 Travel straight ahead the length of the pad.

Step 8 Maintain a low engine speed in order to provide proper speed and avoid engine lugging.

Step 9 Approach the end of the pad and gradually make a turn. Care must be taken during the turn to avoid kicking out the material on the right side of the tracks, which could cause irregularities on the pad.

Using Specialty Blades

Dozer manufacturers provide a number of different blade designs for special applications. These blades are designed to increase production while performing certain tasks. The most common types of specialty blades are angling blades, cushion blades, landfill U blades, and rake blades. Other specialty blades include KG blades, which are used for land clearing, and V-tree cutter blades, which are designed for shearing trees and stumps at ground level.

In some cases, such as landfill operations or open pit quarries, the dozer is designed specifically for use with a specialty blade. Design features such as track configuration, pushing power, and operator protection allow the blade to be used effectively while protecting the equipment and operator. Before using any specialty blade on the dozer, make sure the machine is designed to handle the blade and its intended purpose. Check the operator's manual to see if the blade is specified for the model of dozer you are operating.

(A)

(B)

CAT D7R TRACK-TYPE TRACTOR. REPRINTED COURTESY OF CATERPILLAR INC.

(C)

CAT D10T TRACK-TYPE TRACTOR. REPRINTED COURTESY OF CATERPILLAR INC.

(D)

22302-14_SA05.EPS

Step 10 Align the roller with the right edge of the pad end.

Step 11 Select the first reverse gear and slowly move backward over the area that was missed when the turn was made. When backing, turn in the seat and observe the backward movement. Observe when the bottom of the roller drum reaches the edge of the pad.

Step 12 Select the first forward gear and slowly move forward.

Step 13 Continue this procedure completely around the pad, and then proceed to the next pass, overlapping one-half of the first pass.

Discing loosens and blends surface material so that it can be re-compacted. If the work area is large, a dozer towing a large disc and working with a motor grader is an efficient and effective way to get the job done. While the disc continuously cuts and mixes the soil, the motor grader forms windrows and moves the material back and forth across the subgrade.

The initial procedures for hooking up the disc and traveling to the work site are the same as with the roller. Use the following procedures for using the disc after reaching the work site:

Step 1 Travel forward over the area, raise the disc's travel wheels, and proceed forward. When discing to blend materials required by specifications, it may be necessary to travel parallel to raised objects in order to ensure full coverage and blending in areas that would be missed by raising the disc out of the ground to travel over the objects.

Step 2 Upon completion, stop forward travel and lower the disc's travel wheels to the ground, raising the disc out of the ground.

Step 3 Travel to the equipment parking area, lower the blade to the ground, place the transmission in neutral, and set the transmission lock. Raise the disc travel wheels in order to lower the weight of the disc onto the ground.

Step 4 Shut the engine off, then operate the disc control until all movement is out of the system, with disc and travels both resting on the ground.

3.3.14 Using a Side Boom

Side booms or pipe booms are used extensively for heavy hoisting and carrying, particularly in pipeline work (*Figure 48*). When operating a dozer with a side boom, operators must make sure that they do not exceed the load rating for the equipment. The load rating is determined using factors such as the distance between the load and the dozer, and the size and position of the counterweight. Exceeding the load rating will cause the side of the dozer that is opposite the load to tip. Information about the load rating for a particular piece of equipment can be found in the operator's manual.

The following procedure summarizes the steps for handling a load with a side boom:

Step 1 Make sure the weight of the load does not exceed the maximum load capacity.

Step 2 Ensure that the hoist rope is not kinked or twisted on itself.

Step 3 Ensure that the load is secured and balanced in the sling or lifting device before it is lifted more than a few inches off the ground.

Step 4 When two or more side boom tractors are used to lift one load, one designated person should be responsible for the operation. This person should analyze the operation and instruct all personnel involved in the proper positioning, rigging of the load, and the movements to be made.

Step 5 When operating at a fixed boom overhang, the bottom hoist pawl or other positive locking device should be engaged.

3.3.15 Moving the Dozer

When dozer operations in one area are completed, the dozer must be moved to a new work location.

REPRINTED COURTESY OF CATERPILLAR INC.

22302-14_F48.EPS

Figure 48 Dozer with side boom laying pipe.

Before loading the equipment for transport, make sure the following tasks have been completed:

- Check the operator's manual to determine if the loaded equipment complies with height, width, and weight limitations for over-the-road hauling.
- Check the operator's manual to identify the correct tie-down points on the equipment.

- Be sure to get the proper permits, if required.
- Plan the loading operation so that the loading angle is at a minimum.

Once the above tasks are completed and the loading plan determined, carry out the following procedures:

Step 1 Position the trailer or transporting vehicle. Always block the wheels of the transporter before loading the dozer.

Dozer Transporting

A dozer can usually be driven to a new area on the same job site. However, tracked vehicles cannot be driven over public roads and must be transported. This means the dozer must be loaded onto a trailer or other transport vehicle and transported to the new location. Loading, transporting, and unloading a dozer can be dangerous work. While your involvement may be limited to driving the dozer onto and off of a trailer, you need to always be aware of the hazards and the proper procedures for securing a dozer for transport.

REPRINTED COURTESY OF CATERPILLAR INC.

22302-14_SA06.EPS

Step 2 Raise the blade high enough to clear obstacles and drive the dozer onto the transporter. Whether the dozer is facing forward or backward will depend on the recommendation of the manufacturer.

Step 3 Lower the blade to the floor of the transporter.

Step 4 Move the transmission lever to neutral and engage the parking brake and the transmission neutral lock. Turn off the engine.

Step 5 Manipulate the blade and operate all controls to remove any remaining hydraulic pressure.

Step 6 Remove the engine start key and secure vandal guards. Activate any disconnect switches. Place the fuel shutoff valve in the OFF position.

Step 7 Lock the door to the cab and any access covers. Attach any vandalism protection.

Step 8 Secure the machine with the proper tie-down equipment as specified by the manufacturer. The tie-down points are specified in the operator's manual.

WARNING!

The machine may shift while in transit if it is not properly tied down. If the machine shifts in transport, it could cause personal injury or death. Follow the manufacturer's safety procedures.

Step 9 Cover the exhaust and air intake openings with tape or a plastic cover.

Step 10 Place appropriate flags or markers on the equipment if needed for height and width restrictions.

Unloading the equipment from the transporter is accomplished by reversing the steps of the loading operation.

3.0.0 Section Review

1. During the startup of a dozer, the blade and accessory controls should be _____.

   a. de-activated with the disconnect switch
   b. locked out using the joystick button
   c. in the hold position
   d. set to the down position

2. A dozer's blade position can be changed by adjusting the _____.

   a. lift, angle, tilt, and pitch
   b. lift, rip, forward, and backward
   c. lift, left, right, and down
   d. push, shift, float, and down

3. The most common blade angle to use for side-hill cutting is _____.

   a. 5 to 10 degrees
   b. 10 to 20 degrees
   c. 25 to 30 degrees
   d. 45 to 60 degrees

SUMMARY

A dozer operator controls some of the largest and most powerful heavy equipment used in the industry and must therefore be able to operate the equipment safely, efficiently, and effectively. Basic dozer operation involves excavating and moving various types of earth and manufactured materials along the ground. There are also many attachments that can be used on a dozer to do special jobs such as land clearing, tree harvesting, and pipeline construction.

The main tasks that are performed by a dozer include cutting, sidehill cutting, ditch digging, slot dozing, backfilling, clearing and grubbing, and finishing.

Because dozers are large and powerful machines, you must constantly be aware of what is going on around the work area. Avoid situations that may produce accidents. Know the operational limits of the dozer, such as the maximum sidehill slope operation and maximum traction power.

The dozer needs to be well-maintained in order to be safe and fully operational. You should be familiar with the routine daily maintenance checks, as well as with any other special requirements for the specific equipment. Follow the daily inspection checklist.

Before attempting to move the dozer, you should develop a complete understanding of the controls. You or other personnel could be injured as a result of operating the controls in an inappropriate manner. As you develop skill with the controls, ease of operation and the application of the dozer in different tasks will be realized.

A safety-conscious operator is the greatest safety device. Most dozers have been designed to incorporate every safety device possible. Even with all these safety features, there are still dangers. You must be extremely careful in the operation of this machine.

Review Questions

1. One common name that is often used for a dozer is _____.

   a. crawler loader
   b. digger
   c. blade scraper
   d. track-type tractor

2. A high-track crawler dozer is used primarily for _____.

   a. pushing
   b. grading
   c. demolition
   d. rocky terrain

3. What do many wheel dozers have that makes them easier to operate and maneuver than crawler dozers?

   a. Separate brake pedals
   b. Rubber tracks
   c. A steering wheel
   d. A guidance display

4. The governor on a dozer is typically used to control _____.

   a. machine direction
   b. engine speed
   c. the rate of the hydraulic system response
   d. the movement of any attachments

5. If the engine key switch is located on the steering control joystick of a dozer, it is likely that the switch will also _____.

   a. lock the parking brake
   b. toggle digital readouts
   c. lock the decelerator pedal
   d. activate the PAT system

6. On some dozers, a single joystick is used to control the _____.

   a. steering and the engine RPM
   b. blade position and attachments
   c. transmission and steering
   d. blade and the steering

7. If a dozer has a ripper attachment and the ripper control lever is released, what is most likely to occur?

   a. An alarm will sound and the decelerator will stop any forward movement of the dozer.
   b. The lever will stay at its present position and the ripper will rise to its highest position.
   c. A warning light will flash and a disconnect switch will deactivate the ripper control.
   d. The lever will return to the central hold position and the ripper will stop moving.

8. PAT blades, variable radius blades, and straight blades are all considered to be _____.

   a. landfill attachments
   b. general-purpose blades
   c. push-plate attachments
   d. specialty blades

9. A type of dozer blade that has two sides angled toward the center is known as a(n) _____.

   a. straight blade
   b. S blade
   c. universal blade
   d. PAT blade

10. The shank and the tip are two parts of a _____.

    a. ripper tooth
    b. V-tree cutter blade
    c. winch
    d. drill attachment

11. What is ordinarily used to wind the wire rope around the drum of a rear-mounted winch attachment?

    a. A pusher plate
    b. A motor
    c. Gears driven by the dozer tracks
    d. A cable clamp

12. A basic safety rule for operating a dozer while going downhill is _____.

    a. shift to a lower gear and allow the dozer to coast
    b. lower the ripper to reduce speed
    c. shift the transmission to a higher gear
    d. do not coast or shift gears

13. A dozer's sprockets should be checked during the daily inspection because worn or broken sprocket teeth can cause the dozer to _____.

    a. throw a track
    b. tip over
    c. lose power
    d. slip gears

14. All of the following are causes of premature equipment failure *except* _____.

    a. operating the equipment in less-than-ideal conditions
    b. using aftermarket replacement parts and fluids
    c. performing preventive maintenance on a regular basis
    d. ignoring onboard diagnostic recommendations

15. One of the first steps to perform during the startup of a dozer is to _____.

    a. take the blade controls out of the hold position
    b. engage the parking brake
    c. activate the battery disconnect switch
    d. purge the fuel lines

16. To reduce engine wear and damage during the shutdown of a dozer, it is important to _____.

    a. shut down the engine immediately to allow cooling
    b. park the dozer so that the engine is facing downhill
    c. engage the engine compartment shrouds
    d. let the engine run on low idle for approximately five minutes

17. A dozer blade adjustment that changes the slope of the blade from top to bottom is _____.

    a. pitch
    b. lift
    c. tilt
    d. angle

18. The process of moving material straight forward with a dozer is called _____.

    a. discing
    b. drafting
    c. dozing
    d. ripping

19. The best way to clear medium-sized trees of about 10 to 20 inches (25.4 to 50.8 centimeters) in diameter with a dozer is to start by _____.

    a. winching the trees to break up the roots before pushing with the blade
    b. cutting down the trees and dozing only the stumps and roots
    c. creating a dirt ramp and riding down the trees from up high
    d. raising the blade as high as possible and slowly applying pressure

20. When performing finish work with a dozer, an operator should _____.

    a. maintain an empty blade that is perpendicular to the grade
    b. avoid reverse floating unless forward operation is not feasible
    c. use the ripper attachment to help with leveling
    d. use reverse floating to decrease wear on the blade

Trade Terms Introduced in This Module

Bank state: A term indicating that material, such as soil, is in its natural state. Also referred to as virgin ground.

Blade: The main attachment on the front end of the machine that is used for moving material.

Blade float: When the blade is allowed to float over the surface for a smooth finish.

Blade pitch: The angle of the blade from the vertical.

Blade tilt: The angle of the blade from the horizontal.

Charging hoppers: The process of filling a hopper or temporary storage container with a material, such as coal, that will be dispersed later.

Dozing: Using a blade to scrape or excavate material and move it to another place.

Grouser: A ridge or cleat across a track shoe that improves the track's grip on the ground.

Grubbing: Digging out roots and other buried material.

KG blade: A type of dozer-mounted blade used in forestry and land clearing operations. On some blades, a single spike called a stinger splits and shears stumps at the base.

Pawl: Pivoted lever which has a free end to engage with the teeth of a cogwheel or ratchet so that the wheel or ratchet can only move one way.

Ripping: Loosening hard soil, concrete, asphalt, or soft rock.

Slot dozing: A method that creates a trench for controlling spillage from the blade ends.

Stockpile: Material that is dug and piled for future use.

Winch: An attachment commonly mounted on the rear of a dozer that uses an electric motor and wire rope to pull equipment, trees, and other objects.

Typical Daily Inspection Check Sheet

DAILY CHECKS AND LUBRICATION

Check the crankcase oil level:

☐ Check the OIL LEVEL on the DIPSTICK. Oil must be **between the marks.** Some dipsticks have a cold check side and a hot check side. Be sure to read the cold side.

☐ If oil is to be added, remove the FILLER CAP and add the required amount of oil to the CRANKCASE. Replace the FILLER CAP.

☐ REPLACE THE DIPSTICK when the oil level is correct.

Check the coolant level:

☐ REMOVE the RADIATOR FILLER CAP. Coolant level must be within three inches of the bottom of the FILLER NECK.

☐ If COOLANT is low, add sufficient coolant to restore to the correct level.

☐ Replace the FILLER CAP.

Drain fuel filter sediment:

☐ Loosen the DRAIN SCREW and drain fuel for several seconds.

☐ Tighten the DRAIN SCREW.

Check the transmission/hydraulic oil level:

☐ Be sure all equipment is lowered.

☐ A hydraulic check may be done with a DIPSTICK or SIGHT GLASS on the TANK.

☐ REMOVE the DIPSTICK. Oil must be in the crosshatched area on the dipstick. If necessary, remove the FILLER CAP and add the required oil.

☐ Replace the DIPSTICK.

Lubricate the lift arms and blade:

☐ Lubricate the LIFT ARM CYLINDERS with two shots of grease each.

☐ Lubricate the BLADE SUPPORTS.

Check the tracks:

☐ Check the TRACK ADJUSTMENT for correctness.

☐ Check the GROUSER PADS.

☐ Check the ROLLERS for leaks.

22302-14 A01.EPS

Figure Credits

Deere & Company, Module opener, Figures 1A, 1D, 2, 16–18, 21, 23, 26, 29, 33, 34, 39–41, 47, SA03, SA05A, SA05B

Reprinted Courtesy of Caterpillar Inc., Figures 1B, 1C, 3–15, 19, 22, 24, 27, 28, 31, 35–38, 42–46, 48, SA01, SA04, SA05C, SA05D, SA06

Topaz Publications, Inc., Figure 25

Tim Davis/NCCER, Figure 30

Answer	Section Reference	Objective
Section One		
1. b	1.1.1	1a
2. d	1.2.0	1b
3. a	1.3.2	1c
4. c	1.4.2	1d
5. b	1.5.1	1e
6. c	1.6.1	1f
Section Two		
1. a	2.1.1	2a
2. d	2.2.0	2b
3. b	2.3.0	2c
Section Three		
1. c	3.1.1	3a
2. a	3.2.4	3b
3. c	3.3.5	3c

NCCER CURRICULA — USER UPDATE

NCCER makes every effort to keep its textbooks up-to-date and free of technical errors. We appreciate your help in this process. If you find an error, a typographical mistake, or an inaccuracy in NCCER's curricula, please fill out this form (or a photocopy), or complete the online form at **www.nccer.org/olf**. Be sure to include the exact module ID number, page number, a detailed description, and your recommended correction. Your input will be brought to the attention of the Authoring Team. Thank you for your assistance.

Instructors – If you have an idea for improving this textbook, or have found that additional materials were necessary to teach this module effectively, please let us know so that we may present your suggestions to the Authoring Team.

NCCER Product Development and Revision

13614 Progress Blvd., Alachua, FL 32615

Email: curriculum@nccer.org
Online: www.nccer.org/olf

❏ Trainee Guide   ❏ Lesson Plans   ❏ Exam   ❏ PowerPoints   Other _____

Craft / Level: _____   Copyright Date: _____

Module ID Number / Title: _____

Section Number(s): _____

Description: _____

_____

_____

_____

_____

Recommended Correction: _____

_____

_____

_____

_____

Your Name: _____

Address: _____

_____

Email: _____   Phone: _____

22304-14

Excavators

OVERVIEW

Excavators are available in many sizes and in different styles, and they are produced by numerous manufacturers. Two of the most common types are hydraulic excavators and telescoping-boom excavators. As a heavy equipment operator, you need to understand how an excavator functions and how to use the bucket and other accessories so that the equipment works effectively. You must also be aware of potential dangers that exist with excavators, know how to conduct daily preventive maintenance on the machine, and practice safety at all times. Finally, you need to know how to perform common work activities associated with excavators.

Module Six

Trainees with successful module completions may be eligible for credentialing through NCCER's National Registry. To learn more, go to **www.nccer.org** or contact us at **1.888.622.3720**. Our website has information on the latest product releases and training, as well as online versions of our **Cornerstone** magazine and Pearson's product catalog.

Your feedback is welcome. You may email your comments to **curriculum@nccer.org**, send general comments and inquiries to **info@nccer.org**, or fill in the User Update form at the back of this module.

This information is general in nature and intended for training purposes only. Actual performance of activities described in this manual requires compliance with all applicable operating, service, maintenance, and safety procedures under the direction of qualified personnel. References in this manual to patented or proprietary devices do not constitute a recommendation of their use.

Objectives

When you have completed this module, you will be able to do the following:

1. Identify and describe types, uses, and components of excavators.
   a. Identify and describe common types of excavators.
   b. Identify and describe common uses of excavators.
   c. Identify and describe major parts of excavators.
   d. Identify and describe excavator instrumentation and controls.
   e. Identify and describe common excavator buckets and attachments.
2. Identify and describe safety, inspection, and service guidelines associated with an excavator.
   a. Describe guidelines associated with excavator safety.
   b. Describe prestart inspection procedures.
   c. Describe preventive maintenance requirements.
3. Describe basic startup and operating procedures for a track-mounted hydraulic excavator.
   a. Describe startup, warm-up, and shutdown procedures.
   b. Describe basic maneuvers and operations.
   c. Describe common work activities.
   d. Describe activities involving special attachments.

Performance Tasks

Under the supervision of your instructor, you should be able to do the following:

1. Demonstrate a proper prestart inspection and maintenance on an excavator.
2. Perform proper startup, warm-up, and shutdown procedures.
3. Perform basic maneuvers with a hydraulic excavator, including moving forward, moving backward, making a pivot turn, and making a spot turn.
4. Demonstrate basic excavator operation by creating a 10 feet by 10 feet (3 meter by 3 meter) excavation at least 3 feet (1 meter) deep to grade.

Trade Terms

Boom	Hydraulic breaker (hammer)	Spoil
Breakout force	Inslope (foreslope)	Stick
Counterweight	Joystick	Undercarriage
Demolition	Knuckle boom	Undercutting
Grapple	Platen	Upper carriage
Haunches	Shooting-boom excavator	Upper structure
Hoe	Slope	

Industry Recognized Credentials

If you are training through an NCCER-accredited sponsor, you may be eligible for credentials from NCCER's Registry. The ID number for this module is 22304-14. Note that this module may have been used in other NCCER curricula and may apply to other level completions. Contact NCCER's Registry at 888.622.3720 or go to **www.nccer.org** for more information.

Contents

Topics to be presented in this module include:

1.0.0 Types, Uses, and Components of Excavators 1
  1.1.0 Common Types of Excavators 1
    1.1.1 Hydraulic Excavators 2
    1.1.2 Telescoping-Boom Excavators 3
  1.2.0 Common Uses of Excavators 5
    1.2.1 General Excavation and Trenching 5
    1.2.2 Ditch Cutting, Shaping, and Cleaning 6
    1.2.3 Side Sloping 6
    1.2.4 Finish Grading 7
    1.2.5 Placing Riprap 7
    1.2.6 Ripping Pavement 7
    1.2.7 Lifting 7
    1.2.8 Demolition 7
    1.2.9 Scrap Handling 8
  1.3.0 Major Parts of Excavators 8
    1.3.1 Track-Mounted Hydraulic Excavator Components 8
    1.3.2 Truck-Mounted Telescoping-Boom Excavator Components 10
  1.4.0 Instrumentation and Controls 10
    1.4.1 Track-Mounted Excavator Instruments and Controls 10
    1.4.2 Telescoping-Boom Excavator Instruments and Controls 11
    1.4.3 Telescoping-Boom Excavator Braking System 11
  1.5.0 Excavator Buckets and Attachments 13
    1.5.1 Buckets 13
    1.5.2 Attachments 15
2.0.0 Safety, Inspection, and Maintenance 18
  2.1.0 Safety Guidelines 18
    2.1.1 Operator Safety 18
    2.1.2 Safety of Co-Workers and the Public 18
    2.1.3 Equipment Safety 19
  2.1.4 Working Around Overhead Power Lines and
        Underground Utilities 19
  2.2.0 Prestart Inspection Procedures 20
  2.3.0 Preventive Maintenance 21
3.0.0 Hydraulic Excavator Startup and Operations 24
  3.1.0 Startup, Warm-Up, and Shutdown 24
    3.1.1 Startup 24
    3.1.2 Warm-Up 25
    3.1.3 Shutdown 25
  3.2.0 Basic Maneuvers and Operations 25
    3.2.1 Vehicle Movement 26
    3.2.2 Operating the Hoe and Bucket 27
    3.2.3 Stability While Operating 30
  3.3.0 Work Activities 30
    3.3.1 General Excavation and Loading 31

3.3.2 Excavating a Foundation 33

3.3.3 Loading Trucks 33

3.3.4 Lifting Objects... 35

3.3.5 Trenching and Laying Pipe 36

3.3.6 Placing Bedding Material... 37

3.3.7 Setting Pipe... 37

3.3.8 Working in Unstable Soils 38

3.4.0 Special Attachments and Activities 39

3.4.1 Demolition 39

3.4.2 Blading 40

3.4.3 Digging with the Shovel 40

3.4.4 Material Handling... 41

3.4.5 Setting Up and Calibrating a Laser Guidance System 41

3.4.6 Transporting the Excavator... 42

Appendix   CAT 336E Hydraulic Excavator Lifting Capacities Chart 48

Figures

Figure 1 Standard hydraulic excavator 2

Figure 2 Telescoping-boom excavator 2

Figure 3 Track-mounted hydraulic excavator 3

Figure 4 Wheel-mounted excavator 3

Figure 5 Mini excavator 4

Figure 6 Truck-mounted telescoping-boom excavator 4

Figure 7 Wheel-mounted telescoping-boom excavator 5

Figure 8 Track-mounted telescoping-boom excavator 5

Figure 9 General excavation 6

Figure 10 Trenching excavation 6

Figure 11 Side sloping and riprap placement 7

Figure 12 Demolition shears 8

Figure 13 Breaking up pavement 8

Figure 14 Scrap handling 8

Figure 15 Components of a track-mounted hydraulic excavator 9

Figure 16 Components of a truck-mounted telescoping-boom excavator ... 10

Figure 17 Instruments and controls on a typical track-mounted
          hydraulic excavator 12

Figure 18 Typical joystick layout 13

Figure 19 Excavator buckets 14

Figure 20A Excavator attachments 15

Figure 20B Excavator attachments 16

Figure 21 Daily prestart inspection for a track-mounted hydraulic
          excavator 21

Figure 22 Check track tension 22

Figure 23 Lubricate boom and stick linkage fittings
          (for severe service only) 22

Figure 24 Tracked excavator steering 26

Figure 25 Excavator hoe and bucket controls 27

Figure 26 Excavator stick and swing control 28

Figure 27 Excavator bucket and boom control 29

Figure 28 Location of hydraulic activation control lever 30

Figure 29 Starting position for digging 31

Figure 30 Bucket angle for maximum breakout 32

Figure 31 Full bucket 33

Figure 32 Basic excavation pattern 34

Figure 33 Correct position for dumping 35

Figure 34 Lifting capacity based on hoe position 36

Figure 35 Sling attached to a bucket 38

Figure 36 Placing mats 39

Figure 37 Using a hydraulic breaker 40

Figure 38 A grapple in use 41

Figure 39 Position of a loaded excavator 43

SECTION ONE

1.0.0 Types, Uses, and Components of Excavators

Objective

Identify and describe types, uses, and components of excavators.

a. Identify and describe common types of excavators.
b. Identify and describe common uses of excavators.
c. Identify and describe major parts of excavators.
d. Identify and describe excavator instrumentation and controls.
e. Identify and describe common excavator buckets and attachments.

Trade Terms

Boom: A part of the digging component (hoe) of an excavator. The boom is attached to the upper carriage and serves as the main lifting stick of the machine.

Breakout force: The amount of digging force that can be exerted by the bucket on material in its natural state.

Counterweight: Dead or nonworking load that is attached to one end or side of a machine to help improve the balance and directly impacts lifting capacity.

Demolition: The destruction of a structure or the scrapping of material that has been torn down.

Grapple: An excavator attachment that is used for lifting. The grapple has two or more hooks that can be closed on each other to pick up solid material, such as logs, by grabbing onto them.

Hoe: The component of an excavator that is made up of the boom, stick, and bucket. Also, a term used to refer to any type of excavator that digs by pulling the bucket from front to back.

Hydraulic breaker (hammer): A hydraulic attachment for an excavator that is used for breaking boulders and other solid objects

Inslope (foreslope): The slope of a drainage ditch that is between the shoulder of the road and the flow line of the ditch.

Joystick: A control mechanism that pivots about a fixed point and is used to control the motion of an object. Joystick controls are used in an excavator to control the movement of the boom, stick, bucket, and upper carriage.

Knuckle boom: A term sometimes used for a gooseneck boom and stick combination that resembles a knuckle at the pivot point of the boom and stick.

Platen: A flat plate that is used to support another structure, and that can be rotated around a pivot point. The platen is the supporting base of a turntable mechanism.

Shooting-boom excavator: A term sometimes used to describe a telescoping-boom excavator.

Slope: Ground that forms a natural or artificial incline.

Stick: The part of the digging component of an excavator between the end of the boom and the bucket. The bucket is attached to the end of the stick. Sometimes referred to as an arm.

Undercarriage: The lower frame of an excavator that supports the turntable and has the tracks or wheels attached.

Upper carriage: The upper frame of an excavator that includes the turntable, engine, cab, controls, and counterweights.

Upper structure: The revolving turntable, swing mechanism, counterweight, boom, and cab of a telescoping-boom excavator.

Excavators are available in a vast array of sizes and configurations. Most of these machines can be fitted with numerous attachments that make the equipment useful for many applications. In order to operate an excavator safely and efficiently, the operator first needs to have a fundamental understanding of the basic types, uses, and components of these machines.

1.1.0 Common Types of Excavators

Generally speaking, the most common types of excavators used in the construction industry can be divided into two basic categories: standard hydraulic excavators and telescoping-boom excavators.

Standard hydraulic excavators, such as the one shown in *Figure 1*, are usually referred to simply as excavators. These excavators are available in various sizes and capacities from mini to very large. This type of excavator has a gooseneck **boom**, a pivoting dipper **stick**, and a bucket. Some manufacturers refer to the dipper stick as a dipper arm. The term *dipper stick*, when used with hydraulic excavators, is a carryover from the days of steam or engine-driven mechanical shovels when the booms were relatively fixed during operation. The dipper stick, with a rigidly attached trap-door shovel, was moved up and down perpendicular to the boom by cables and pulleys or a rack-and-pinion gear system to change the operating depth of the shovel. Sometimes the gooseneck boom with its pivoting dipper stick is called a knuckle boom or hoe.

A less common and more specialized type of hydraulic excavator is a telescoping-boom excavator, sometimes called a shooting-boom excavator. As shown in *Figure 2*, this type of excavator has a straight boom equipped with a pivoting bucket. The boom can be extended, retracted, and rotated. Rotation is used to tilt the bucket from side to side. This type of excavator is generally used for restricted-clearance digging, as well as grading and finishing work. They range in size and capacity from small to large.

Both basic types of excavators are available as track-mounted, wheel-mounted, and truck-mounted machines. Choosing the specific type of excavator to use on a job will depend on the features required. This module focuses primarily on the operation of track-mounted hydraulic excavators and truck-mounted telescoping-boom excavators, since they are among the most commonly used types.

1.1.1 Hydraulic Excavators

Figure 3 shows a typical track-mounted hydraulic excavator of medium size being used to excavate a trench for a stormwater system installation. The excavator can move under its own power using the tracks mounted on the undercarriage. The operator can rotate the excavator 360 degrees. A load-hauling truck is visible in the background behind the excavator. The operator can dig facing one direction and then rotate 180 degrees to load the haul truck behind it. This flexibility is an asset when operating in a tight area.

Unless the work requires extensive travel to, from, and around the job site, a tracked excavator is the best choice. Tracked excavators provide good traction and flotation in almost all types of soil conditions. A tracked undercarriage also provides good overall stability. If frequent positioning is required, a tracked excavator will provide better operating efficiency because raising and lowering outriggers on a wheeled excavator takes extra time.

A wheel-mounted excavator, such as the one shown in *Figure 4*, can perform the same type of

Figure 1 Standard hydraulic excavator.

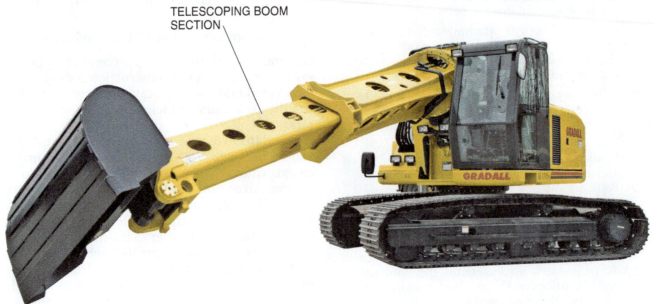

Figure 2 Telescoping-boom excavator.

Figure 3 Track-mounted hydraulic excavator.

operations as the track-mounted excavator. It sits on basically the same type of undercarriage. This undercarriage, however, is attached to several axles with rubber tires that make it suitable for driving on the highway.

A wheel-mounted excavator can be a highly versatile machine that does more than mass excavation and trenching. It combines standard excavator features with the mobility of a wheeled undercarriage. The rubber tires allow the excavator to travel paved roads and work on parking lots and other paved areas without damaging the pavement.

A wheel-mounted excavator is also a good machine to use for material handling. Because of its mobility, it can load and unload trucks and carry loads around the job site. Stabilizers and a push-back (grading) blade can be pinned to the undercarriage to increase the machine's stability during lifting.

Today, mini excavators (*Figure 5*) are increasingly popular for small jobs. Some units are small and light enough to be towed on a small flatbed trailer behind a pickup truck. Many of these excavators have little or no tail swing, which allows them to operate in very restricted spaces. Mini excavators are typically track-mounted machines that are available with a wide array of smaller attachments similar to those used for larger excavators.

1.1.2 Telescoping-Boom Excavators

A telescoping-boom excavator has many uses in the earthmoving and excavation business, and is a popular piece of equipment with highway departments and utility contractors. It is a good machine for performing the final shaping of slopes and spreading topsoil on any area it can reach.

Figure 4 Wheel-mounted excavator.

Figure 5 Mini excavator.

Figure 6 Truck-mounted telescoping-boom excavator.

Truck-mounted or wheel-mounted telescoping-boom excavators can be driven on public highways and do not have to be moved on a trailer. *Figure 6* shows a typical truck-mounted telescoping-boom excavator. The excavator portion of the machine is mounted on the rear part of the truck frame. There are separate engines and operator cabs for the operation of the undercarriage truck and the excavator. However, creeper movement of the undercarriage can be controlled from operator cab foot pedals.

Figure 7 shows a wheel-mounted telescoping-boom excavator. This excavator is designed for rough terrain that is unsuitable for a truck-mounted unit. However, it can travel on roads at speeds under 20 mph (about 32 kilometers per hour) for movement to various nearby work sites. These excavators use four outriggers or two outriggers in combination with a grading blade to stabilize the undercarriage during operation. The grading blade can be used to level an area before positioning the machine for operation. These units can be rotated through 360 degrees because they are not obstructed by a truck cab. One engine is used to power the hydraulic cylinders for excavator operation and the hydraulic motors that drive each of the wheels. These types of units use a steering wheel in the operator cab to control movement of the excavator.

Figure 8 shows a track-mounted telescoping-boom excavator. Like the wheel-mounted unit, this machine is equipped with only one engine that powers both the excavator and the tracks. The excavator is mounted at the front of the tracked unit and can rotate 360 degrees. Track-mounted telescoping-boom excavators are very useful in terrain where wheeled units would have diffi-

Hydraulic Excavator Choices

As with most pieces of heavy equipment, there are pros and cons that must be considered when choosing among different types of hydraulic excavators. Selecting the proper machine for the job can involve questions about size and capacity, traction capabilities, the means of locomotion, maneuverability, stability, and work site conditions. As a general guideline, keep the following characteristics in mind when choosing a particular type of hydraulic excavator:

Track-Mounted Excavator	Wheel-Mounted Excavator
Flotation	Mobility and speed
Traction	Less pavement damage
Maneuverability	Better stability with outriggers
Severe soil conditions	Leveling machine with outriggers
Faster digging	Usable for backfilling and repositioning trench boxes

Figure 7 Wheel-mounted telescoping-boom excavator.

culty getting enough traction, or might get stuck in unstable material.

1.2.0 Common Uses of Excavators

Excavators are used in a variety of earthmoving and trenching activities, as well as in other jobs such as demolition and scrap handling. With the availability of a number of special attachments, the usefulness of excavators can be extended to many operations. This section provides a summary of the main jobs that are performed with a hydraulic excavator.

1.2.1 General Excavation and Trenching

Hydraulic excavators are designed and built mainly to handle excavation jobs. They come in a wide range of sizes based on the type of undercarriage, bucket size, boom length, and digging power. Excavators are versatile pieces of equipment that have many applications such as digging, trenching, laying pipe, and loading material. They can dig deeper than most other earthmoving equipment and are efficient for digging material both above and below grade. *Figure 9* shows a track-mounted excavator performing an excavation operation.

With the excavator's long boom and its ability to rotate 360 degrees, excavating even on rough or sloping terrain is not difficult. Notice that the excavator in *Figure 9* has a small grading blade attached to the front. One use of this blade is to make a smooth working surface for the machine. When the blade is rested firmly on the ground, it acts a stabilizer to keep the machine from tipping over in the front operating quadrant.

The excavator is a good piece of equipment for trenching excavations (*Figure 10*). Use the most efficient bucket size available for digging. Larger

Figure 8 Track-mounted telescoping-boom excavator.

Figure 9 General excavation.

Figure 10 Trenching excavation.

buckets are not always the most efficient. A ditch bucket is also available as an attachment.

Telescoping-boom excavators can also be used for general excavation and lift trenching to the extent that the reach of the boom will get the job done.

The advantage of the telescoping-boom excavator is that it can create a smoother **slope** and finish to the sides or walls of the excavation. Telescoping-boom excavators are extremely useful in low-overhead or close-quarter work areas.

1.2.2 Ditch Cutting, Shaping, and Cleaning

Cutting and shaping new ditch lines and cleaning existing ditches are among the main tasks performed by a telescoping-boom excavator. Because it can travel easily up and down a section of road, it can be set up quickly to clean out and realign the ditch flowline. The long reach of the boom allows the machine to stay on the shoulder or stable **inslope (foreslope)** and work from there without getting stuck in the ditch.

1.2.3 Side Sloping

Telescoping-boom excavators are also very good at grading because of their retractable boom. When fine grading up or down a slope, the bucket is kept lightly in contact with the ground while retracting the boom. The hydraulic controls can help keep the bucket at its most efficient angle so that a smooth slope can be made.

1.2.4 Finish Grading

Finish grading is another task that telescoping-boom excavators perform well. A 24- to 46-foot (about 7- to 14-meter) extension allows it to reach areas that other machines cannot. Pulling the boom inward allows it to make a smooth finish on any surface within reach.

1.2.5 Placing Riprap

As shown in *Figure 11*, telescoping-boom excavators can also be used to place riprap on embankments and slopes. Once the riprap is placed on the ground, it can be moved around by using the bucket and boom to push.

1.2.6 Ripping Pavement

With the special pavement removal bucket, telescoping-boom excavators can rip asphalt and concrete pavements, and load the broken pieces into a waiting dump truck.

1.2.7 Lifting

An excavator will not only dig a trench but it will lift and place a pipe in the correct position. It will also lift and place other items such as trench boxes and shoring for the installation.

A pick point that is welded to the back side of the bucket by the manufacturer allows a rigging device to be attached for lifting purposes. This enables the excavator to lift many different types of objects.

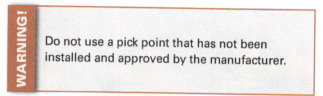

WARNING!

Do not use a pick point that has not been installed and approved by the manufacturer.

Most lifting operations are performed in conjunction with other work, such as excavation of trenches, sewer covers, or footings. Because of the machine's lifting capacity and reach, it can be conveniently used to do this work without bringing in another piece of equipment.

Both tracked and wheeled excavators are also used for material handling. For this type of work, they are equipped with special attachments such as grapples, magnets, and forks that can be used for lifting and placing different materials.

1.2.8 Demolition

Demolition work can be performed by both types of hydraulic excavators using either the bucket or special attachments such as shears (*Figure 12*).

A general bucket or ripper bucket can be used for demolition of buildings or brick walls. The ma-

22304-14_F11.EPS

Figure 11 Side sloping and riprap placement.

Figure 12 Demolition shears.

terial can then be picked up with the bucket and placed in dump trucks for removal from the site. Attachments such as a ripper tooth or hydraulic breaker (hammer) are used to break up rock and concrete. An excavator with a breaker attachment that is being used to break up pavement is shown in *Figure 13*.

1.2.9 *Scrap Handling*

An excavator equipped with a grapple (*Figure 14*) or magnet is a very useful piece of equipment in a scrap-metal handling yard. This type of material is difficult to work with; it requires a machine with good lifting capabilities as well as the durability to withstand the constant scraping and rubbing that occurs as scrap metal is processed. Excavators equipped with magnets are typically used for sorting scrap or for loading it into a processing bin.

Figure 13 Breaking up pavement.

REPRINTED COURTESY OF CATERPILLAR INC.

Figure 14 Scrap handling.

1.3.0 Major Parts of Excavators

The following sections cover the parts and controls of a typical track-mounted hydraulic excavator and a truck-mounted telescoping-boom excavator. In the case of the telescoping-boom excavator, only the controls of the upper structure are covered, because the truck cab controls are the same as those of any standard truck.

1.3.1 *Track-Mounted Hydraulic Excavator Components*

The components of a typical track-mounted hydraulic excavator are labeled in *Figure 15*. The engine and operator's cab are located on the upper structure platform above the frame. The digging system, or hoe, is made up of a boom, a dipper stick, and a dipper (bucket). Each component is controlled by double-acting hydraulic cylinders. The motion of the boom is usually controlled by two hydraulic cylinders mounted on either side of the boom and the upper structure. The motion of the dipper stick is controlled by a cylinder located on top of the boom and attached to the end of the dipper stick.

The dipper (bucket), located at the end of the dipper stick, is controlled by a cylinder attached

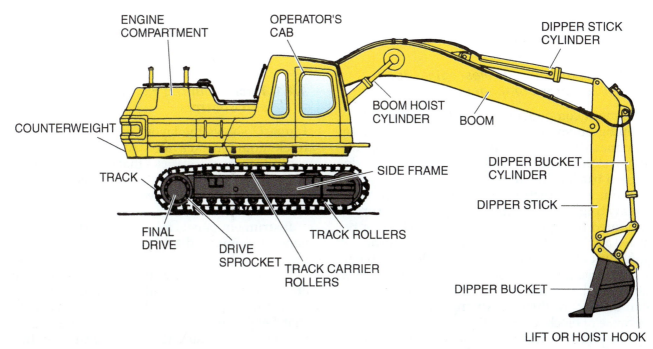

Figure 15 Components of a track-mounted hydraulic excavator.

22304-14_F15.EPS

to the stick and the back of the bucket. The boom, dipper stick, and bucket work together to perform a digging motion that is similar to how a person's arm and hand can be used to scoop material towards the operator. The boom can be compared to the upper portion of a person's arm; the dipper stick is like the portion between the elbow and wrist; and the bucket is similar to a hand, with the pivot point being the wrist.

Buckets and other attachments can be changed by removing several pins and bolts at the end of the dipper stick. Different size counterweights can be placed at the back of the engine compartment to balance the digging and lifting forces on

Watch Your Weight

Counterweights serve a valuable function on excavators. Without them, the weight of the boom and load could cause the excavator to tip over during operation. From a safety standpoint, there are several factors to remember about counterweights. First of all, counterweights may only be used with manufacturer approval. Never install a counterweight that is not approved for the excavator. Operators must also know the rated capacity and never exceed it by adding a heavier-than-approved counterweight. In addition, never attempt to lift an excavator using the counterweight lifting lugs. Those lugs are intended for lifting the counterweight only during installation and removal. Also, pay close attention to warning labels or decals pertaining to the counterweight. Typically, these labels warn about crushing hazards associated with side-to-side and backward movement of the excavator.

REPRINTED COURTESY OF CATERPILLAR INC.

22304-14_SA01.EPS

an extended boom when used according to manufacturer's specifications.

1.3.2 Truck-Mounted Telescoping-Boom Excavator Components

The main components of a truck-mounted telescoping-boom excavator are shown in *Figure 16*. The machine is made up of two separate operating systems. The lower system is a specially designed truck called the carrier. The upper system is the upper structure, which includes all of the excavating components. The upper structure is mounted on a platen that allows it to turn or revolve a full 360 degrees. The right-hand portion of the truck cab is often cut away to allow the boom to be carried forward at a low angle. The upper structure has full rotational mobility. However, it can only work through an angle of 260 degrees because of blockage from the truck cab.

The platen or turntable that is mounted on the bed of the truck supports the excavator engine and hydraulics, the operator's station, the boom, and a counterweight. The boom is made from a hollow box girder that has two sections that hydraulically telescope, one inside the other, to extend or retract the boom. The bucket is attached directly to the end of the boom and can be pivoted hydraulically. The boom can be rotated about 165 degrees in either direction.

1.4.0 Instrumentation and Controls

Controls for excavators have changed over the years. Early models of most heavy equipment used the mechanical action of levers, cranks, and wheels to control the motion of the machine. These mechanical controls have been replaced by hydraulic and electronic controls that are easier to operate and are much more sensitive to move-

ment. Because an excavator has many moving parts, there are numerous foot and hand controls, as well as gauges and switches.

1.4.1 Track-Mounted Excavator Instruments and Controls

Figure 17 shows the layout of a cab on a modern track-mounted excavator. At the center of these controls are two joysticks that operate the motion of the boom, the dipper stick, and the bucket.

The foot pedals and levers in front of the operator control the direction and speed of the tracks.

Instrument panels for gauges and switches are generally mounted to the right of the operator's seat.

One of the most important levers in the operator's cab is the hydraulic activation control (4), shown in *Figure 17*. This lever is a safety feature that keeps operators from leaving the hydraulic controls activated if they leave the cab. In the unlocked position, the lever blocks the operator's exit from the cab. The function of each of the controls shown in *Figure 17* is as follows:

- *Swing lock pin (1)* – Locks and unlocks the position of the upper structure relative to the lower carriage structure.
- *Left console (2)* – Contains controls on the left side of the operator's seat.
- *Left travel pedal (3)* – Operates the speed and direction of the left track.
- *Hydraulic activation control (4)* – Locks and unlocks all hydraulic controls.
- *Right travel pedal (5)* – Operates the speed and direction of the right track.
- *Left joystick (6)* – Controls the extend (reach)/retract (crowd) movement of the dipper stick and the cab swing left and right.
- *Service hour meter (7)* – Keeps track of the hours the machine has operated.

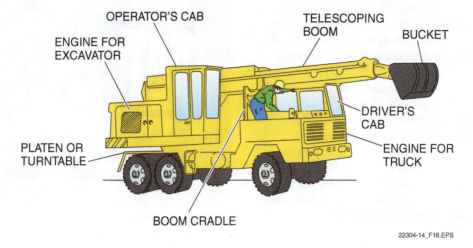

OPERATOR'S CAB
ENGINE FOR EXCAVATOR
TELESCOPING BOOM
BUCKET
DRIVER'S CAB
PLATEN OR TURNTABLE
ENGINE FOR TRUCK
BOOM CRADLE

22304-14_F16.EPS

Figure 16 Components of a truck-mounted telescoping-boom excavator.

- *Left track lever (8)* – Operates the speed and direction of the left track. This lever is attached to the left travel pedal.
- *Right track lever (9)* – Operates the speed and direction of the right track. This lever is attached to the right travel pedal.
- *Instrument panel (10)* – Contains gauges and switches for operation of various systems and accessories.
- *Right joystick (11)* – Controls the up and down motion of the boom and the curl of the bucket.
- *Right console (12)* – Contains controls on the right side of the operator's seat.
- *Operator's seat (13)* – Four-way adjustable seat.
- *Backup system switches (14)* – Console boxes that contain backup switches for electric and hydraulic controls.
- *Engine speed dial (15)* – Controls the engine operating speed (rpm).

The types of controls and their location within the operator's cab vary among models and manufacturers. Always study the operator's manual before attempting to operate an unfamiliar piece of equipment.

1.4.2 Telescoping-Boom Excavator Instruments and Controls

Telescoping-boom excavators have electronic controls similar to those of track-mounted excavators. An example of a typical joystick layout for an excavator is shown in *Figure 18*. The functions of these controls are listed by number:

1. Ignition switch
2. Ignition ON light
3. Engine throttle
4. Engine stop
5. Control shutoff lever
6. Swing lock brake
7. Left-hand control for upper carriage rotation and boom extension
8. Right-hand control for boom up/down and bucket open/close
9. Bucket tilt switch (rotates the boom and bucket)
10. High-speed boom hoist button
11. Parking brake control
12. Hydraulic remote travel pedal
13. Hydraulic remote steering pedal

14. Horn button
15. Electronic system protection warning module
16. Two-speed fan switch
17. Heater temperature control
18. Heater fan control
19. Work lights

Figure 18 shows the two main joystick controls located to the left and right side of the operator's seat. Another common arrangement for these controls is mounted on a movable pedestal in front of the operator.

One of the joysticks controls the operation of the upper carriage rotation and boom extension. Remote operation of the carrier truck has been moved to two floor pedals (12 and 13, *Figure 18*) in front of the operator. The swing lock brake (6, *Figure 18*) is to the side of the operator.

Another important safety feature of the telescoping-boom excavator is the control shutoff lever (5, *Figure 18*). This two-position lever automatically relieves all hydraulic pressure, sets the swing and travel brakes, and turns off remote functions when in the fully raised or locked position. When it is in the unlocked or down position, it is difficult for the operator to leave the cab.

1.4.3 Telescoping-Boom Excavator Braking System

The truck-mounted telescoping-boom excavator is a unique type of machine. Its braking system is a little different from those found in most other pieces of heavy equipment. The telescoping-boom excavator has an air brake system, which includes a service brake, emergency brake, parking brake, and operating brake.

- *Service brake* – The service brake in the carrier truck cab provides braking power to the wheels of the carrier. It includes two reservoirs to store and furnish air pressure for service brake operation. One reservoir supplies pressure to apply brakes to the wheels on the front axle, and the other supplies pressure to apply brakes to the wheels of the rear axles. If the pressure in either portion of the system falls below the safe operating range, the low air indicator light will be activated.
- *Emergency brake* – The emergency brake functions only when air pressure has been lost from some portion of the dual-brake system. Emergency brakes are applied by normal foot pressure on the brake pedal. If air pressure is lost from the front portion of the system, nor-

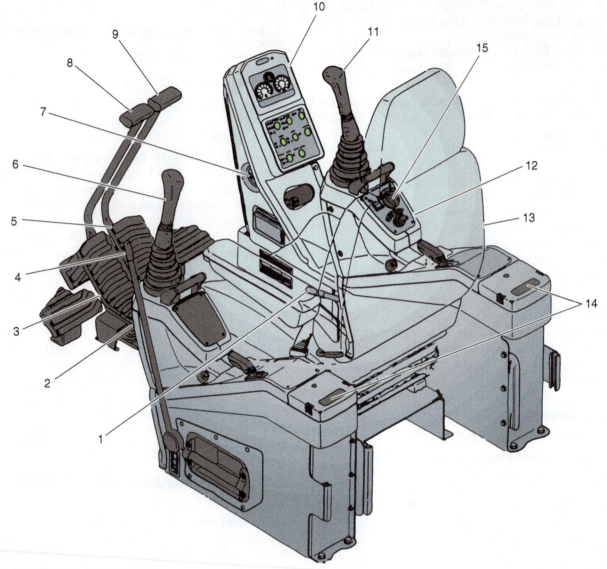

22304-14_F17.EPS

Figure 17 Instruments and controls on a typical track-mounted hydraulic excavator.

mal actuation of the brake pedal valve will still apply service brakes to the wheels on the rear axle. There will be no braking on the wheels of the front axle. If air pressure is lost from the rear portion of the dual-brake system, normal actuation of the brake pedal will apply service brakes to the wheels of the front axle. It will also cause a controlled application of emergency brakes to the wheels of the rear axle.

> **NOTE**
>
> Because air pressure is required to release the emergency brakes, the brakes will remain on until the air pressure can be restored. Do not try to move the machine under this condition.

- *Parking brake* – The parking brake can be applied by raising the parking brake control knob. This causes air pressure to be vented from the spring chambers, allowing the springs to apply brakes to the wheels on the rear axle. The knob will rise automatically if air pressure is lost from the front and rear portions of the system. The parking brake can be released by depressing the brake control knob. The system must be pressurized to release the parking brake.
- *Operating brake* – When activated, the operating brake is applied to all the wheels to hold the carrier stationary while the excavator is digging. If air pressure is lost after the operating brake is applied, the emergency brake will be

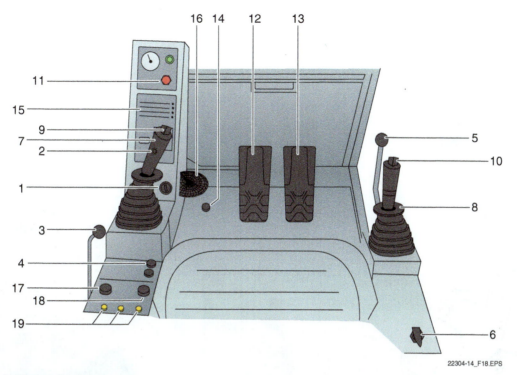

Figure 18 Typical joystick layout.

applied automatically. When the remote switch in the carrier track cab is depressed, the operating brake is automatically applied. Any time the remote signal light is lit, the brake is applied, unless the unit is being driven from the excavator cab.

- *Remote control braking* – The operating brake is applied by the remote switch in the carrier cab. The operating brake will be released and reapplied automatically as the travel pedal in the upper structure operator's cab is actuated and released. A parking brake control is also located in the upper structure operator's cab. This control is to be used in case of failure of the automatic operating brake. If the operating brake fails to apply when the travel pedal is released, raise the parking brake control to apply the parking brake.

1.5.0 Excavator Buckets and Attachments

Excavator manufacturers, as well as third party equipment makers, produce and sell various types of buckets and attachments that can be used on excavators. This variety of options makes it possible to use an excavator for many different applications. Details about the function and operation of these buckets and attachments are described in the manufacturer's operator's manuals.

Always make sure that the bucket or attachment being used is suitable for the equipment, and always follow the recommendations stated in the applicable owner's manual.

1.5.1 Buckets

The bucket is the main attachment for an excavator. A number of different types of buckets are made to accommodate different work requirements. Some of the buckets that are used with excavators are shown in *Figure 19*.

- *Excavating bucket* – The general excavation, or standard, bucket is designed for capacity and **breakout force**. It has a wide profile that makes it easy to heap material and a short tip radius that maximizes the breakout force.
- *Mass excavation bucket* – This bucket has the same profile as the excavation bucket, but is wider for more capacity.
- *Trenching bucket* – Bite width is usually dictated by pipe diameter in trenching applications. The trenching bucket is narrower than most other buckets but deeper than the excavation bucket. It has a large tip radius.
- *Utility bucket* – This bucket is used for excavating, loading, and easy digging in light material. It can also be used as a finishing and cleanup bucket for utility work. It is generally equipped with a weld-on cutting edge.

(A) STANDARD BUCKET

(B) MASS EXCAVATION BUCKET

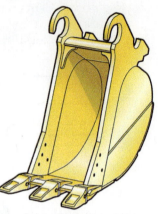

(C) TRENCHING BUCKET

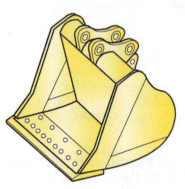

(D) UTILITY BUCKET

(E) ROCK BUCKET

(F) PUSH-TYPE REHANDLING BUCKET

(G) DITCH CLEANING BUCKET

(H) FORMED DITCH BUCKET

A, C–H: REPRINTED COURTESY OF CATERPILLAR INC.

22304-14_F19.EPS

Figure 19 Excavator buckets.

- *Rock bucket* – This bucket is specially designed for extreme digging and working conditions. A stepped tooth design allows the center tooth to enter the ground first at a 45-degree angle. The two teeth on each side of the center tooth enter the ground next. The outer teeth, which project outward, enter last and slice the trench wall, leaving the wall clean and straight. Since the main purpose of this bucket is to break rock, capacity is secondary.

- *Dredging bucket* (not shown) – The dredging bucket is a wide bucket. It has holes or grates in it to allow water to drain from wet materials. It tilts hydraulically, but a fixed link is available for applications where tilting is not required.
- *Push-type rehandling bucket* – Push-type buckets (sometimes called front shovels) for excavators come in several different sizes and configurations. The bucket can be either a normal front loading type or a bottom dump.

- *Ditch cleaning bucket* – This bucket has a smooth edge and wide mouth that allows scraping large quantities of material from ditches. It has a shallow rounded back with several holes that allow water to drain when working with wet materials.
- *Pavement removal bucket* (not shown) – The pavement removal bucket is strong and specially designed for ripping up asphalt, but it can also be used on slab concrete and other hard surfaces. The wide side-less mouth and longer teeth of the bucket allow it to be maneuvered under the edge of pavement and to rip up large chunks of material. The bucket can be pivoted to tightly clamp the material between the bucket and a boom or dipper stick for loading into trucks.
- *Formed ditch bucket* – Formed ditch buckets are available in various sizes of V-shaped or trapezoidal-shaped buckets. One pass of the bucket in soft material will create the desired shape.

1.5.2 Attachments

Special attachments enable an excavator to do jobs that are well beyond normal trenching and excavating tasks. These jobs range from loosening compacted soil to grinding and cutting wood and metal objects. In order to use attachments safely and effectively, always read and follow the operator's manual for the attachment. *Figure 20* identifies some common excavator attachments.

- *Ripper* – Rippers are available in several versions. One is a heavy-duty type that can be pivoted by the hydraulic cylinder that is normally used for the bucket. The other is a set of ripper shanks with teeth that are welded to the back of an extreme-service trenching bucket. (This attachment should not be confused with the rock ripping bucket that has a stepped tooth design with the teeth at the front edge of the bucket.) The ripper breaks up material by using the dipper stick and/or boom as a lever to pull the ripper teeth through hard soil or soft rock.
- *Grading blade* (not shown) – A grading blade can be attached to the stick or boom of an excavator and used for smoothing and sloping tasks.
- *Grapple* – A grapple can be attached at the end of the dipper stick. With this attachment, the excavator can be used to grab, pick up, and place material. Different designs of the grapple are used for picking up logs in forestry work or scrap metal in scrap yard work.
- *Thumb* – A thumb, attached to the underside of a dipper stick and fixed in position for operation or retracted for storage, is used in conjunction with a bucket to perform tasks similar to those accomplished with a grapple.
- *Cutter head* (not shown) – An excavator-powered cutter head can be attached to the end of a dipper stick and used to grind stumps, trees, and brush. This attachment uses the excavator's hydraulic system to run a hydraulic motor that drives the cutting teeth. It is available in numerous sizes to match excavators of different capacities.
- *Pile driver* – A pile driver is an attachment that can be used to drive posts or pilings into the ground.

(A) THUMB   (B) GRAPPLE   (C) DEMOLITION SHEARS

C: REPRINTED COURTESY OF CATERPILLAR INC.

22304-14_F20A.EPS

Figure 20A Excavator attachments. (1 of 2)

REPRINTED COURTESY OF CATERPILLAR INC.

(A) RIPPER

(C) BOOM EXTENSION

(B) HYDRAULIC
BREAKER (HAMMER)

(D) PILE DRIVER

22304-14_F20B.EPS

Figure 20B Excavator attachments. (2 of 2)

- *Hydraulic breaker* – The primary function of the hydraulic breaker, sometimes called a hammer, is to break boulders, thick slabs, and other solid objects. Selecting the proper breaker size is very important. Using the wrong size could result in structural damage to the excavator.
- *Boom extensions* – Boom extensions can be attached to extend a boom or dipper stick for a longer reach. Actuators for extended-boom attachments are extended from the end of the boom or dipper stick through or beside the boom extension to the attachment.

- *Demolition shears* – Shears can be used to cut through steel beams and rebar during the demolition of buildings or other structures. The shears are available in a large range of capacities and must be matched to the excavator to prevent damage to the excavator.

WARNING!

When using a powered cutter head, a hydraulic breaker, or demolition shears, there must be a steel cage in front of the operator.

1.0.0 Section Review

1. One of the advantages of a wheel-mounted hydraulic excavator is that it _____.

   a. has a much greater lifting capacity than a track-mounted unit
   b. can travel on paved surfaces without damaging the pavement
   c. is able to rotate further than any other kind of excavator
   d. has special tires that prevent it from sinking in any kind of terrain

2. Because a truck-mounted telescoping-boom excavator can easily travel along a section of road, it is commonly used for _____.

   a. clearing snow and ice from roadways
   b. backfilling low spots on shoulders
   c. removing brush along road banks
   d. cleaning out and realigning ditches

3. On a track-mounted hydraulic excavator, the motion of the dipper stick is usually controlled by a hydraulic cylinder located _____.

   a. between the boom and the frame
   b. under the dipper
   c. on top of the boom
   d. within the telescoping section

4. How does the braking system of a truck-mounted telescoping-boom excavator differ from that of a standard hydraulic excavator?

   a. It has independent hydraulic brakes for each wheel.
   b. It is an air brake system that powers all of the brakes.
   c. It does not require emergency or parking brakes.
   d. It requires a special brake fluid approved by the DOT.

5. An excavator attachment that connects to the underside of a dipper stick and works in conjunction with a bucket to perform tasks similar to those accomplished with a grapple is called a _____.

   a. thumb
   b. shear
   c. breaker
   d. ripper

2.0.0 SAFETY, INSPECTION, AND MAINTENANCE

Objective

Identify and describe safety, inspection, and service guidelines associated with an excavator.

a. Describe guidelines associated with excavator safety.
b. Describe prestart inspection procedures.
c. Describe preventive maintenance requirements.

Performance Task 1

Demonstrate a proper prestart inspection and maintenance on an excavator.

The responsibility for preventing accidents involving powered equipment rests on the operator of the machine. Manufacturers have designed excavators to incorporate many new safety devices, but even with all these safety features, there are still dangers. Most excavators, for example, have long booms that can easily come into contact with overhead power lines. A safety-conscious operator has much greater value than any safety device. The operator of an excavator must be extremely careful, making sure the area is clear of any obstructions and that no one is near the machine before swinging the boom around.

It is also the operator's responsibility to ensure that the equipment itself is functioning properly. This means operators must know how to perform prestart inspections, recognize potential problems with the equipment, and understand how to perform routine maintenance.

Most companies have policies and guidelines about safety, pre-startup procedures, and machine maintenance. It is important for operators to be familiar with those policies and follow them.

2.1.0 Safety Guidelines

Nobody wants to have an accident or be hurt. There are a number of things that can be done to protect yourself and others around you from getting hurt on the job. Always be alert and do your best to avoid accidents.

Safety can be divided into three basic areas: safety of the operator, safety of other personnel, and safety of the equipment. Operators are re-sponsible for performing the work safely, protecting the public and co-workers from harm, and protecting the equipment from damage.

2.1.1 Operator Safety

Operators must know and follow their employer's safety rules. The foreman, supervisor, or company safety officer should provide you with the requirements for proper dress and safety equipment. The following are recommended general safety procedures:

- Keep the steps, grab irons, and the operator's compartment of the excavator clean and free of debris.
- Mount and dismount the equipment only where steps and handholds are provided. Always use three-point contact to avoid slips and falls.
- Always wear and use appropriate personal protective equipment.
- Keep the windshield, windows, and mirrors on the cab clean at all times.
- Do not wear loose clothing or jewelry that could catch on controls or moving parts.
- Never operate equipment while under the influence of alcohol or drugs.
- Never smoke while refueling, or while checking batteries or fluids.
- Never remove protective guards or panels from the equipment while in operation.
- Hydraulic and cooling systems operate at high pressure. Never attempt to search for leaks with your hands. Fluids under high pressure can cause serious injury.
- Always lower the bucket or other attachment to the ground before performing any service or before leaving the equipment unattended.
- Observe extreme caution when operating under low overhead obstructions such as power lines, or when digging where there may be underground wires and pipes.
- Do not refuel with the engine running.

Each piece of equipment will have a number of warning signs and labels mounted at specific locations. Take time to read and become familiar with these safety signs. Make sure all of them can be easily read. Clean or replace any that cannot be read or understood.

2.1.2 Safety of Co-Workers and the Public

You are responsible not only for your own personal safety, but also for the safety of other individuals who may be working nearby. It will often be necessary to work in areas that are very close to pedestrians or motor vehicles. Remember these safety points when working around other people:

- Make sure you have a clear view in all directions. If working in tight areas, have a spotter on the ground alert you to any people or equipment that you may not be able to see.
- Be sure everyone is clear of the equipment before starting and moving.
- Before working in a new area, take a walk to determine if there are any steep banks, holes, power or gas lines, or other obstacles that would create a hazard to safe operation.
- Never allow riders on the equipment.
- If working in traffic, find out what kind of warning devices are required. Make sure you know the rules and the meaning of all flags, hand signals, signs, and markers that will be used.

2.1.3 Equipment Safety

All excavators have been designed with certain safety features to protect the operator as well as the equipment; for example, they have guards, cabs, shields, and seat belts. Know the equipment's safety devices and be sure they are in working order.

- Make sure clearance flags, lights, and other required warnings are on the equipment when it is transported on a trailer.
- Check all components, hinge pins, and their securing devices often.
- When loading, transporting, and unloading equipment, know and follow the manufacturer's recommendations for the proper sequence and tie-down procedures.
- Never park on an incline.
- Always travel with the bucket in the travel position, low to the ground, and close to the cab.
- Never exceed the manufacturer's limits for operating on slopes and inclines.
- Always turn the engine off and secure the controls before leaving the machine.
- Never exceed the manufacturer's recommendations for the lifting capacity of the machine. Refer to the load rating label in the operator's cab.
- Never exceed the manufacturer's recommended bucket size.
- Be careful when operating the bucket or other attachment close to the machine. They can strike and damage the undercarriage or cab.

WARNING!

Serious injury could result from tipping of the machine. Never exceed the rated load capacity of the machine. If the machine is not on level ground, load capacities will be reduced.

The basic rule is know the equipment and its limitations. Learn the purpose and use of all gauges and controls. Never operate the machine if it is not in good working order.

Finally, never operate the equipment from any position other than the operator's seat. Whenever possible, avoid obstacles such as rocks, fallen trees, curbs, or ditches.

2.1.4 Working Around Overhead Power Lines and Underground Utilities

Excavators are very popular for digging and trenching. Much of this work is done in populated areas. This creates additional maneuvering problems for the operator. The operator not only needs to put the machine into the most effective position for digging and loading, but must also make sure to keep a safe distance from overhead power lines and underground utilities. It is very important to know the maximum height and reach of the machine. Refer to the operator's manual for these specifications.

If there are any underground utilities where work is being performed, operators must make sure they are marked or staked so that the location is clearly visible. If the location of an underground utility is not known, then a call must be placed to the local One Call Center. The nationwide "Call Before You Dig" telephone number is 8-1-1. Calling 8-1-1 enables call center operators to dispatch the appropriate people to locate and mark the buried utilities. Typically, the service must be called at least 48 hours before excavation begins. This gives the utility companies time to come to the job site and mark their lines.

The legal safe distance for operating around overhead power lines varies, depending on the voltage of the lines. OSHA minimum requirements are as follows:

- For lines rated 50kV or below, minimum clearance between the line and any part of the equipment or load is 10 feet (3 meters).
- For lines rated over 50kV, minimum clearance between the lines and any part of the equipment or load is 10 feet plus 0.4 inches (3 meters plus 1 centimeter) for each 1kV over 50kV, or

twice the length of the line insulator but never less than 10 feet (3 meters).

- In transit with no load and boom lowered, the clearance is a minimum of 4 feet (1.2 meters).

Always check the clearance of the boom and stick before operating around or beneath overhead power lines. If you feel that they are too close to risk working in the area, ask about getting the lines temporarily moved. If this is not possible, then maybe a smaller piece of equipment could be used that will not come as close to the lines.

Be very careful when working near any power lines. Be sure to follow all instructions given by the spotter. If some part of the excavator should touch a power line, follow these procedures to reduce the risk of injury to yourself and others:

- Warn other workers not to touch the machine.
- If you are in no immediate danger, reverse the operation that caused contact with the line and move a safe distance away before leaving the machine.
- If it is necessary to leave the machine, do not step off the platform. Jump off as far from the machine as possible, landing with feet together. Hop away with both feet touching the ground together until you are a safe distance from the machine and the electrical current. Do not touch anything.
- Have someone call the proper authorities or the utility company to see if the line can be shut down immediately.

2.2.0 Prestart Inspection Procedures

Routine inspection and maintenance can greatly reduce equipment breakdowns and improve performance. The first thing that should be done before starting any work is to conduct a walk-around inspection. This should be done before starting the engine. The inspection will identify any potential problems that could cause a breakdown and will indicate whether the machine can be operated. It is also a good idea to inspect the equipment during and after operation.

The operator's manual for the excavator must be consulted for the specific checks and the procedures used to accomplish them. The following list describes typical prestart inspections and actions that should be performed on hydraulic excavators.

- Perform a walk-around inspection, including checking for any fuel, engine oil, transmission, and hydraulic system leaks and for any loose, worn, or missing parts (*Figure 21*).
- Check and fill the cab tilt hydraulic system, if applicable.
- Check and fill the cooling system as needed.
- Check and fill the engine oil as needed.
- Drain water and sediment from the fuel filter/separator.
- For wheeled excavators, check for tire wear and proper inflation.
- Drain water and sediment from the fuel tank.
- Check and fill hydraulic system oil.

The Shocking Truth

Electrical shock is the fourth leading cause of death in the construction industry. Did you ever stop to consider what happens if you strike a power line with an excavator? Well, as you probably know, electricity always tries to flow to ground. It will travel through water, metal, and many other materials—including you—to get there. If the equipment touches an energized power line, avoid touching anything on the excavator that you are not already touching. The last thing you want to do is to create another path for electrical current flow. The resistance in your body will cause an enormous amount of heat to be generated, which will burn body tissue immediately. Current flow of as little as 60 milliamps can cause your heart to stop. Many ordinary power tools operate at 2,000 to 4,000 milliamps, so you can imagine what would occur if you were to become the circuit for a power line. Following the so-called ten-foot (3 meter) rule is a good habit. You should keep a clearance of at least ten feet (3 meters) or follow the equipment manufacturer's recommendation, whichever is greater, around any power lines—and even farther away if you are around high-voltage lines. Do not assume that a power line is de-energized or that a line is grounded. Always ask.

22304-14_SA02.EPS

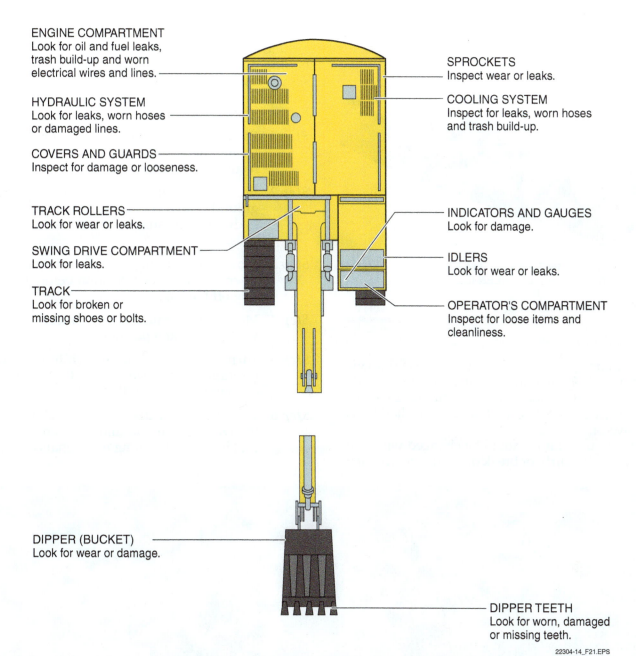

ENGINE COMPARTMENT
Look for oil and fuel leaks, trash build-up and worn electrical wires and lines.

HYDRAULIC SYSTEM
Look for leaks, worn hoses or damaged lines.

COVERS AND GUARDS
Inspect for damage or looseness.

TRACK ROLLERS
Look for wear or leaks.

SWING DRIVE COMPARTMENT
Look for leaks.

TRACK
Look for broken or missing shoes or bolts.

SPROCKETS
Inspect wear or leaks.

COOLING SYSTEM
Inspect for leaks, worn hoses and trash build-up.

INDICATORS AND GAUGES
Look for damage.

IDLERS
Look for wear or leaks.

OPERATOR'S COMPARTMENT
Inspect for loose items and cleanliness.

DIPPER (BUCKET)
Look for wear or damage.

DIPPER TEETH
Look for worn, damaged or missing teeth.

22304-14_F21.EPS

Figure 21 Daily prestart inspection for a track-mounted hydraulic excavator.

- For tracked excavators, check track tension (*Figure 22*).
- Lubricate boom and stick linkage fittings (*Figure 23*).

2.3.0 Preventive Maintenance

Preventive maintenance involves an organized effort to regularly perform periodic lubrication and other service work on the equipment to avoid poor performance and breakdowns at critical times. Performing preventive maintenance on the excavator will keep the equipment operating efficiently and safely, and reduce the possibility of costly failures in the future.

Preventive maintenance of equipment is essential and it is not difficult with the right tools. The leading cause of premature equipment failure is putting things off. Preventive maintenance should become a habit, and be performed on a regular basis.

Maintenance time intervals for most heavy equipment are established by the Society of Automotive Engineers (SAE) and adopted by most equipment manufacturers. Examples of time intervals are 10 hours (daily), 50 hours (weekly), 100 hours, 250 hours (monthly), 500 hours, 1,000 hours, and 2,000 hours. Use the hour meter on the equipment to determine when inspection and service are required. Instructions for inspections

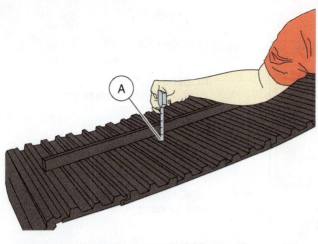

A – TRACK SAG 1⅝ TO 2⅛ INCH (ABOUT 4 TO
6 CENTIMETERS) BETWEEN IDLER AND
CARRIER ROLLER

REPRINTED COURTESY OF CATERPILLAR INC.

22304-14_F22.EPS

Figure 22 Check track tension.

and maintenance, including any special servicing requirements, are usually provided in the operator's manual. Some information may also be provided on labels and notices attached to the excavator.

Hydraulic fluids should be changed whenever they become dirty or break down due to overheating. Continuous and hard operation of the hydraulic system can heat the hydraulic fluid to the boiling point and cause it to break down. Filters should also be replaced during regular servicing.

Before performing maintenance, always go through these procedures:

Step 1 Park the machine on a level surface to ensure that fluid levels are indicated correctly.

Step 2 Be sure the bucket is lowered to the ground or to a cradle and that there is no hydraulic pressure on the machine controls.

Step 3 Move any hydraulic and directional lock levers to the locked position before shutting off the engine.

Step 4 Move the hydraulic activation control lever to the locked position.

Step 5 Turn the engine start switch to the Off position and remove the key. Place a lockout/tagout notice on the switch.

Step 6 Turn the battery disconnect switch key to the Off position and allow the engine to cool before performing maintenance.

22304-14_F23.EPS

Figure 23 Lubricate boom and stick linkage fittings.

Preventive Maintenance Records

Accurate, up-to-date maintenance records are essential for knowing the history of a piece of equipment. Each machine should have a record that describes any inspection or service that is to be performed and the corresponding time intervals. Typically, an operator's manual and some sort of inspection sheet are kept with the equipment at all times. Operators should update maintenance records as part of their daily routine.

2.0.0 Section Review

1. When using an excavator to trench in an area where there might be underground utilities, the best thing to do is _____.

   a. make numerous shallow trenches until the utilities are spotted
   b. use whatever GPS equipment is available to locate the utilities
   c. phone the local One Call Center to have the utilities marked
   d. look for disturbed soil or backfill material that would indicate utilities

2. A walk-around inspection for an excavator typically involves _____.

   a. examining other equipment used at the work site for compatibility
   b. identifying and locating overhead and underground utilities
   c. asking a field mechanic to analyze your performance on the machine
   d. checking the equipment and fluid levels prior to startup and operation

3. An operator can generally identify when preventive maintenance is needed on the excavator by _____.

   a. taking pressure readings for the hydraulic system
   b. using the hour meter to determine hours of equipment use
   c. using a tachometer to read the monitoring system
   d. examining the amount and color of smoke from the equipment exhaust

SECTION THREE

3.0.0 HYDRAULIC EXCAVATOR STARTUP AND OPERATIONS

Objective

Describe basic startup and operating procedures for a track-mounted hydraulic excavator.

   a. Describe startup, warm-up, and shutdown procedures.
   b. Describe basic maneuvers and operations.
   c. Describe common work activities.
   d. Describe activities involving special attachments.

Performance Tasks 2, 3, and 4

Perform proper startup, warm-up, and shutdown procedures.

Perform basic maneuvers with a hydraulic excavator, including moving forward, moving backward, making a pivot turn, and making a spot turn.

Demonstrate basic excavator operation by creating a 10 feet by 10 feet (3 meter by 3 meter) excavation at least 3 feet (1 meter) deep to grade.

Trade Terms

Haunches: The portions of a pipe extending from the bottom to the springline, which is line of maximum diameter.

Spoil: Soil, rock, or other material that is removed from a ditch or other excavation.

Undercutting: The digging of material from beneath the machine. Also, the digging of material from a bank or vertical face of a trench below the grade.

O peration of a hydraulic excavator requires constant attention to the use of the controls while remaining alert to the surrounding environment. The operator must plan the work and movements in advance, and be aware of other operations going on around the machine. If there is any doubt about the capability of the excavator to do a particular job, do not take risks. Stop the equipment, investigate the situation, and discuss it with the site competent person. Whether it is a slope that may be too steep, or an area that looks like the material is too unstable, know the limitations of the equipment and decide how to do the job before starting. Once you find yourself in the middle of a situation, it may be too late to get out of it; the result might be that an unsafe working condition will be created, or that extra work will be necessary.

3.1.0 Startup, Warm-Up, and Shutdown

The proper startup and shutdown of an engine are very important. Proper starting lengthens the life of the engine and other components. A slow warm-up is essential for proper operation of the machine under load. Similarly, the shutdown of the machine is critical because of all the hot fluids circulating through the system. These fluids require cooling in order to cool the metal parts before the engine is shut down.

3.1.1 Startup

There may be specific startup procedures for the piece of equipment being operated, but in general, the startup procedure should follow this sequence:

Step 1 Adjust the seat to a comfortable and safe operating position. Make sure you can reach all controls and pedals.

Step 2 Make sure all the controls are in the neutral position.

Step 3 Move the hydraulic activation control lever to the locked position.

Step 4 Make sure there is no one close to the machine.

Step 5 Place the engine throttle control lever in the low idle position.

Step 6 Turn the key switch to the On position.

Step 7 Adjust the engine speed dial or knob to the Start position.

Step 8 Start the engine and allow it to warm up for five minutes.

> **NOTE**
>
> Never operate the starter for more than 30 seconds at a time. If the engine fails to start, refer to the operator's manual before activating the starter again.

Step 9 After the engine starts, check all the gauges and instruments to make sure they are working properly.

Step 10 Release the hydraulic activation lever to the unlocked position.

Step 11 Manipulate the controls to move forward, turn, and back up.

Step 12 Manipulate the boom controls to raise, swing, and lower the boom.

Step 13 Manipulate the dipper stick control to move the stick in and out.

Step 14 Manipulate the dipper (bucket) cylinder control to curl the bucket.

Step 15 Place all the controls in neutral.

Step 16 Reset the hydraulic activation control lever to the locked position.

Step 17 Make a final check of the gauges and indicators. Check for leaks, unusual noises, and vibrations.

Some machines have special procedures for starting the engine in cold temperatures. Some units are equipped with glow plugs or starting aids. Review the operator's manual to fully understand the procedures for using these aids.

> **WARNING!**
>
> Ether is used in some older equipment to start cold engines. Ether is a highly flammable gas and should only be used under strict supervision in accordance with the manufacturer's instructions.

As soon as the engine starts, release the starter switch and adjust the engine speed to approximately half throttle. Let the engine warm up to operating temperature before moving the excavator.

Let the machine warm up for a longer period of time when it is cold. If the temperature is at or slightly above freezing (32°F or 0°C), let the engine warm up for 15 minutes. If the temperature is between 32°F (0°C) and 0°F (–18°C), warm the engine for 30 minutes. If the temperature is less than 0°F (–18°C), or hydraulic operations are sluggish, additional time will be needed. Follow the manufacturer's procedures for cold starting.

3.1.2 Warm-Up

Gauges and other indicators provide information about the operating condition of the equipment. One of the most important gauges to monitor during startup and operation is the oil pressure gauge. Check the operator's manual to find out the proper operating range for the oil pressure. Keep the engine speed low until the oil pressure registers. If oil pressure does not register within ten seconds, stop the engine and investigate.

Check the other gauges and indicators to see that the engine is operating normally. Check that the water temperature, ammeter, and hydraulic pressure are in the normal range. If there are any problems, shut down the machine and investigate, or get a field mechanic to look at the problem.

3.1.3 Shutdown

Shutting down the engine too quickly after it has been working under load can result in overheating and accelerated wear of the engine's components. A gradual shutdown is needed to allow the engine to cool. This will prevent excessive temperatures in the turbocharger, which could cause a malfunction or result in damage to the lubrication system.

Shutting down the equipment is easy and straightforward. Follow these steps every time the machine is shut down:

Step 1 Make sure the bucket or other attachment is lowered to the ground.

Step 2 Let the engine run for approximately five minutes at low idle.

Step 3 Push or turn the engine throttle to its lowest setting.

Step 4 Push and hold the engine stop button until the engine stops or put the toggle switch to the Off position.

Step 5 Operate the controls to make sure all the hydraulic pressure has been released.

Step 6 Turn the key to the Off position and remove it.

Step 7 Clean up the cab before dismounting.

Step 8 Give the machine a walk-around inspection to make sure there are no leaks or damaged components.

3.2.0 Basic Maneuvers and Operations

There are a number of general guidelines that must be considered before and during excavator operation. Some of these factors affect safety; others affect the operating efficiency. Before using an excavator, it is necessary to understand these general guidelines.

The appropriate bucket or attachment to be used for an operation is determined by the type of work that is to be performed. If a bucket is used inappropriately for a task, there will be increased wear on the machine. Exceeding the operating limits will also reduce the service life of the machine. The following are some suggestions for general operation and maintenance of the excavator:

- Keep the excavator clean.
- Service the excavator regularly.
- Keep the bucket teeth sharp.
- Load the bucket quickly and easily with thin slices of material.
- Take even cuts and heap the bucket as much as possible.
- Hoist the bucket as soon as it is filled.
- When loading haul units, keep productive while waiting for the next truck; make necessary adjustments, loosen dirt, or clean up debris from the bottom and corners of the excavation.
- Make cleanup at final depth to minimize hand labor.
- Set up the overall work cycle so that it will be as short as possible with a minimum amount of swing (optimally 45 degrees).
- Keep travel distances as short as possible.
- Curl the bucket and keep it level to reduce spillage when moving from the dig position to the dump position.

When preparing to work, mount the excavator using the grab rails and footrests while maintaining three points of contact. Adjust the seat to a comfortable operating position, making sure that you can manipulate all the pedals and reach all the levers and controls easily. If you are not in a comfortable position you will not be able to give full attention to the operation. Preparing to work involves getting organized in the cab, starting the machine, and fastening the seat belt.

Determine how much visibility there is to the sides and to the rear of the machine. Since the excavator rotates 360 degrees, make sure the boom will not swing around and slam into a tree or another piece of equipment because you weren't able to see it.

The layout and features of the operator's cab vary with the manufacturer, size, and age of the equipment. However, all stations have gauges, indicators, switches, levers/joysticks, and pedals. Gauges and displays show the status of critical items such as water temperature, oil pressure, engine speed, battery voltage, and fuel level. The indicators will alert you to problems such as low oil pressure, engine overheating, clogged air and oil filters, and electrical system malfunctions. There are also switches for disconnecting the electrical system, activating the glow plugs and starting the engine, as well as turning accessories, such as lights, on and off. The specific controls on the machine may be different from the controls shown

in this section. However, the basic procedure for controlling the movement and direction of all track-mounted excavators is the same. Be sure to study the operator's manual for the equipment to learn the specific location of the controls and how they are used.

Because the excavator's upper carriage can rotate 360 degrees, the orientation for normal steering is when the final drive sprockets are under the rear of the machine and the idlers are under the front of the cab. The reverse steering position places the cab over the sprockets and the directional and steering functions will both be reversed.

> **NOTE**
>
> Always travel with the sprockets under the rear of the machine.

3.2.1 Vehicle Movement

The excavator is moved forward and backward and is turned left and right by using pedals and levers. The farther a lever or pedal is moved forward or pulled backward, the faster the excavator will travel. *Figure 24* provides an operator's view of the pedal and lever arrangement in a tracked excavator. Levers are attached to the backside of the floor pedals, and are used for pulling the pedals back to reverse the direction of the tracks.

The left and right travel levers/pedals operate their respective tracks. Refer to *Figure 24* for these specific positions:

- *Reverse (1)* – Move the travel lever/pedal backward to operate the track in a reverse direction.
- *Stop (2)* – Release the travel lever/pedal to stop the track.

REPRINTED COURTESY OF CATERPILLAR INC.

22304-14_F24.EPS

Figure 24 Tracked excavator steering.

- *Forward (3)* – Move the travel lever/pedal forward to operate the track in a forward direction.

Move both travel levers/pedals the same distance in the same direction, either forward or reverse, for straight line travel. When the travel levers/pedals are moved in the forward direction, the machine will always travel toward the idlers. When the travel levers/pedals are moved in the backward direction, the machine will always travel toward the sprockets.

There are two types of turning maneuvers that can be performed to reposition the excavator. The spot turn allows for a quick turn to the right or left. The pivot turn is a more gradual turn around either the right or left track.

To make a spot left turn, move the left travel lever/pedal backward and move the right travel lever/pedal forward at the same time. This motion causes counter-rotation of the tracks, resulting in a quick left-turn movement. Reversing the position of each travel lever/pedal causes the tracks to rotate in the opposite direction, resulting in a quick turn to the right.

Pivot turns are made by moving either the right or left travel lever/pedal forward and allowing the machine to turn right or left while pivoting on the opposite track.

Remember the following points when maneuvering the excavator:

- The farther the travel levers/pedals are moved, the faster the machine will travel.
- Reduce the engine speed when driving in tight areas or when coming out of a steep grade to help maintain control over the steering.
- When making many turns in soft material, travel in a straight direction once in a while to clean the tracks.
- When traveling all but very short distances, the cab should be lined up correctly with the tracks, with the front of the cab facing the direction of the idlers.

Most tracked excavators have some type of speed control switch or lever that regulates the travel speed of the machine. Select the lower speed setting when traveling on rough or soft surfaces. Low speed should also be used when loading onto a transporter, or when unloading. Use the high-speed position when traveling on a hard, even surface.

3.2.2 Operating the Hoe and Bucket

The operation of the hoe (boom and dipper stick) and bucket of a hydraulic excavator is controlled by two joysticks on either side of the operator's seat. A top view of this arrangement is shown in *Figure 25*. The joystick to the left of the operator usually controls the swing of the upper structure and the in-and-out motion of the dipper stick. The joystick on the right controls the raising and lowering of the boom and the curling of the bucket. Both joysticks can be moved right and left as well as forward and backward.

For the model of excavator shown, each joystick has eight different positions. The digging motion of the hoe is controlled by smoothly manipulating the two joysticks at the same time. To perform a complete digging cycle efficiently, the operator must be able to work both joysticks at the same time. Hand coordination is very important to the smooth operation of the hoe and bucket.

The eight operating positions of the left-hand joystick shown in *Figure 26* control specific motions of the upper carriage and stick components. Moving the joystick forward and backward controls the raising and lowering of the stick only. Moving the joystick to the left or right controls the swing of the carriage only. Diagonal positions, for example Position 2, control movement of both the upper carriage and stick simultaneously.

The right-hand joystick also has eight operating positions, as shown in *Figure 27*. The movement of this joystick directly forward and backward moves the boom up and down only. Movement of

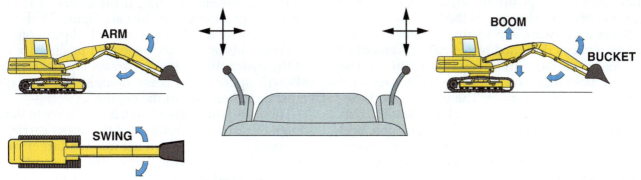

Figure 25 Excavator hoe and bucket controls.

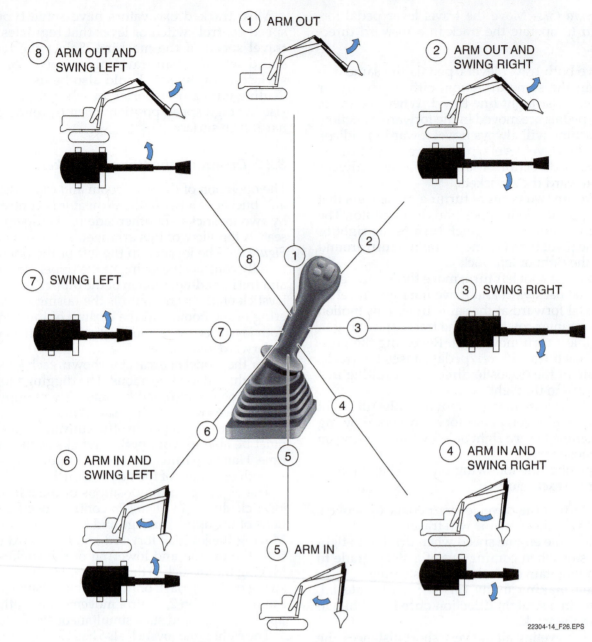

ARM OUT

ARM OUT AND SWING LEFT

ARM OUT AND SWING RIGHT

SWING LEFT

SWING RIGHT

ARM IN AND SWING LEFT

ARM IN AND SWING RIGHT

ARM IN

22304-14_F26.EPS

Figure 26 Excavator stick and swing control.

the joystick directly to the right and left operates the curl of the bucket only. Movement of the joystick to diagonal positions will cause combined movements of the boom and the bucket.

Most excavators are equipped with several safety devices designed to restrict movement of the control surfaces under certain conditions. The two joysticks that control the hoe and bucket have a hold or neutral position. The mechanism is designed so that when the joystick is released from any position, it returns to the hold position, and the swing, stick, boom, and bucket movement stop and remain where they were when the joystick was released.

Another safety feature is the hydraulic activation control lever, designed to keep the machine from accidentally moving if the operator is not properly positioned in the cab. *Figure 28* shows the location of this lever in the locked position.

When this lever is in the locked position, none of the hydraulic controls or travel joysticks/pedals will operate. On some models, the controls must be locked to start the engine.

There are situations when it is necessary to slow down or restrict the rotation of the upper carriage to avoid damaging the machine. On larger machines, there is a swing brake control used to reduce the speed and force of rotation of the upper carriage and boom in order to keep the machine

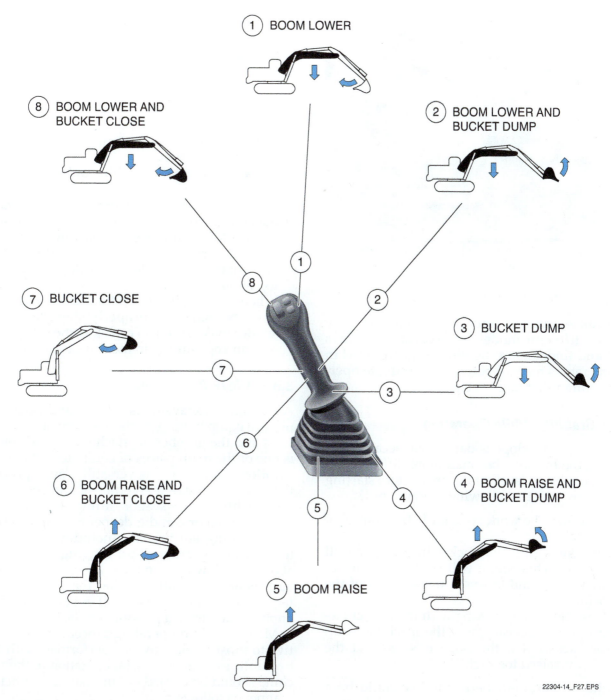

Figure 27 Excavator bucket and boom control.

from turning over. Smaller machines have an automatic feature that controls the momentum of the swing. If there is a heavy load at the end of the hoe and the upper carriage is rotated too quickly, there is a possibility that the force will create an unstable condition and pull one side of the machine off the ground.

The swing lock pin is a manually activated device that locks the upper and lower carriages together and prevents all rotation of the machine. Engage the swing lock pin only when the upper structure is aligned with the lower carriage structure. Do not attempt to engage the pin while the upper carriage is rotating.

REPRINTED COURTESY OF CATERPILLAR INC.

22304-14_F28.EPS

Figure 28 Location of hydraulic activation control lever.

Because controls and control layouts vary among different models of excavators, always read and understand the directions provided in the operator's manual before attempting to operate an unfamiliar piece of equipment.

3.2.3 *Stability While Operating*

Excavating on a slope is dangerous because stabilizing the machine becomes more difficult. If it is not possible to level off the area before starting excavation, take the following precautions:

- Slow down the work cycle for better control of the machine.
- Avoid swinging the bucket in the downhill direction. This action will shift the center of gravity and tend to lessen the stability of the machine.
- If possible, avoid working with the bucket facing uphill. This can cause the machine to tip over backward if the slope is steep and the bucket is raised too high.

Other situations where stability could be a problem include working in soft material, excavating or **undercutting** an embankment, digging under an overhang, or working at the bottom of a pit or excavation. Remember the following points when facing a situation where the stability of the excavator is a concern:

- When excavating a trench or other opening, keep the machine as far away from the edge of the excavation as possible. Always provide adequate shoring to keep the machine from falling into the excavation.
- Do not undercut the machine unless absolutely necessary.
- Be careful when working in areas with unstable material or standing water. Check out the situation before you move the machine into these areas. Determine if mats are needed for stable operation. Do not enter water deeper than the top of the track.
- When working inside a pit or excavation, make sure the walls are properly shored. Have a spotter make sure you can clear the sides without causing damage to the walls or shoring.

3.3.0 Work Activities

A hydraulic excavator is a large and powerful piece of equipment. With the assortment of bucket sizes and the number of attachments available, it is one of the main pieces of equipment of any excavation contractor. The versatility and speed of the excavator make it an efficient and cost-effective machine for excavation, lifting, and loading.

The excavator can dig deeper than most other earthmoving equipment, including front-end loaders, and is efficient for digging material at, above, or below its own level. In some operations, such as mud excavations, it has the added advantage of being able to raise itself out of the mud by applying downward pressure on the bucket.

Many standard operating procedures are relative to most work activities performed with the hydraulic excavator. General operation of the machine should be carried out in a safe and efficient manner as follows:

Momentum

A 2009 safety report from the state of Washington cites a situation in which an excavator operator was moving a 6000-pound (2,722-kilogram) load on a bridge with the excavator boom arm almost fully extended. As the load was being swung toward the center of the bridge, a worker walked in front of it. The operator quickly swung the load in the other direction toward the bridge rail and extended the boom even more in order to avoid hitting the rail. As the load swung over the rail, the combination of the fully extended boom and the momentum of the load caused the machine's capacity to be exceeded. The excavator tipped sideways and toppled over the edge into the water 60 feet (over 18 meters) below. The operator was killed.

- For efficient operation, use more than one control at a time when possible.
- Use smooth, comfortable speeds while operating. Do not use fast, jerking motions that put extra stress on the equipment.
- Keep the machine under control. Know the maximum height, reach, and lifting capacity of the machine and do not work above its capacity.
- Before you begin to operate or move the excavator, make sure there is no one in the area around the machine.
- To work into and out of close places, use the bucket to push or pull the machine or to lift the tracks.
- Watch boom clearance when moving the machine. Uneven ground can cause the boom to bounce from side to side or up and down, increasing clearance requirements.
- Because of the combination of boom, stick, and bucket movement, the bucket can hit the cab and tracks. Use extreme care when working with the bucket close to the cab and undercarriage.
- Do not work the hoe at the end stroke of a cylinder. Repeated use of the components under this condition will damage the cylinder.
- When working with a spotter, always review and agree on the signaling technique to be used before beginning operation. Signaling is commonly done using hand signals, but flag signals, radio communication, and voice are also used.

3.3.1 General Excavation and Loading

The hydraulic excavator is mainly designed to perform excavation work below grade. It can also work at grade level and above grade with equal effectiveness. Different types of excavation work include digging foundations and footings, cutting and smoothing slopes, trenching, cleaning ditches, backfilling, and loading. Each of these jobs requires knowledge of the machine's capabilities, and skill in operating the controls smoothly and effectively.

The operation of digging at or below grade is the major activity performed by the excavator operator. Use of the machine's hoe for digging requires movement of the boom, the stick, and the bucket. The following steps describe the operation of these components to perform basic digging:

Step 1 Plan the work so you do not have to continually move the excavator. Position the machine so that the bucket can reach the area to be excavated without fully extending the boom and stick. Look at the point on the ground where you want the bucket to penetrate. Plan the cuts in advance so there will be no time lost at the beginning of each cycle deciding what to do.

Step 2 Position the boom, stick, and bucket as shown in *Figure 29* so that the stick is at an angle of approximately 70 degrees to the ground.

REPRINTED COURTESY OF CATERPILLAR INC.

22304-14_F29.EPS

Figure 29 Starting position for digging.

Step 3 Extend the stick and position the bucket cutting edge at a 120-degree angle to the ground (*Figure 30*). This is the position of maximum breakout force of the bucket.

Step 4 Move the boom control forward to lower the bucket to the ground.

Step 5 Move the stick in and raise the boom to keep the bucket parallel to the ground. As the bucket enters the material, manipulate the bucket cylinder control so that the cutting edge is working parallel to the ground. As the stick moves past the vertical point, it may be necessary to lower the boom again in order to keep the bucket in the digging position.

Step 6 Raise or lower the boom control slightly if the stick slows or completely stops due to the load. To apply the greatest force at the cutting edge, down pressure must be decreased as the stick moves in toward the boom.

Step 7 Maintain a bucket position that ensures a continuous flow of material into the bucket.

Step 8 When the bucket is full, curl the bucket and raise the boom as shown in *Figure 31*. As soon as the bucket clears the ground it should be swung to the side and dumped.

Step 9 Move the loaded bucket to the dumping point. This may be either a **spoils** pile or a haul unit. To dump, move the stick out and open the bucket in one smooth motion.

Step 10 Return to the cut and continue the digging cycle until the excavation is complete. It may be necessary to reposition the machine from time to time to complete the job.

Performing excavation work efficiently and smoothly requires good coordination and constant attention to the operation of the controls. Because the machine is so quick and powerful, a careless operator can quickly lose control. Keep the following points in mind when operating the excavator:

- Look at the point on the ground where you are trying to make the bucket penetrate. Do not watch the bucket until it gets to the cut. Know where the next cut needs to be while dumping the current load; do not wait until you get back to the cut to decide what to do next.
- Use a short digging pass rather than a long pass; a short pass causes less wear on the bucket.
- Take immediate action if the bucket stalls while digging, either because of ground resistance or load weight. Reduce the resistance by moving the boom up and curling the bucket. Failure to relieve the force could result in overheating of the hydraulic oil and cylinder, or the machine itself may be dragged toward the bucket.

REPRINTED COURTESY OF CATERPILLAR INC.

22304-14_F30.EPS

Figure 30 Bucket angle for maximum breakout.

- Move the machine whenever the digging position is not efficient. It can be moved forward or backward anytime during the operating cycle.

3.3.2 Excavating a Foundation

To excavate a foundation or other opening efficiently, a pattern is usually established. This pattern will vary, depending on such factors as the size of the excavation, the type of soil, and the available area for dumping the spoil. *Figure 32* shows one pattern that can be used for excavating a basic rectangular hole with enough area to dump the spoil close to the side.

Beginning at the south edge of the excavation, the machine is lined up as shown in *Figure 32(A)*. Accurate alignment along the stakes is essential for a clean job. The boom and tracks should be parallel with the digging line. The first cut is made in the corner and a ditch is dug to bottom grade with the left edge on the digging line. The spoil should be dumped on the left side. After digging as much of the south wall as possible from this position, begin to excavate a face on the west wall by reaching to the center and digging a trench back from there.

Back up the excavator a few feet to the position shown in *Figure 32(B)*. Now cut the west end of the ditch vertically because the position of the bucket is more extended. The ditch along the south wall is excavated as near to the machine as possible, with the spoil being dumped along the south outside edge.

Unless the spoils pile is pushed back periodically, it will build up sharply and roll back into the excavation. The material is pushed back using the bucket as it swings around during the dump. Knocking the material back should be started before the pile gets too high.

Continue to dig until the east end of the south wall *Figure 32(C)* is reached. Begin to cut the east wall as close as possible to vertical before moving the machine to the northeast corner, facing south along the east wall as shown in *Figure 32(D)*. Trim up the southern portion of the east wall while moving backward towards the northeast corner.

The northern half of the foundation is excavated in the same way as the southern half, moving from east to west. The west edge can be cleaned up, if necessary, by turning the excavator parallel to the west edge so the bucket can dig straight up.

3.3.3 Loading Trucks

Instead of dumping excavated material on the ground, it can be loaded directly into trucks for removal from the site. The loading arrangement will depend on the site access, as well as the area around the actual excavation. Since the excavator can rotate 360 degrees, the position of the truck can be anywhere the excavator can reach without having to move the tracks. The most efficient way to load is to position the truck on the side of the excavator; this way the machine only needs to turn 90 degrees for the dump. If possible, a load-

REPRINTED COURTESY OF CATERPILLAR INC.

22304-14_F31.EPS

Figure 31 Full bucket.

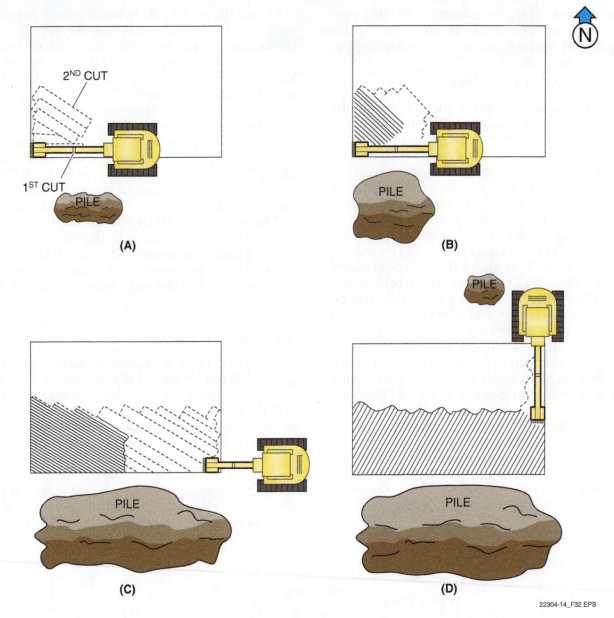

Figure 32 Basic excavation pattern.

22304-14_F32.EPS

ing area should be set up on each side of the excavator so that when one truck is full, a truck on the other side can be loaded. In this way, if the trucks keep lining up properly, the excavator never has idle time.

Follow these steps to load a truck:

Step 1 Raise the boom and bucket clear of the ground. Keep the bucket parallel to the ground so material will not spill out.

Step 2 Operate the swing control when the bucket is clear of the ground. Begin rotation of the upper carriage toward the truck.

Step 3 Raise the boom so the bucket can clear the side of the truck as it swings around over

the truck bed. Do not swing the bucket over the cab of the truck.

Step 4 Make sure the stick and bucket are out over the truck bed, as shown in *Figure 33*.

Step 5 To dump, move the stick out and open the bucket in one smooth motion.

Step 6 Close the bucket and move the stick back in toward the boom. Return the hoe to the digging position for the next cycle.

3.3.4 Lifting Objects

The excavator's design allows it to be used for lifting as well as digging. Some excavators are specifically outfitted to be used as material handling

Figure 33 Correct position for dumping.

machines for logging operations, scrap metal handling, and manufactured material storage such as precast concrete pipe. These machines are usually fitted with special booms, sticks, and handling attachments such as grapples or magnets.

Standard tracked and wheeled excavators primarily used for digging are also capable of safely lifting objects to be moved or placed as part of the construction process. On many sewer jobs, an excavator must lift and swing heavy pipe and trench boxes in and out of the trench and unload material from trucks. This is normally done with a lifting device attached to the bucket.

Lifting is done with the use of a lift eye or hook welded to the back side of the bucket by the bucket manufacturer, or with a special lifting bar on the linkage. The configuration depends on the model of the excavator. All lifting should be done

with the proper slings, wire ropes, and shackles as described in previous modules.

An excavator's lifting capacity depends on its weight, center of gravity, lift point position, and hydraulic capability. The lifting capacity for any given lift position is limited by the excavator's tipping stability or hydraulic capability. *Figure 34* is an example of how lifting capacity varies with the horizontal and vertical distance from the center point of the machine. Changes in boom, stick, and bucket position affect hoe geometry and can drastically change a machine's lifting capacity.

The greatest lifting capacity shown in the figure is 26,900 pounds (12,202 kilograms); notice that this is the closest point to the machine and has the shortest load radius of any position in the figure. The position that has the largest load radius and still has some hydraulic lifting capacity

Work Efficiently

Keep in mind that it is not necessary to begin an excavation in a certain corner based on mapping directions. The actual starting corner will depend on such factors as site layout, the use of spoils, and the construction requirements. The type of material being excavated and the size of the spoils pile will also affect the location. Make sure there is enough area available to accommodate the spoil before starting to dig; having to handle the material twice is expensive and takes time.

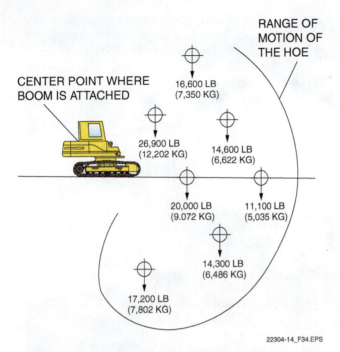

CENTER POINT WHERE BOOM IS ATTACHED

RANGE OF MOTION OF THE HOE

16,600 LB (7,350 KG)

26,900 LB (12,202 KG)

14,600 LB (6,622 KG)

20,000 LB (9.072 KG)

11,100 LB (5,035 KG)

14,300 LB (6,486 KG)

17,200 LB (7,802 KG)

22304-14_F34.EPS

Figure 34 Lifting capacity based on hoe position.

can only lift 11,100 pounds (5,035 kilograms). As a rule, lift capacity decreases as the distance from the swing center line is increased. Always check the operator's manual or the load rating chart attached to the machine to determine the maximum lifting capacities for different positions of the hoe. An example of a lifting capacities chart is included in the *Appendix*.

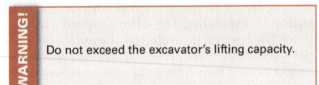

WARNING!

Do not exceed the excavator's lifting capacity.

3.3.5 *Trenching and Laying Pipe*

The hydraulic excavator is an ideal machine for trenching and ditching operations. Its capacity and reach provide effective digging. Its mobility allows the operator to position the machine and begin the work without taking time for setup. Digging trenches is similar to general excavation except that the machine is positioned parallel to the trench and travels backward as it works along the trench line. The excavator can dig the trench, move and position the trench box, lift and place the pipe, and then place backfill.

NOTE

Pipe is normally laid from the lowest elevation to the highest elevation.

Trenching with an excavator is easiest and neatest when the trench width is the same width as the bucket cut. This allows the machine to stand over the center line of the trench with the tracks parallel to it, digging a straight-sided ditch by peeling the material off in layers. When the correct depth is obtained along the area within reach of the bucket, the excavator is walked backward, and an adjoining section is excavated. Short moves should be made for deep excavations, cutting the bottom to exact grades, or cutting curves. Longer moves are used for rough, shallow work.

Curves are dug as a succession of short, straight ditches with beveled edges to produce a smooth curve. The excavator is positioned with its center a little outside the trench center line, and digging is done with the outer half of the bucket.

Many kinds of pipe are laid in straight lines or at angles, rather than in curves. Angles are made by digging slightly past the angle point, then shifting the excavator to straddle the new center line at the required angle.

Spoils from a trench are usually piled only on one side, far enough away to allow a footpath or working space between the spoil and the trench. If a large volume of material is being moved, the pile must be pushed back by the bucket as it is built. It may be necessary to load some material into trucks for removal from the area. Make sure enough material is retained along the trench for adequate backfill.

Before beginning a trench excavation, understand the digging requirements and pay attention to the following items:

- Be aware of the location of any buried cable or other utilities. Have their location and depth marked before excavation begins.
- Understand the alignment and grade requirements. Check the plans and make sure the staking is correct.
- Decide where to begin. If one end of the trench is blocked, always try to work away from that end.
- If haul trucks are being used, make sure everyone knows the loading pattern. Never swing the excavator bucket over the truck cab.

Unless working in extremely stable material such as rock, some type of shoring is required for all trenches beyond a certain depth. In some states, the depth is 4 feet (1.2 meters), but the exact amount can vary. Always check local regulations for shoring requirements. The excavator can be used to lift shoring material into the trench, and then remove it when the work is complete. For many jobs, a prefabricated trench box is used if space is limited and if it is not possible to slope the bank back from the trench bottom.

A simple trench box consists of two sheets of steel separated by welded steel bracing. The open area in the center between the braces allows pipe to be lowered through the shield to the trench bottom. More elaborate shields have a hopper at the front that feeds gravel to the correct grade. As the shield moves forward, the pipe is lowered through the center opening and a hopper at the back of the shield backfills gravel over the pipe. The excavated material is then backfilled up to the back of the shield.

The excavator can be used to position and move the trench box. The initial positioning will depend on the site and the trench depth. In some cases, the trench box can be pushed into the trench, while other conditions will require lifting the box with a sling and placing it in the required position.

If there is the option of sloping the sides of the trench instead of using a trench box or shoring, it must be done according to OSHA regulations. The slope of the trench walls will depend on the depth of the trench and the type of soil being excavated. Walls can either be smooth or benched, but must conform to the maximum allowable slope for the specific conditions. For OSHA Class A type soils, the slope must be at least ¾:1 for trenches up to 20 feet deep (6 meters). This means that for every foot of trench that rises vertically, it must slope out ¾ of a foot (.23 meters) horizontally. For example, if a trench is 5 feet (1.5 meters) deep, the edge of the cut must be 3¾ feet (1.1 meters) out from the edge of the trench bottom.

3.3.6 Placing Bedding Material

In trenching operations, the excavator is also able to undercut the trench bottom and place the bedding material. The purpose of the bedding is to give the pipe uniform support and a smooth foundation. The type of bedding material used will depend on the type of pipe being placed and the local soil conditions.

When undercutting is necessary to remove unstable material, dig deep enough below the grade line to remove all of the material or to provide enough depth to backfill with suitable material to form a firm bed for the pipe. The actual depth and width of the bedding will be given in the plans or specifications.

The undercutting operation and the placement of the bedding material is guided by a worker in the trench. Because this is fine grading work, the excavator needs to be closer to the cut so there will be greater control over the position of the bucket as it digs. It is important for operators and ground personnel to understand and agree on the signals being used.

Final placement of the bedding material in the bottom of the trench to the required width, depth, and shape is the responsibility of the work crew in the trench. The bucket is filled with bedding material from a stockpile, truck, or material box and is then positioned over the trench. Following the signals of the spotter, the bucket is lowered into the trench and positioned so that the tip will clear the bottom of the trench when the bucket is uncurled. While raising the boom slightly, the bucket is uncurled to cast out the material. The dumped material is then smoothed out by dragging the bucket through the material with the teeth down. Final grading to required elevations must be done by a finish crew working in the trench.

> **WARNING!**
>
> When working with trench crews, make sure all personnel have left the trench.

3.3.7 Setting Pipe

Once the bedding material has been placed and compacted, the pipe can be set. The excavator can be used to unload the pipe from the delivery truck, move the pipe into position along the trench, and then lift and set the pipe into the trench. The method of lifting the pipe will depend on the type and shape of the pipe as well as the type of lifting devices used. Contractors in different geographical regions will use different methods. However, all lifting equipment must conform to basic OSHA standards.

> **WARNING!**
>
> Only lifting eyes provided by the equipment manufacturer should be used for lifting activities. Never attempt to lift objects using lifting devices that have been added to the equipment by welders or other personnel.

The following procedure is used to lift and set pipe into a trench:

Step 1 Make sure the pipe is within easy reach of the excavator. It should be in line with the side of the trench or loaded on a truck positioned to the side of the excavator.

Step 2 Connect the sling or lifting devices to the pipe first. Make sure it is secured before swinging the bucket over the pipe.

Step 3 Swing the bucket over the pipe and lower the boom so the sling can be attached, as shown in *Figure 35*.

Step 4 Raise the boom until the sling is tight and the pipe is just off the ground. At this point, you will be able to tell whether the pipe is balanced or unstable. If the pipe is not balanced, lower the boom and allow the ground crew to reposition the sling.

Step 5 Continue raising the boom until the pipe clears all obstructions along the side of the trench. Swing the boom around and position the pipe over the trench above the point where it will be set.

Step 6 Begin lowering the pipe into the trench, following the spotter's signals. Be careful not to hit any shoring along the sides of the trench.

Step 7 As the pipe is lowered, line up the end of the pipe with the last pipe section placed. The workers in the trench will maneuver the pipe into place as it is lowered the final 1 to 2 feet (30 to 60 centimeters). At this point, it may be necessary to wait until the trench crew prepares the end of the pipe for fitting to the previous section. This will involve attaching any gaskets or pipe joint compound.

Step 8 Once the ends of the pipe are lined up, roll the bucket out slightly and move the pipe into place.

Step 9 The trench crew can now disconnect the sling from the pipe. The bucket is raised out of the trench with the loose sling attached and repositioned to lift and set the next section of pipe.

Once the pipe is properly placed, backfilling can be done using the excavator. This requires the close attention of the operator, because too much material placed on one side of the pipe will create a side force capable of knocking the pipe out of alignment. The initial backfill material is spread on either side of the pipe and compacted under the **haunches**. After the haunch material is placed, backfill can be dumped and compacted on alternating sides of the pipe until it is covered.

3.3.8 Working in Unstable Soils

Unstable soil or mud may be encountered on some jobs. The tracks of the excavator may not be able to get through the material. When this happens, some type of supporting timbers or mats must be placed under the tracks. Depending on the local conditions and the type of work in progress, the excavator can use prefabricated metal or timber mats as a stable platform for traveling and digging.

The following procedure is used for laying mats:

Step 1 Before beginning to move the excavator, determine exactly where the mats should go. If there are a limited number of mats, use them as efficiently as possible by picking them up and replacing them as you travel.

Step 2 Position the excavator to pick up the first mat. Make sure you are close enough to position the bucket over the area where the sling will be attached to the mat.

22304-14_F35.EPS

Figure 35 Sling attached to a bucket.

Step 3 Swing the boom over the first mat and lower the bucket into position. Have the ground crew attach the sling to the lifting eye or bar on the excavator. Depending on the size of the mat, it may be possible to pick the mat up and position it using the bucket and the stick. In this maneuver, the mat is clamped in the middle between the bucket and stick, picked up, and placed using the bucket.

Step 4 Raise the boom until the mat is clear of the ground and swing the excavator into position for setting the first mat.

Step 5 Lower the boom until the end of the mat is in the correct position in front of the excavator.

Step 6 Continue to lower the boom until the mat is flat on the ground and there is enough slack to remove the sling.

Step 7 Raise the boom and swing around to pick up the next mat. Repeat the lifting and swinging procedures until enough mats are in place to support the excavator, allowing it to travel forward.

Step 8 Prepare to walk the excavator onto the mats by positioning the tracks to engage the mats evenly at the edge.

Step 9 Walk the excavator forward onto the mats, moving forward until the front of the tracks are on the last mat laid.

Step 10 Swing the excavator around until the boom is over the end of the first mat set. Lower the boom until the sling can be connected to this mat.

Step 11 Raise the boom until the mat is high enough to clear the ground and any other obstructions. Swing the excavator around to the front and lower the boom so the edge of the mat is resting on the ground in line with the other mats, as shown in *Figure 36*.

Step 12 Repeat the procedure of traveling and moving the mats forward until the excavator is in the proper position to begin operation.

During the setup, mats must be positioned to allow for a series of forward and backward movements. During the course of travel, if it becomes necessary to make turns, the mats can be laid at a right angles to each other.

3.4.0 Special Attachments and Activities

The hydraulic excavator has many special attachments that can be used to do different jobs. The buckets and digging attachments are used for earthmoving and excavation tasks. Other attachments are used for specialty work, such as materials handling. The following attachments and activities are commonly used with track-mounted and wheel-mounted hydraulic excavators.

3.4.1 Demolition

Effective demolition using a hydraulic breaker requires selecting the proper size and bit type to do the job. Bits come in several different shapes and are designed for specific types of work. Using the wrong size bit or breaker could result in structural damage to the boom and stick. The hydraulic breaker should only be used for breaking up material. Before starting the operation, place the machine on a level, stable surface and position the breaker against the material as shown in *Figure 37*.

WARNING!

Be sure to close the front window or place a shield on the front of the cab for protection against flying debris.

Before the breaking operation begins, put the upper structure in the correct position, which is approximately 30 degrees right or left of either the front or back of the undercarriage. Do not operate the breaker with the upper structure sideways to the undercarriage and tracks, and do not try to operate the breaker with the boom perpendicular to the tracks.

Do not attempt to break rocks or concrete by using the force of the excavator in addition to the in-out motion of the hydraulic breaker; this could cause structural damage to the machine. Allow the hydraulics of the breaker to do the work.

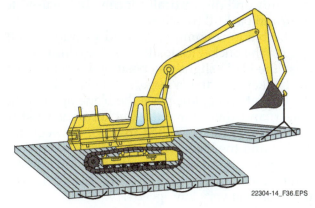

22304-14_F36.EPS

Figure 36 Placing mats.

Figure 37 Using a hydraulic breaker.

> **CAUTION**
>
> Do not allow the breaker to operate continuously for more than one minute at any one point, or overheated hydraulic oil could result in damage to the breaker mechanism. Change the breaking location and repeat the procedure. Stop the breaking operation immediately if any of the hydraulic hoses are wriggling or twisting rapidly. This indicates a punctured accumulator in the breaker. Contact the field mechanic or repair shop to have any necessary repairs made.

3.4.2 Blading

On some smaller models of wheeled and tracked excavators, manufacturers have added the option of installing a dozer-type blade on the front of the undercarriage. This attachment can be used to level the work space so that the excavator will sit firmly and dig vertically. It can also be used to doze and backfill trenches.

The blade is commonly used to trap objects against the bucket for loading. To use this feature, raise the blade slightly to position it evenly with the bucket. To carry a large, solid object between the bucket and blade, place the blade against the object, and then move the bucket in against it. The object is then pinched in place between the bucket and the blade. Raise the blade and bucket slightly to lift the object off the ground so that it can be moved. When performing this maneuver, travel only at very low speeds.

3.4.3 Digging with the Shovel

The shovel excavator is configured differently than the normal hoe excavator. The shovel is designed to dig and load with an outward motion rather than curling inward. This puts the bucket operating cylinder behind the stick instead of in front of it. The design is similar to a tracked front-end loader with an extra pivot point. The advantage of this configuration is the ability to load narrow haul units more accurately with less spilling of material.

Operation of the bucket on a shovel excavator is as follows:

Step 1 Position the excavator in front of the material.

Step 2 Move the stick forward so that the shovel will crowd into the material.

Step 3 Pull the bucket up into the material until it is full.

Step 4 Move the bucket up and out of the material.

Step 5 Swing the boom around to position the bucket over the truck bed or dump position.

Step 6 Center the bucket over the truck bed and raise the bucket to clear the side of the truck.

Step 7 Move the bucket out over the truck bed and simultaneously dump the material by uncurling the bucket.

Step 8 Curl the empty bucket and pull the boom up to move away from the truck.

Step 9 Swing back to the material pile or excavation and lower the bucket to the digging position.

Step 10 Repeat the cycle until the operation is complete.

As the truck fills, the load will need to be pushed across the truck bed to even it out. As the leading edge of the bucket passes the sideboard of the truck, the bucket is rolled down quickly and the load is dumped, then pushed across the truck bed. By raising the bucket and pulling the boom up slowly, the load will be distributed evenly across the bed. Be careful to watch the bucket, making sure it clears the sideboards of the truck, and be careful not to hit the truck with the bucket or stick.

3.4.4 Material Handling

The grapple (*Figure 38*) is used mainly in the logging and scrap-handling business. It attaches to the end of the stick, which allows it to drop down in a vertical position and swivel about the stick. The grapple has paddle-style tines that can grip logs and irregular-shaped objects and pick them up. To pick up a log, position the grapple over the log. Open the tines wide enough to get them around the log, then close the tines and lift the log.

Be very careful when lifting material with the grapple. Make sure no one is in the area where the lifting and placing are being done. Material can slip out of the grasp of the tines and fall on workers and other equipment.

The magnet is a special attachment that is used in scrap metal handling operations. Like the grapple, the magnet is attached to a swivel bracket at the end of the stick.

Magnet operation is essentially the same as grapple operation, except that the magnet will only pick up material when the magnetic head is activated. The magnet head is placed over the material to be picked up and the magnet is activated using the switch located on the cab controls. To release the material, the power switch for the magnet is turned off.

REPRINTED COURTESY OF CATERPILLAR INC.

22304-14_F38.EPS

Figure 38 A grapple in use.

3.4.5 Setting Up and Calibrating a Laser Guidance System

The hydraulic excavator is one of several pieces of heavy equipment that can make use of a laser guidance system for maintaining constant slope and elevation while digging. Before the use of lasers, equipment operators followed the grade and slope stakes put in the ground by a survey crew, or were given directions by a grade checker. Using the laser guidance system, the boom and bucket follow signals from a laser receiver unit. The receiver captures directional information from a laser level sending unit set up at a distance from the actual operation.

The laser guidance system consists of the laser level sending unit, the receiving unit, and the controls. The level is usually mounted on a tripod that is in line of sight with the area being excavated. The sending unit transmits a beam of light that can be picked up by the receiver attached to the boom of the excavator. The receiver senses the laser beam and keeps it centered in the receiver's target area. This signal operates the excavator controls to raise or lower the bucket according to the distance needed to keep the laser in the center mark of the target area.

The laser guidance system on the excavator can be used in several different modes, including slope control, fixed depth, constant elevation, and blind applications. Slope control provides automatic control of sloped cuts from flat to 100 percent. Fixed-depth mode can control excavation to a designated depth from a surface reference. The constant elevation mode is used to provide a constant elevation reference when excavating trenches for laying sewer and drainage pipe.

Blind applications involve the use of a computer screen mounted in the operator's cab to show the bucket's progress toward final grade with an animated picture. In this mode, the operator does not have to watch the bucket to adjust the controls. The diagram or picture on the screen is observed and the controls are manipulated based on what is shown by the computer.

Before the laser guidance system can be put into operation, it must be set up and calibrated for the project. The first task is to position the level and sending unit and establish the height of the laser beam. After this is completed, the receiver must be mounted on the excavator and calibrated so that the digging operation can be carried out according to the elevation and grade requirements. The calibration usually involves sliding the receiver up or down the mast until a level reading is indicated on the panel in the cab. Whenever the receiver intercepts the laser beam from the level,

it sends a reading to the operator indicating cut, fill, or on-grade. This makes checking the grade a continuous process as the excavator continues to dig.

Laser leveling equipment is manufactured by several companies. Each system has slightly different characteristics and capabilities. Before beginning operations with a laser guidance system, review the requirements and instructions in the manufacturer's manual and make sure the equipment is operating properly.

3.4.6 Transporting the Excavator

The tracked excavator cannot be driven on public highways; it must be moved using a transporter. This is usually a flatbed trailer that is pulled by a truck. Before transporting the excavator, investigate the travel route to make sure there will be adequate clearance for the machine.

Observe all state and local laws governing the weight, width, height, and length of a load. Follow

GPS Technology

Excavator control can also involve the use of Global Positioning System (GPS) technology. This GPS is similar to that is used for automobile navigation. Instead of providing travel directions, however, an excavator system uses satellite guidance to relate the bucket position to a computerized model of the site. The onboard computer then signals the operator or the excavator's hydraulic system to adjust the bucket position to achieve the design requirements. This system increases excavation accuracy and speed, and removes some of the burden on the operator. The result is higher productivity and lower operating costs.

GPS RECEIVERS

REPRINTED COURTESY OF CATERPILLAR INC.

22304-14_SA03.EPS

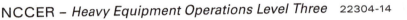

all load regulations governing wide loads and obtain the necessary permits before moving the equipment.

To prevent the machine from slipping during loading, or from shifting while in transit, remove ice, snow, or other slippery material from the loading dock and the trailer bed before loading. Clean the tracks of any mud or debris that could fly off during travel.

The excavator is typically loaded by crawling up a ramp at the front end of the trailer. Some types of trailers can be loaded from the back. Before starting to load, check the operator's manual for any specific instructions regarding activities such as blocking the tracks, tie-down points, and position of the controls.

General loading procedures are as follows:

Step 1 Position the trailer on a level surface and block the wheels.

Step 2 Place the boom, stick, and bucket in the travel position and walk the excavator onto the trailer. Use a spotter on both sides of the excavator. After it has been moved onto the trailer, the excavator should be facing backward with the boom lowered and the stick and bucket curled underneath, as shown in *Figure 39*.

Step 3 Lower the bucket to the bed of the trailer and fold the boom, stick, and bucket up as tightly as possible. This will reduce the height of the boom and create the lowest profile.

Step 4 Align the upper carriage with the tracks and engage the swing lock pin.

Step 5 Manipulate the hydraulic controls to remove any remaining pressure.

Step 6 Move the hydraulic activation control to the locked position.

Step 7 Turn the engine start switch to the Off position and remove the key.

Step 8 Turn off any disconnect switches and remove the key.

Step 9 Lock the door and access covers. Attach any vandalism protection shields.

Step 10 Cover the exhaust and air intake openings.

Step 11 Tie down the machine according to directions provided in the operator's manual. Use only approved chains and load binders.

Step 12 Place appropriate flags or markers on the equipment if needed for height and width restrictions.

Unloading the excavator from the trailer is basically the reverse of the loading operation. When unloading the machine, make sure there is enough room to maneuver once the machine is off the trailer and on the ground. Sometimes there will be enough room to unload, but the excavator will not be able to move away from the trailer, and the tractor may not be able to move into position to pull the trailer out of the way.

22304-14_F39.EPS

Figure 39 Position of a loaded excavator.

3.0.0 Section Review

1. During startup, which of these controls should be placed in the locked position initially, then unlocked so that the boom and bucket can be manipulated, and finally reset to the locked position while the excavator warms up?

   a. The engine throttle control.
   b. The differential lock mechanism.
   c. The hydraulic activation lever.
   d. The independent braking system joystick.

2. Because a hydraulic excavator's upper carriage can rotate 360 degrees, the orientation for normal steering is when the final drive sprockets are _____.

   a. under the rear of the machine
   b. on the left side of the operator's seat
   c. under the front of the cab
   d. on the right side of the boom mounts

3. Which of these choices best describes a standard procedure that can help improve the safe and efficient operation of an excavator?

   a. Try to perform all excavation work with the hoe or other attachment at the end stroke of its hydraulic cylinder.
   b. When working into and out of close places, use the bucket to push or pull the excavator.
   c. Before operating or moving an excavator, position at least three co-workers around the machine.
   d. During normal operation, never use more than one control at a time.

4. Two excavator attachments that are most likely to be used during scrap-handling operations are the _____.

   a. hydraulic breaker and vise
   b. thumb and bucket hoe
   c. telescoping forks and wedge
   d. grapple and magnet

SUMMARY

Excavators are most commonly available as crawler-type hydraulic excavators. They are also available as rough-terrain wheel-mounted units. A special type of excavator that is available is a telescoping straight-boom excavator that is usually truck-mounted. Telescoping-boom excavators are also available as wheel- and crawler-mounted units. Excavators are available with a wide variety of attachments. Besides the standard buckets, hydraulic shears, grapples, and breakers along with specialized buckets, rippers, and lifting equipment are available.

Excavators are large, powerful machines, so safety is always a primary concern during operation. As an operator, you are responsible for the safety of the machine, the safety of co-workers and other people in the area, as well as your own safety. Know and follow your employer's safety rules. Read the operator's manual for the equipment and follow the safety and operating procedures provided. Know the working limits of the machine and do not attempt operations the machine is not designed to perform.

Excavators have specific preventive maintenance requirements that are described in the operator's manual. To extend machine life and safety, always perform walk-around checks as well as all prescribed periodic checks during pre-start and operation inspections.

Being a good excavator operator takes patience and skill. Much damage can result if the boom and any attachments are not controlled properly. Good hand and eye coordination is essential. Learn to operate the excavator smoothly and efficiently using both hand controllers at the same time.

Review Questions

1. Hydraulic excavators and telescoping-boom excavators are both available as track-mounted, truck-mounted, and _____.
   a. stationary-point machines
   b. wheel-mounted machines
   c. skid-mounted machines
   d. grader-mounted machines

2. What type of hydraulic excavator is generally best suited for work that does *not* require a lot of travel to, from, and around a job site?
   a. A truck-mounted excavator.
   b. A telescoping-boom excavator.
   c. A wheeled mini excavator.
   d. A track-mounted excavator.

3. The type of telescoping-boom excavator that has two engines is a _____.
   a. truck-mounted type
   b. track-mounted type
   c. wheel-mounted type
   d. hydraulic-track type

4. One distinct advantage of telescoping-boom excavators is that they _____.
   a. have a built-in serrated blade that can be used for dozing the work area
   b. have very few moving parts to inspect and maintain
   c. use twin hydraulic cylinders to operate the boom, stick, and bucket
   d. are extremely useful in low-overhead or close-quarter work areas

5. A pick point on the back side of a bucket enables an excavator to be used for _____.
   a. lifting different types of objects
   b. finish grading jobs
   c. placing riprap on roadways
   d. soil ripping applications

6. On a track-mounted hydraulic excavator, the component that connects the bucket to the boom is typically called the _____.
   a. telescoping sleeve
   b. boom extender
   c. dipper stick
   d. articulating pivot

7. On a truck-mounted telescoping-boom excavator, the engine, hydraulics, operator's cab, boom, and counterweight are supported by the _____.
   a. carriage or riser
   b. hydraulic cradle
   c. cross member
   d. platen or turntable

8. What control device, if left unlocked, will block the operator's exit from the cab of a track-mounted hydraulic excavator?
   a. Left vehicle movement joystick.
   b. Hydraulic activation control lever.
   c. Main braking system control lever.
   d. Right boom movement joystick.

9. The basic purpose of the operating brake on a truck-mounted telescoping-boom excavator is to _____.
   a. provide hydraulic fluid pressure to the brakes during roading
   b. slow or dampen the movement of the boom and platen
   c. hold the carrier stationary while the excavator is digging
   d. function as an emergency backup if air pressure is lost

10. What type of excavator bucket has a stepped-tooth design that allows the center tooth to enter the ground first at a 45-degree angle, while two teeth on each side of the center tooth enter the ground next?
    a. Rock bucket.
    b. Grapple bucket.
    c. Dredging bucket.
    d. Utility bucket.

11. An excavator attachment that is commonly used to break boulders and other solid objects is called a _____.
    a. breaker bar
    b. hydraulic breaker
    c. grapple
    d. ripper

12. The minimum safe working distance between an excavator and electrical power lines is determined by _____.

    a. the equipment operation team
    b. the engineer in charge of the project
    c. a One Call Center dispatcher
    d. local codes and working conditions

13. A daily inspection of an excavator should be made _____.

    a. before starting the engine
    b. once the equipment is in the shop
    c. after the engine reaches normal operating temperature
    d. only when a problem is suspected

14. Before any type of preventive maintenance is performed on an excavator, be sure to move the hydraulic activation control lever to the _____.

    a. release position
    b. unlocked position
    c. locked position
    d. travel position

15. When starting the engine of a hydraulic excavator, the throttle control lever should be placed in the _____.

    a. choke position
    b. low idle position
    c. open draft position
    d. high idle position

16. When making frequent turns in soft material during the operation of a track-mounted excavator, travel in a straight direction occasionally to _____.

    a. self-lubricate the tracks
    b. allow the tracks to cool
    c. allow the track tensioners to activate
    d. clean the tracks

17. The position of maximum breakout force for the cutting edge of a hydraulic excavator bucket is at a _____.

    a. 45-degree angle to the ground
    b. 90-degree angle to the horizon
    c. 120-degree angle to the ground
    d. 180-degree angle to the undercarriage

18. When using a track-mounted hydraulic excavator in unstable soil or mud, _____.

    a. drive in the same tracks every time you move the machine
    b. place supporting timbers or mats under the tracks
    c. operate in the highest gear to avoid spinning the tracks
    d. use the partial brake lock mechanism to gain traction

19. Before using a laser guidance system for excavating or grading, first _____.

    a. measure the exact distance from the sending unit to the bucket
    b. re-survey the work site and install marker spikes
    c. set up and calibrate the laser sending and receiving units
    d. make sure that no other electronic equipment is operating nearby

20. Before transporting a track-mounted hydraulic excavator, the tracks of the excavator should be cleaned to _____.

    a. prevent the machine from damaging the pavement
    b. be in compliance with OSHA regulations
    c. prevent mud or debris from flying off during travel
    d. enable the track lubrication system to function

Trade Terms Introduced in This Module

Boom: A part of the digging component (hoe) of an excavator. The boom is attached to the upper carriage and serves as the main lifting stick of the machine.

Breakout force: The amount of digging force that can be exerted by the bucket on material in its natural state.

Counterweight: Dead or nonworking load that is attached to one end or side of a machine to help improve the balance.

Demolition: The tearing down of buildings and other structures.

Grapple: An excavator attachment that is used for lifting. The grapple has two or more hooks that can be closed on each other to pick up solid material, such as logs, by grabbing onto them.

Haunches: The portions of a pipe extending from the bottom to the springline, which is line of maximum diameter.

Hoe: The component of an excavator that is made up of the boom, stick, and bucket. Also, a term used to refer to any type of excavator that digs by pulling the bucket from front to back.

Hydraulic breaker (hammer): A hydraulic attachment for an excavator that is used for breaking boulders and other solid objects.

Inslope (foreslope): The slope of a drainage ditch that is between the shoulder of the road and the flow line of the ditch.

Joystick: A control mechanism that pivots about a fixed point and is used to control the motion of an object. Joystick controls are used in an excavator to control the movement of the boom, stick, bucket, and upper carriage.

Knuckle boom: A term sometimes used for a gooseneck boom and stick combination that resembles a knuckle at the pivot point of the boom and stick.

Platen: A flat plate that is used to support another structure, and that can be rotated around a pivot point. The platen is the supporting base of a turntable mechanism.

Shooting-boom excavator: A term sometimes used to describe a telescoping-boom excavator.

Slope: Ground that forms a natural or artificial incline.

Spoil: Soil, rock, or other material that is removed from a ditch or other excavation.

Stick: The part of the digging component of an excavator between the end of the boom and the bucket. The bucket is attached to the end of the stick.

Undercarriage: The lower frame of an excavator that supports the turntable and has the tracks or wheels attached.

Undercutting: The digging of material from beneath the machine. Also, the digging of material from a bank or vertical face of a trench below the grade.

Upper carriage: The upper frame of an excavator that includes the turntable, engine, cab, controls, and counterweights.

Upper structure: The revolving turntable, swing mechanism, counterweight, boom, and cab of a telescoping-boom excavator.

CAT 336E HYDRAULIC EXCAVATOR LIFTING CAPACITIES CHART

336E L Heavy Duty Reach Boom Lift Capacities – Americas

Load Point Height Load at Maximum Reach Load Radius Over Front Load Radius Over Side

Boom – 6.50 m (21'4") **Counterweight** – 6.0 mt (6.6 t) **Bucket** – None
Stick – R3.9DB (12'10") **Shoes** – 850 mm (34") triple grouser

		1.5 m/5.0 ft		3.0 m/10.0 ft		4.5 m/15.0 ft		6.0 m/20.0 ft		7.5 m/25.0 ft		9.0 m/30.0 ft		Max reach		m / ft
		front	side	front	side	front	side	front	side	front	side	front	side	front	side	
9.0 m / 30.0 ft	kg													*6250	*6250	7.35
	lb													*13,950	*13,950	23.64
7.5 m / 25.0 ft	kg									*7700	*7700			*5800	*5800	8.53
	lb									*17,050	16,550			*12,850	*12,850	27.74
6.0 m / 20.0 ft	kg									*8000	7600	*7500	5600	*5650	5200	9.33
	lb									*17,550	16,300	*14,550	11,900	*12,450	11,550	30.48
4.5 m / 15.0 ft	kg							*9800	*9800	*8750	7350	*8200	5450	*5650	4700	9.84
	lb							*21,200	*21,200	*19,100	15,750	*17,950	11,750	*12,450	10,350	32.22
3.0 m / 10.0 ft	kg					*15 300	14 950	*11 600	9750	*9750	7000	8300	5300	*5850	4400	10.10
	lb					*32,900	32,200	*25,100	21,000	*21,150	15,100	17,850	11,400	*12,850	9,650	33.11
1.5 m / 5.0 ft	kg					*18 450	13 750	*13 300	9150	10 650	6700	8100	5100	*6200	4300	10.12
	lb					*39,800	29,700	*28,800	19,750	22,950	14,400	17,450	11,000	*13,600	9,400	33.22
Ground Line	kg			*8550	*8550	*20 100	13 150	*14 500	8750	10 400	6450	7950	5000	*6750	4350	9.93
	lb			*19,400	*19,400	*43,450	28,250	31,250	18,850	22,300	13,850	17,100	10,700	*14,850	9,500	32.56
−1.5 m / −5.0 ft	kg	*8900	*8900	*13 300	*13 300	*20 350	12 900	14 300	8500	10 200	6250	7850	4900	7300	4600	9.48
	lb	*19,900	*19,900	*30,050	*30,050	*44,100	27,700	30,700	18,350	21,950	13,500	16,950	10,550	16,150	10,050	31.09
−3.0 m / −10.0 ft	kg	*14 100	*14 100	*19 400	*19 400	*19 500	12 900	14 250	8450	10 150	6250			8200	5100	8.76
	lb	*31,550	*31,550	*43,850	*43,850	*42,150	27,750	30,550	18,250	21,900	13,450			18,100	11,300	28.66
−4.5 m / −15.0 ft	kg	*20 200	*20 200	*24 050	*24 050	*17 350	13 100	*13 200	8600	*9900	6400			*9450	6200	7.69
	lb	*45,400	*45,400	*51,900	*51,900	*37,450	28,250	*28,300	18,550	*20,850	13,800			*20,850	13,800	25.01
−6.0 m / −20.0 ft	kg					*13 250	*13 250	*9400	9000					*9250	8900	6.06
	lb					*27,950	*27,950							*20,250	*20,250	19.44

*Indicates that the load is limited by hydraulic lifting capacity rather than tipping load. The above loads are in compliance with hydraulic excavator lift capacity standard ISO 10567:2007. They do not exceed 87% of hydraulic lifting capacity or 75% of tipping load. Weight of all lifting accessories must be deducted from the above lifting capacities. Lifting capacities are based on the machine standing on a firm, uniform supporting surface.

Always refer to the appropriate Operation and Maintenance Manual for specific product information.

REPRINTED COURTESY OF CATERPILLAR INC.

22304-14_A01.EPS

Figure Credits

Deere & Company, Module opener, Figures 1, 3, 4, 9, 10, 13, 19(B), 23, 33, 35, 37, 39

Gradall Industries, Inc., Figures 2, 6–8, 11, 20B(C)

Stanley LaBounty, Figures 12, 20B(B)

Reprinted Courtesy of Caterpillar Inc., Figures 14, 17, 19(A, C–H), 20A(C), 20B(A), 22, 24, 28–31, 38, SA01, SA03, Appendix

Badger Construction Equipment Company, Figure 16

Woods Equipment Company, Figures 20A(A, B)

Topaz Publications, Inc., Figure 20B(D)

Associated General Contractors, Figures 15, 36

Answer	Section Reference	Objective
Section One		
1. b	1.1.1	1a
2. d	1.2.2	1b
3. c	1.3.1	1c
4. b	1.4.3	1d
5. a	1.5.2	1e
Section Two		
1. c	2.1.4	2a
2. d	2.2.0	2b
3. b	2.3.0	2c
Section Three		
1. c	3.1.1	3a
2. a	3.2.0	3b
3. b	3.3.0	3c
4. d	3.4.4	3d

NCCER CURRICULA — USER UPDATE

NCCER makes every effort to keep its textbooks up-to-date and free of technical errors. We appreciate your help in this process. If you find an error, a typographical mistake, or an inaccuracy in NCCER's curricula, please fill out this form (or a photocopy), or complete the online form at **www.nccer.org/olf**. Be sure to include the exact module ID number, page number, a detailed description, and your recommended correction. Your input will be brought to the attention of the Authoring Team. Thank you for your assistance.

Instructors – If you have an idea for improving this textbook, or have found that additional materials were necessary to teach this module effectively, please let us know so that we may present your suggestions to the Authoring Team.

NCCER Product Development and Revision

13614 Progress Blvd., Alachua, FL 32615

Email: curriculum@nccer.org
Online: www.nccer.org/olf

❏ Trainee Guide   ❏ Lesson Plans   ❏ Exam   ❏ PowerPoints   Other _____

Craft / Level: _____ Copyright Date: _____

Module ID Number / Title: _____

Section Number(s): _____

Description: _____

_____

_____

_____

_____

Recommended Correction: _____

_____

_____

_____

_____

Your Name: _____

Address: _____

_____

Email: _____ Phone: _____

22305-14

Motor Graders

OVERVIEW

A motor grader is a rubber-tired, hydraulically operated, single-engine machine used to shape and finish materials. As a heavy equipment operator, you need to understand how a motor grader functions and how to control the blade and other attachments to ensure that the equipment works effectively. You must be aware of hazards associated with motor graders, operate the equipment safely at all times, and know how to conduct daily inspections and preventive maintenance on the machine. You must be able to maneuver the motor grader under various conditions and perform common earthmoving activities.

Module Seven

Trainees with successful module completions may be eligible for credentialing through NCCER's National Registry. To learn more, go to **www.nccer.org** or contact us at **1.888.622.3720**. Our website has information on the latest product releases and training, as well as online versions of our *Cornerstone* magazine and Pearson's product catalog.

Your feedback is welcome. You may email your comments to **curriculum@nccer.org**, send general comments and inquiries to **info@nccer.org**, or fill in the User Update form at the back of this module.

This information is general in nature and intended for training purposes only. Actual performance of activities described in this manual requires compliance with all applicable operating, service, maintenance, and safety procedures under the direction of qualified personnel. References in this manual to patented or proprietary devices do not constitute a recommendation of their use.

Objectives

When you have completed this module, you will be able to do the following:

1. Identify and describe uses and components of a motor grader.
   a. Identify and describe common uses and types of motor graders.
   b. Identify and describe major parts of a motor grader.
   c. Identify and describe motor grader instrumentation.
   d. Identify and describe motor grader controls.
   e. Identify and describe common motor grader attachments.
2. Identify and describe safety, inspection, and service guidelines associated with a motor grader.
   a. Describe guidelines associated with motor grader safety.
   b. Describe prestart inspection procedures.
   c. Describe preventive maintenance requirements.
3. Describe basic startup and operating procedures for a motor grader.
   a. Describe startup, warm-up, and shutdown procedures.
   b. Describe basic maneuvers and operations.
   c. Describe common work activities.

Performance Tasks

Under the supervision of your instructor, you should be able to do the following:

1. Demonstrate a proper prestart inspection and maintenance on a motor grader.
2. Perform proper startup, warm-up, and shutdown procedures.
3. Perform basic maneuvers on a motor grader, including moving forward, moving backward, and turning.
4. Demonstrate basic motor grader operation by:
   - Grading a rough grade by following grade stakes placed along a 300-foot (91-meter) section that is at least double the width of the machine.
   - Demonstrating rotation of the blade for high-bank grading.
   - Cutting a V-ditch with a 3-to-1 slope.

Trade Terms

Articulated-frame motor grader	Pitch
Blade	Rigid-frame motor grader
Blending	Scarifying
Cutting edge	Side-casting
Heel of the blade	Toe of the blade
Moldboard	

Industry Recognized Credentials

If you are training through an NCCER-accredited sponsor, you may be eligible for credentials from NCCER's Registry. The ID number for this module is 22305-14. Note that this module may have been used in other NCCER curricula and may apply to other level completions. Contact NCCER's Registry at 888.622.3720 or go to **www.nccer.org** for more information.

Contents

Topics to be presented in this module include:

1.0.0 Uses and Components of Motor Graders 1
  1.1.0 Common Uses and Types of Motor Graders 1
    1.1.1 Rough Grading 1
    1.1.2 Windrowing and Mixing 1
    1.1.3 Spreading New Material 1
    1.1.4 Finish Grading 2
    1.1.5 Ditch Cutting and Cleaning 3
    1.1.6 Snow Plowing.. 3
    1.1.7 Motor Grader Types and Configurations 4
  1.2.0 Major Parts of a Motor Grader 4
    1.2.1 Operator's Cab 7
  1.3.0 Instrumentation... 8
    1.3.1 Voltmeter... 9
    1.3.2 Fuel Level Gauge 10
    1.3.3 Hydraulic Oil Temperature Gauge 10
    1.3.4 Tachometer 10
  1.4.0 Controls.. 10
    1.4.1 Vehicle Movement Controls 11
    1.4.2 Blade Controls 13
  1.5.0 Special Blades and Attachments 14
    1.5.1 Snow Plow Blades... 14
    1.5.2 Serrated Blades 15
    1.5.3 Straight Blades 16
    1.5.4 Scarifiers 16
    1.5.5 Rippers 16
    1.5.6 Automated Guidance Systems 16
2.0.0 Safety, Inspection, and Maintenance 20
  2.1.0 Safety Guidelines... 20
    2.1.1 Operator Safety 20
    2.1.2 Safety of Co-Workers and the Public 21
    2.1.3 Equipment Safety 22
  2.2.0 Prestart Inspection Procedures 22
  2.3.0 Preventive Maintenance 24
3.0.0 Basic Startup and Operations 27
  3.1.0 Startup, Warm-Up, and Shutdown... 27
    3.1.1 Startup.. 27
    3.1.2 Warm-Up 28
    3.1.3 Shutdown.. 28
  3.2.0 Basic Maneuvers and Operations.. 29
    3.2.1 Moving Forward 30
    3.2.2 Reverse and Backing 30
    3.2.3 Steering and Turning 30
    3.2.4 Operating the Blade 31
    3.2.5 Calibrating the Laser Setup 32

3.3.0 Work Activities..33
   3.3.1 Windrowing 34
   3.3.2 Blending and Spreading Material 35
   3.3.3 Cutting Ditches...35
   3.3.4 Ditching in Wet Areas 37
   3.3.5 Ripping and Scarifying..37
   3.3.6 Grading Unpaved Roads...38
   3.3.7 Grading Around an Object...38
   3.3.8 Grading Slopes...39
   3.3.9 Finish Grading 39
   3.3.10 Snow Plowing...40
   3.3.11 Roading a Motor Grader...41
Appendix Daily Checks and Lubrication....................................48

Figures

Figure 1 Motor grader..2
Figure 2 Rough grading an unpaved road 2
Figure 3 Windrowing 2
Figure 4 Spreading gravel 2
Figure 5 Finish grading..3
Figure 6 Cleaning a ditch 3
Figure 7 Clearing snow with a V-plow blade.......................................4
Figure 8 Rigid-frame motor grader..4
Figure 9 Articulated-frame motor grader 4
Figure 10 Major parts of a typical motor grader...................................6
Figure 11 Pivot point in an articulated-frame grader 6
Figure 12 Articulated-frame grader operating modes 7
Figure 13 Typical blade components..7
Figure 14 Drawbar, circle rail, and blade..8
Figure 15 Common operator cab layouts..8
Figure 16 Instrument panel 9
Figure 17 Lever controls 10
Figure 18 Joystick controls 11
Figure 19 Fingertip armrest controls 11
Figure 20 Transmission and engine speed controls...............................12
Figure 21 Tilt front wheels for better traction 12
Figure 22 Circle rail 13
Figure 23 Hydraulic cylinders lift the circle rail...................................13
Figure 24 Blade control levers 14
Figure 25 Grader with snow blades..14
Figure 26 Snow wing 15
Figure 27 V-plow blade 15
Figure 28 One-way plow blade 15
Figure 29 Serrated blades..15
Figure 30 Straight blade 16

Figures (continued)

Figure 31 Scarifier 16
Figure 32 Ripper 17
Figure 33 Combination scarifier/ripper............................... 17
Figure 34 Laser leveling system 18
Figure 35 GPS guidance system 18
Figure 36 Radio mounted on grader.................................... 18
Figure 37 On-board GPS display 19
Figure 38 Prepare for reflected light in snow conditions............................ 21
Figure 39 Be aware of other workers and equipment on the job site 22
Figure 40 Daily inspection spots for motor graders.................... 23
Figure 41 Check the air filter.. 24
Figure 42 Report maintenance problems to foreman or field mechanic...... 24
Figure 43 Joystick controls for the blade and attachments 31
Figure 44 Blade operating positions................................... 33
Figure 45 Angle of the blade 34
Figure 46 Common blade pitch adjustments...................... 34
Figure 47 Adjust the blade to avoid driving over the windrow.................... 35
Figure 48 Ditch profiles.. 36
Figure 49 Ripper attached to the rear of a motor grader 37
Figure 50 Scarifier attached to the front of a motor grader.................... 38
Figure 51 Grading a rural road.. 38
Figure 52 Finish grade stakes... 40
Figure 53 Roading a motor grader....................................... 41

1.0.0 USES AND COMPONENTS OF MOTOR GRADERS

Objective

Identify and describe uses and components of a motor grader.

   a. Identify and describe common uses and types of motor graders.
   b. Identify and describe major parts of a motor grader.
   c. Identify and describe motor grader instrumentation.
   d. Identify and describe motor grader controls.
   e. Identify and describe common motor grader attachments.

Trade Terms

Articulated-frame motor grader: Frames that have a hinge or joint in the middle and are able to pivot for better traction and handling.

Blade: The blade comprises the moldboard, cutting edge, and end bits.

Blending: The mixing of two or more materials together until they are uniformly distributed throughout the soil mass.

Cutting edge: The sharp steel bar attached to the bottom of the moldboard used for cutting into the ground.

Moldboard: The long concave strip of metal that is used to push the dirt. The cutting edge is fastened to the bottom of the moldboard.

Rigid-frame motor grader: A single metal frame that extends from the front to the back of a motor grader.

Scarifying: To loosen the top surface of material using a set of metal shanks (teeth).

Motor graders come in many different sizes and are manufactured by several companies in the United States and abroad. These machines are easy to recognize; no other piece of heavy equipment resembles a grader. *Figure 1* shows a modern motor grader being used to grade a level surface. There are, however, numerous other uses for motor graders.

1.1.0 Common Uses and Types of Motor Graders

A motor grader's primary purpose is cutting, filling, and moving soil or other material using a **blade** for final shaping and finishing. There are many types, sizes, and manufacturers of motor graders, but the basic features and operations are essentially the same.

There are seven primary operations that a motor grader performs using its blade, including rough grading, windrowing, mixing, spreading new material, finish grading, ditch cutting and clearing, and snow plowing. Other operations such as ripping and **scarifying** can also be performed using attachments.

1.1.1 Rough Grading

Rough grading is the process of taking an existing surface and removing its irregularities, then shaping it into the desired surface at a specific elevation. This requires cutting, pushing, and spreading the material. *Figure 2* shows a motor grader doing rough grading on a rural unpaved road.

Motor graders are used by most state highway departments and county public works agencies for building and maintaining shoulders and unpaved roads. Private contractors use motor graders for grading roads, building pads, and landscaping on industrial, commercial, and residential projects.

1.1.2 Windrowing and Mixing

A motor grader is capable of making windrows of material for either removal or mixing (*Figure 3*). The creation of a windrow allows other equipment, such as scrapers or front-end loaders, to pick up the material without having to do all the preparation work.

Motor graders can also build windrows for the purpose of mixing or **blending** two or more materials together. This is done on the ground very close to the final location. Large quantities of material can be mixed effectively and efficiently using this method.

1.1.3 Spreading New Material

When new material has been dumped or piled in a general location, a motor grader can be used to spread that material to create a smooth surface. As in the grading operation, the material will be spread over an area to a specified depth. *Figure 4* shows a motor grader spreading gravel that has been delivered to a commercial site.

Figure 1 Motor grader.

Figure 2 Rough grading an unpaved road.

Figure 3 Windrowing.

Motor graders are used by many highway and public works agencies for maintaining unpaved roads by smoothing and placing new material on the surface. Grading contractors use motor graders to spread new material to build up low areas. New materials are also spread to raise pads and subgrades to final elevations.

1.1.4 Finish Grading

Finish grading is one of the most important uses of the motor grader. The purpose of this activity is to finish the surface to the final elevation or slope required by the specifications. Motor graders are very good machines to use for finish grading if there are large enough areas for the equipment to maneuver properly. If the operator can precisely control the blade height and position, the required

Figure 4 Spreading gravel.

final elevation of a surface may be achieved with great accuracy.

Figure 5 shows a motor grader trimming a base course for a driveway. Note how the end of the blade is up against the curb at the lower left side of the picture.

Computerized equipment is now used with graders for finish grading. Electronic equipment is attached to masts at each end of the **moldboard**. These devices receive signals from laser levels or Global Positioning System (GPS) satellites. This information is used to control blade elevation for precise grading. Additional detail about automated guidance systems is provided later in this module.

1.1.5 Ditch Cutting and Cleaning

Another job of the motor grader is ditch cutting and cleaning. Ditches are normally designed as long, narrow excavations used to drain water away from a roadway, parking area, or other structures. The grader's blade can be extended out to the side of the machine. The end of the blade can be used to cut a V-shape in the ground, or used to create a ditch line parallel to other structures in long distances.

The same procedure can be used to clean existing ditches that have been clogged with silt or other eroded material from the sides of the ditch. *Figure 6* shows a grader cleaning a ditch on the side of a road. Note how the front wheels can be tilted to increase the stability of the machine.

REPRINTED COURTESY OF CATERPILLAR INC.

22305-14_F06.EPS

Figure 6 Cleaning a ditch.

1.1.6 Snow Plowing

Many public works agencies use motor graders to clear snow from rural roads and local streets. This can be done with the blade if the snowfall is not very deep. In areas where large amounts of snow accumulate, special snow blade and snow blower attachments can be mounted on the front of the motor grader. *Figure 7* shows a motor grader with a front V-plow blade clearing snow. Chains are commonly fitted onto the grader's tires for better traction. Motor graders equipped with an appropriate snow plow blade can be used on drifts up to five feet high.

22305-14_F05.EPS

Figure 5 Finish grading.

Figure 7 Clearing snow with a V-plow blade.

1.1.7 Motor Grader Types and Configurations

The basic design of the motor grader has not changed significantly over the past 50 years. Early models did not have their own power and had to be pulled by another piece of equipment. These machines used mechanical wheels and levers to adjust the position of the blade. Today, motor graders are equipped with 6- to 12-cylinder diesel engines that drive the wheels and provide power for the hydraulic control system.

Historically, there have been two basic types of motor graders: rigid-frame motor graders and articulated-frame motor graders. A **rigid-frame motor grader** (*Figure 8*) has a single metal frame that extends from the front to the back of the motor grader. Rigid-frame motor graders are, for the most part, smaller than articulated-frame graders.

Most modern graders are articulated-frame graders. An **articulated-frame motor grader** (*Fig-*

ure 9) has a frame that is hinged at the center. The hinge point is either in front of the cab or behind it, depending on the manufacturer's design. This allows the front of the motor grader to work at an offset angle to the back tandem wheels for such work as cleaning out ditches. It also allows for tighter turning and greater maneuverability.

1.2.0 Major Parts of a Motor Grader

A motor grader can have many individual parts, but most of them can be grouped into four major categories:

- The frame
- The power unit, including the engine, transmission, and differential
- The steering and other controls
- The circle rail, drawbar, and blade (moldboard)

Both rigid-frame and articulated-frame motor graders have these same basic components. In addition, graders are designed so that various attachments can be mounted at the rear, the front, or underneath the frame behind the front wheels.

The major parts of the motor grader shown in *Figure 10* include the frame; the drawbar and circle rail; the blade, or moldboard; the lift cylinders; the operator's cab; and the engine. This particular motor grader also has a ripper attachment that is mounted onto the rear of the grader and a scarifier that is mounted between the front wheels and the blade.

The main difference between the frame of a rigid-frame motor grader and an articulated-frame grader is the hinge or pivot point in the frame (*Figure 11*). This point is located either in front of or behind the operator's cab, and it enables an articulated-frame grader to operate in different configurations, or modes.

REPRINTED COURTESY OF CATERPILLAR INC.

22305-14_F08.EPS

Figure 8 Rigid-frame motor grader.

22305-14_F09.EPS

Figure 9 Articulated-frame motor grader.

Big and Bigger

It is hard to think of a machine that weighs over 15,000 pounds (6,803 kilograms) and has a 100-horsepower engine as being small, but in the world of motor graders that can be the case. Graders can range a great deal in size and power. As you would expect, smaller graders tend to be used for jobs that require less brute power or have less space in which the grader can maneuver. Larger graders are more likely to be found in harsh environments that demand large grading capabilities, such as mining applications. The table provides some approximate specifications for various sizes of motor graders.

REPRINTED COURTESY OF CATERPILLAR INC.

22305-14_SA01A.EPS

	Small	Medium	Large
Grader Weight	8,000 lbs (3,629 kg)	38,000 lbs (17,237 kg)	137,000 lbs (62,142 kg)
Engine Power	50 hp	125 hp	500 hp
Blade Length	8 ft (244 cm)	12 ft (366 cm)	24 ft (732 cm)
Lift Above Ground	11 in (28 cm)	24 in (61 cm)	42 in (107 cm)
Fuel Tank Capacity	15 gal (57 L)	75 gal (284 L)	300 gal (1,136 L)

22305-13_SA01B.EPS

Figure 10 Major parts of a typical motor grader.

Figure 12 shows the top view of an articulated motor grader with the pivot point between the operator's cab and the engine. Because of this pivot point, the grader can operate in a straight mode, an articulated mode, and an offset, or crab mode.

The articulation is controlled by hydraulic cylinders on either side of the frame. Operating controls in the cab actuate the cylinders, causing the front part of the frame to rotate sideways. This process allows the blade to be offset to either side of the motor grader, which provides more maneuverability for cutting ditches and finishing slopes.

All motor graders are designed with the motor in the rear of the machine, over the drive wheels. This arrangement provides extra weight over the wheels for better traction. The operator's cab is forward of the engine and behind the position of the blade. This elevated position allows for clear visibility in all directions and gives the operator a clear view of the front wheels and the ends of the

22305-14_F11.EPS

Figure 11 Pivot point in an articulated-frame grader.

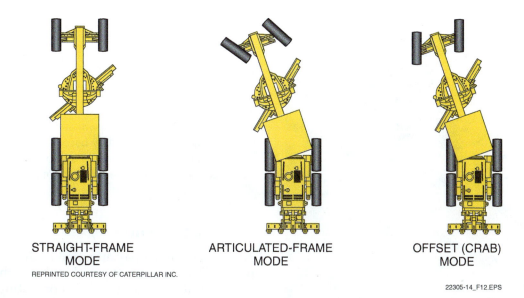

STRAIGHT-FRAME MODE   ARTICULATED-FRAME MODE   OFFSET (CRAB) MODE

REPRINTED COURTESY OF CATERPILLAR INC.

22305-14_F12.EPS

Figure 12 Articulated-frame grader operating modes.

blade. The steering, blade, and attachment controls are all located within the operator's cab.

The terms *blade* and *moldboard* are often used interchangeably to describe the part of the motor grader that cuts and moves material. Actually, the blade is made up of several different components, which include the moldboard, a **cutting edge**, and end bits (see *Figure 13*). For the purposes of this training module, the more generic term *blade* will be used when describing the assembly. The term *moldboard* will be used when describing the steel component onto which the cutting edge and end bits are attached.

The moldboard is a curved piece of steel with a height between 24 and 31 inches (61 to 79 centimeters). The moldboard is typically 12, 14, or 16 feet (3.66, 4.27, or 4.88 meters) long. On some small machines, it is 10 feet (3.05 meters) long, and on the largest machines it is 24 feet (7.32 meters) long. As the grader moves forward, material is lifted by the cutting edge at the bottom of the moldboard and falls forward at the top in a rotary motion.

The cutting edge is bolted to the bottom of the moldboard and provides the primary digging action. Since it can wear quickly in hard soils, the cutting edge can be reversed or replaced when it becomes dull or worn. End bits are attached on either end and protect the moldboard from wear and damage.

The blade is attached to the grader by the drawbar and the circle rail (*Figure 14*). The drawbar and circle rail are located underneath the frame between the operator's cab and the front wheels. They are operated by a series of hydraulic cylinders that allow the operator to maneuver the blade horizontally, vertically, and at an angle to the frame.

1.2.1 Operator's Cab

The grader is operated from the operator's cab. The cab contains the controls for vehicle movement, blade operation, and auxiliary functions. The layout and controls of an operator's cab can vary between different makes and models of graders. For example, *Figure 15* shows two common operator cab layouts. The left cab (A) has a steering wheel and lever controls. The right cab (B) has joystick controls. An operator must study the operator's manual to become familiar with the controls, instruments, and their functions before operating a grader.

The cab is designed for maximum visibility. Most cabs have an open design or have large wind-

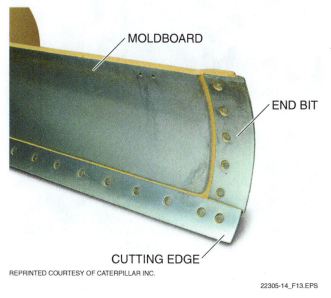

MOLDBOARD

END BIT

CUTTING EDGE

REPRINTED COURTESY OF CATERPILLAR INC.

22305-14_F13.EPS

Figure 13 Typical blade components.

CIRCLE RAIL

DRAWBAR   BLADE

REPRINTED COURTESY OF CATERPILLAR INC.

22305-14_F14.EPS

Figure 14 Drawbar, circle rail, and blade.

shields that extend around the front of the cab. This allows the operator to see the area to the front and sides of the machine. Wipers help keep the windshield clean. Lights aid vision at night or in poor lighting conditions, such as in excavations.

1.3.0 Instrumentation

An operator must pay attention to the instrument panel. The instrument panel includes the gauges that indicate engine speed and operating temperatures. There are also warning lights and indicators that must be monitored for safe operations. Serious damage to the equipment can result if the instrument panel is not monitored closely.

The instrument panel varies on different makes and models of graders. Generally, they include the instruments and indicators described in the sections that follow. One type of instrument panel is shown in *Figure 16*. It contains a quad-gauge cluster, alert indicators, and gauges. Some machines may include additional instruments and displays. Read the operator's manual to be familiar with model-specific instruments.

(A) STEERING WHEEL AND LEVER CONTROLS

REPRINTED COURTESY OF CATERPILLAR INC.

(B) JOYSTICK CONTROLS

22305-14_F15.EPS

Figure 15 Common operator cab layouts.

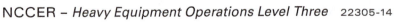

The grader has sensors in various parts of the engine and its related systems. These sensors are connected to a monitoring system. The monitoring system display has alert indicators that light when the machine is not functioning properly. If the situation needs immediate attention, the indicators may flash, remain lit, and/or an audible alarm will sound. The number and arrangement of indicator lights will vary from one machine to another. Always read the operator's manual to understand all warning lights before operating the equipment.

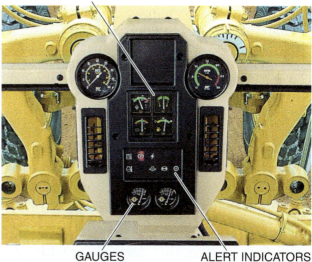

QUAD-GAUGE CLUSTER

GAUGES   ALERT INDICATORS

REPRINTED COURTESY OF CATERPILLAR INC.

22305-14_F16.EPS

Figure 16 Instrument panel.

<table>
<tr><td>CAUTION</td><td>A flashing indicator light may require immediate action. If you need to stop and look up the meaning of an indicator after it is flashing, you may cause serious damage to the equipment. Know the equipment before you begin operating it.</td></tr>
</table>

Most of these instruments and indicators are similar to those in other machines. All motor graders also have indicator lights and an alert system similar to tractors and loaders.

The gauges on the instrument panel in *Figure 16* include a voltmeter, a fuel level gauge, a hydraulic oil temperature gauge, and a tachometer. These instruments are explained in the following sections.

1.3.1 Voltmeter

The voltmeter shows the voltage in the electrical system. A needle shows the voltage over a scale, with low on the left and high on the right. There is a red warning range on both sides, because a voltage that is too high or too low can cause operational problems. This instrument is optional in some models. If the voltmeter is in the warning range, shut the machine down and have it checked by a mechanic.

Seeing and Feeling

While seat and mirror adjustments are not directly involved in grader operation, correct position of these items can aid in safe operation. You should adjust the seat and fasten the seat belt before operating the grader. Position the seat so you can reach all of the controls easily. Your legs must be able to fully depress the foot pedals when your back is flat against the back of the seat. This will permit maximum force to be used on the brake pedals. Adjust the mirrors so you are able to see clearly in all directions.

REPRINTED COURTESY OF CATERPILLAR INC.

22305-14_SA02.EPS

1.3.2 Fuel Level Gauge

The fuel level gauge indicates the amount of fuel in the tank. On diesel engines, the gauge may contain a low fuel warning zone. Some models have a low fuel warning light. Avoid running out of fuel on diesel engines. If this happens, fuel lines and injectors must be bled of air before the engine can be restarted.

1.3.3 Hydraulic Oil Temperature Gauge

The hydraulic oil temperature gauge indicates the temperature of the oil flowing through the hydraulic system. This gauge reads left to right in increasing temperature. It has a red zone that indicates excessive temperature. If the temperature is in the red zone, shut down the machine and have it checked by a mechanic.

1.3.4 Tachometer

The tachometer indicates the engine speed in revolutions per minute (rpm). Most tachometers are marked in hundreds on the meter face and read left to right in an increasing scale. Again, there is a red zone on the high end of the scale that indicates that the engine is overspeeding. On some

graders, there is also an indicator light that will warn the operator if the engine speed is too high.

1.4.0 Controls

While all motor graders operate in basically the same way, the specific type and location of a grader's controls can vary from one manufacturer to another. For that reason, operators should always review the operator's manual for the equipment being used to become familiar with the layout and function of all the controls and instruments.

One common control arrangement is shown in *Figure 17*. It consists of a steering wheel, levers, foot pedals, and switches. The levers that operate the moldboard, circle rail, articulation, and wheel lean are located in front, and to either side of, the steering pedestal. The levers that control the engine speed and transmission are located on the right side of the operator's seat. Switches and gauges are located on the steering pedestal. Foot pedals are used for acceleration, braking, and inching. Other controls, such as the ones used for the moldboard and wheel positioning, will vary depending on the specific manufacturer.

Another common control arrangement is shown in *Figure 18*. In this case, vehicle movement and blade positioning are controlled using two

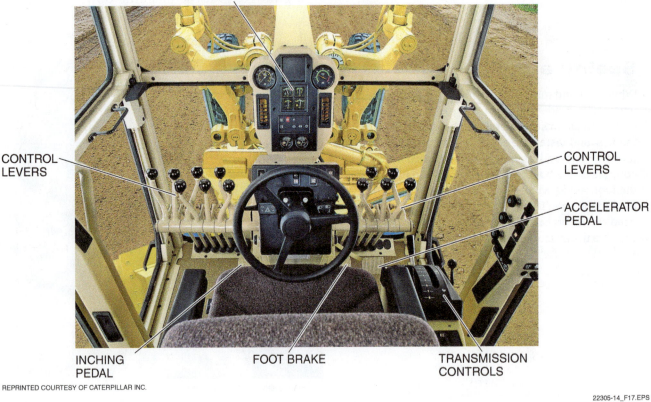

22305-14_F17.EPS

Figure 17 Lever controls.

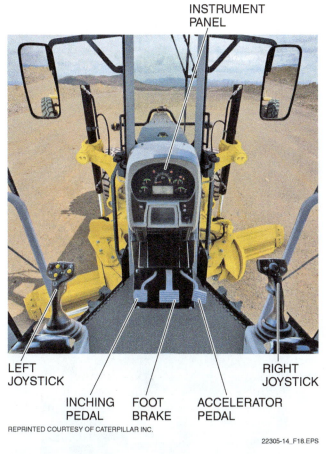

LEFT JOYSTICK

RIGHT JOYSTICK

INCHING PEDAL FOOT BRAKE ACCELERATOR PEDAL

REPRINTED COURTESY OF CATERPILLAR INC.

22305-14_F18.EPS

Figure 18 Joystick controls.

joysticks, one on each side of the operator's seat. Switches and gauge displays are located directly in front of the operator. Pedals are used for accelerating, braking, and inching.

Some motor graders use fingertip armrest controls, such as those shown in *Figure 19*. This particular system is used on a John Deere Grade Pro (GP) model. This control mechanism has knob-integrated push buttons for vehicle movement and blade control. No additional levers are needed with this control arrangement. There is also a steering wheel and pedals for accelerating, braking, and inching.

In addition to these control arrangements, many motor graders are equipped with automated guidance systems. These systems can either display information that an operator can use for blade positioning, or the system can control the blade positioning automatically. Automated guidance systems will be described later in this module.

1.4.1 Vehicle Movement Controls

On many graders, a steering wheel is used to control the directional movement of the machine. The accelerator, brake, and inching pedals are used

22305-14_F19.EPS

Figure 19 Fingertip armrest controls.

to control the vehicle's speed. Most graders also have some type of hand throttle to control engine speed for grader operations. The transmission controls are used to select the gear and direction of movement.

On graders with joystick controls, directional movement is often controlled using the left joystick, while blade movement is controlled using the right joystick.

The functions of the controls are as follows:

- *Accelerator pedal* – Increases the engine's rpm and the speed of travel. When the accelerator is released, the engine's speed will return to the hand-throttle setting.
- *Articulation control*– Moves the rear portion of the motor grader from the center to the right or to the left of center.
- *Brake pedal* – Stops or slows the machine. Some models have right and left pedals that assist in turning.
- *Inching pedal* – Disengages the clutch gradually to provide precise control.

- *Differential lock* – Used to improve pulling power and traction. When the differential lock is engaged, the right and left side wheels are basically locked together so that they spin at the same speed, as if they are on a single, solid axle. When disengaged, the wheels can rotate independently, at different speeds. The differential lock should be disengaged before turning the grader. Many newer model graders have an automatic differential lock, which unlocks the differential during turns and re-locks the differential when the grader is traveling straight.
- *Engine-speed control (hand throttle)* – Governs the speed of the engine.
- *Forward/reverse control lever* – Moves the motor grader forward or backward. The midpoint of the lever's travel is neutral.
- *Transmission control* – Selects the gear range. On some models a lever is combined with a forward/reverse control lever as shown in *Figure 20A*. On models that use joysticks, the transmission control is often grouped with the steering and speed controls on the left joystick, as shown in *Figure 20B*.
- *Parking brake lever* – Secures the motor grader when it is stopped or parked.

The accelerator and hand throttle control the engine speed. The hand throttle is pushed forward or the accelerator is depressed to increase engine speed. The hand throttle is pulled toward the rear or the accelerator is released to decrease engine speed. The hand throttle may have a detent position. This will keep the machine in low idle for warm-up and shutdown. Depressing the inching pedal gradually disengages the clutch to provide precise control of the grader's movement. This feature is particularly helpful during finish grading or close quarter work.

The service brakes are used to slow and stop the machine. Depress the pedal to apply the service brakes; release the pedal to allow the machine to move. Use the service brakes on a downhill slope to prevent the engine from overspeeding.

On most graders, the front wheels can be tilted as much as 20 degrees. The wheels are attached to a lean bar which is moved hydraulically from side to side causing the wheels to lean. This increases the resistance to sliding sideways due to blade load or steering stresses. Note in *Figure 21* how the grader's front wheels are angled for better traction.

TRANSMISSION CONTROL

ENGINE SPEED CONTROL

REPRINTED COURTESY OF CATERPILLAR INC.

22305-14_F20.EPS

Figure 20 Transmission and engine speed controls.

22305-14_F21.EPS

Figure 21 Tilt front wheels for better traction.

1.4.2 Blade Controls

The primary operating component of the grader is the blade. It can be moved in a wide range of directions. The blade can be raised and lowered or angled to create windrows. It can be extended to the side and tilted downwards for ditching, or upwards for embankments.

The blade is mounted on a circle rail. The circle rail is a toothed ring as shown in *Figure 22*. This ring allows the blade to be positioned precisely and remain in place under load. The blade can be rotated 360 degrees in both directions. However, the blade must be positioned so that it will not hit the ground or the machine as it is turned.

The circle rail is lifted, shifted, and tilted by hydraulic cylinders. The blade can be raised up and down to control the depth of cut. Rotation of the blade controls the angle of the cut. The hydraulics are commonly controlled with levers. Two of the hydraulic cylinders used to move the circle rail and blade are shown in *Figure 23*. Other configurations include joystick and armrest mounted controls.

Lever controls are typically arranged in two sets, one at each side of the steering wheel. The handles are offset as shown in *Figure 24*. This makes operating one or more levers with one hand easier. All of the levers have a central hold position and can be moved forward and back to actuate the hydraulics.

REPRINTED COURTESY OF CATERPILLAR INC.

22305-14_F22.EPS

Figure 22 Circle rail.

Each lever is marked with its function. However, the markings and lever position are not standardized. Always review the operator's manual for the machine being operated to understand the function of each control. Do not assume that the levers are in the same position on different machines. Although locations may vary, all graders have the following controls:

- *Blade shift lever* – Moves the blade to the right or left.
- *Blade tilt lever* – Moves the blade rearward or forward.

HYDRAULIC LIFT CYLINDERS

22305-14_F23.EPS

Figure 23 Hydraulic cylinders lift the circle rail.

REPRINTED COURTESY OF CATERPILLAR INC.

22305-14_F24.EPS

Figure 24 Blade control levers.

- *Circle rail shift lever* – Moves the circle rail to the right or left.
- *Circle rail rotation lever* – Moves the circle rail clockwise or counterclockwise.
- *Right and left blade lift levers* – These levers raise and lower the right or left end of the blade. They may be operated one at a time to raise one end of the blade, or both levers may be used together to raise and lower the entire blade.
- *Scarifier/accessory control lever* – Raises or lowers the scarifier or other accessory.
- *Wheel-lean lever* – Tilts front wheels to the right or left.

1.5.0 Special Blades and Attachments

There are several different blades and attachments that can be used with a motor grader. Some attachments, such as snow blades and scarifiers, are mounted on the front of the grader. Other attachments, such as rippers, are mounted onto the rear of the grader. Some attachments can remain attached to the grader and be stowed away while the grader is doing other work. Others must be removed when the work is completed.

1.5.1 Snow Plow Blades

The grader is frequently used to clear snow. Snow plow blades can be fitted to the front or the side of the grader, depending on the model (*Figure 25*). Snow plow blades come in many different sizes and are supplied by several manufacturers. Always check the operator's manual for proper fitting and operation of these attachments.

Snow blades fitted on the side of the grader are often called snow wings. The outboard portion of the blade is suspended by a chain or hydraulic arm attached to a side mast (*Figure 26*). It can be extended hydraulically or manually, depending on the model.

There are also several specialty snow plow blades, including the V-plow blade and the one-way plow blade. The V-plow blade or scraper blade (*Figure 27*) is mounted directly to the front of the grader's frame. It is used to break up deep snow and large snowdrifts. The two sides of the

REPRINTED COURTESY OF CATERPILLAR INC.

22305-14_F25.EPS

Figure 25 Grader with snow blades.

REPRINTED COURTESY OF CATERPILLAR INC.

22305-14_F26.EPS

Figure 26 Snow wing.

large blade are angled toward the back in a V-shape, giving the blade its name.

The one-way plow blade (*Figure 28*) can be fitted on the front or side of the grader. It is used to remove snow and cast it away from the roadway. The blade is angled to one side. The taller side is open and the shorter side has an end cap. This design is more efficient for moving the snow off of the roadway.

1.5.2 Serrated Blades

Some models of motor graders can be fitted with a serrated blade, which is used for placing and smoothing special material. The cutting edge of the blade has serrations that allow small aggregate to pass through the blade instead of being cast to the side. This way, the blade works like a large rake.

22305-14_F27.EPS

Figure 27 V-plow blade.

REPRINTED COURTESY OF CATERPILLAR INC.

22305-14_F28.EPS

Figure 28 One-way plow blade.

Figure 29 shows the two types of serrated blades. The serrated cutting edge is bolted to the bottom of the moldboard. Some serrated blades are made of hard tungsten carbide to withstand

REPRINTED COURTESY OF CATERPILLAR INC.

22305-14_F29.EPS

Figure 29 Serrated blades.

tough conditions, like maintaining mine roads. These blades allow fine matter to pass while larger aggregate is carried forward. The smaller the gaps between the serrations, the smoother the final surface.

1.5.3 Straight Blades

A straight blade is shown in *Figure 30*. It can be attached to the front of a grader equipped with the appropriate front mounting connectors. Straight blades are used for light dozing, backfilling, and spreading gravel or fill material.

1.5.4 Scarifiers

A scarifier is a set of thin teeth mounted on shanks. It is used to scrape through soft soil and loosen it for mixing or scraping. The teeth can be in a straight line or mounted in a V-formation. The teeth on a scarifier tend to be smaller than those on a ripper. There are five or more teeth on a scarifier. *Figure 31* shows a five-shank scarifier. The teeth are attached to a metal bar mounted on the front of the frame. The scarifier can be lowered to different depths, depending on the soil conditions and grading needs.

Scarifiers are attached to the motor grader at the front of the machine or behind the front wheels. The placement depends on the manufacturer and

REPRINTED COURTESY OF CATERPILLAR INC.

22305-14_F30.EPS

Figure 30 Straight blade.

22305-14_F31.EPS

Figure 31 Scarifier.

model. They are attached to the underside of the frame and stowed away until needed.

1.5.5 Rippers

Rippers are also a set of teeth mounted on shanks. Unlike scarifiers, rippers are used to break up hard soil. Ripper teeth are larger, which makes them more versatile than scarifiers. They are also used for ripping up asphalt pavement. Rippers are usually mounted on the back of the machine, where the most ripping power can be applied to the shanks.

Ripper attachments come in different configurations. The usual arrangement is to have a ripper with three to seven teeth that can be raised and lowered in groups. The operator uses all the teeth for regular ripping. When working with harder ground surfaces, fewer teeth are used. In severe conditions, only the middle tooth is used. *Figure 32* shows a five-tooth ripper mounted on the back of a grader.

Figure 33 shows a combination scarifier/ripper. The ripper teeth are raised when the scarifier is in use. Note the difference in the size of the teeth. The teeth on a ripper are replaceable; scarifier teeth generally are not.

1.5.6 Automated Guidance Systems

Some graders are equipped with automated guidance systems, which are particularly useful for finish grading. One type of automated guidance

Figure 32 Ripper.

Figure 33 Combination scarifier/ripper.

system is a laser leveling system *(Figure 34)*. In this system, a laser level that is positioned at the work site sends high-intensity light beams to a receiver unit mounted on the grader's moldboard. The receiver transmits information to an onboard computer, which automatically positions the moldboard to maintain the specified grade.

> **NOTE**
>
> The laser guidance system will not work if the line of sight to the transmitter is blocked.

> **WARNING!**
>
> Lasers must be operated by a trained and authorized person. Inform all people working within the operating range of the instrument that a laser will be in use. Never look directly into a laser beam or point the beam into the eyes of others. Set the laser transmitter so that the height of the emitted beam is above or below normal eye level.

A more common type of automated guidance system is a Global Positioning System (GPS). As *Figure 35* shows, this type of guidance system uses space-based satellites, a base station, receivers, radios, and a computer onboard the grader.

In the GPS system, satellites transmit radio signals that are picked up by receivers at a base station near the job site and on the motor grader. The base station makes corrections to the satellite signals to determine its exact position. The base station then sends corrected data by radio to the grader *(Figure 36)*.

An on-board computer uses the satellite signals and the corrected signals from the base station to determine how to position the blade to meet the grade specifications in the site plan *(Figure 37)*.

Depending on the system, controls in the operator's cab either maneuver the blade automatically or the system displays information that an operator can use to move the blade.

LASER LEVEL   RECEIVER

REPRINTED COURTESY OF CATERPILLAR INC.

22305-14_F34.EPS

Figure 34 Laser leveling system.

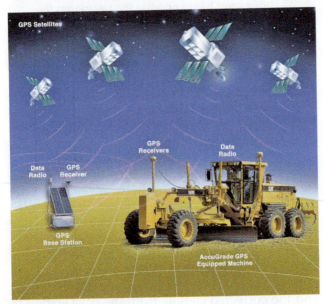

REPRINTED COURTESY OF CATERPILLAR INC.

22305-14_F35.EPS

Figure 35 GPS guidance system.

22305-14_F36.EPS

Figure 36 Radio mounted on grader.

22305-14_F37.EPS

Figure 37 On-board GPS display.

Additional Resources

Basic Equipment Operator, NAVEDTRA 14081, 1994 Edition. Morris, John T. (preparer), Naval Education and Training Professional Development and Technology Center.

Operator's Manual Grader, Heavy Road, Motorized Caterpillar Model 130G, (NSN 3805-01-150-4795), TM 5-3805-261-10, 1992. Washington, DC: Department of the Army.

1.0.0 Section Review

1. The process of using a motor grader to take an existing surface and remove its irregularities, then shape it into the desired surface at a specific elevation is known as _____.

   a. windrowing
   b. finish grading
   c. rough grading
   d. clearing

2. All motor graders are designed with the motor situated _____.

   a. in the rear of the machine
   b. directly below the operator's cab
   c. over the front wheels
   d. in front of the rear wheels

3. A typical motor grader has sensors in various parts of the engine and its related systems that are connected to a _____.

   a. controller pedestal
   b. laser-guidance system
   c. circle rail rotator
   d. monitoring system

4. When the differential lock on a motor grader is engaged, the speed of the right and left wheels will be _____.

   a. identical to the engine's rpm
   b. equal
   c. different
   d. proportional to the wheel tilt

5. An attachment that is typically mounted onto the front of a grader and used to loosen soft soil for mixing or scraping is called a(n) _____.

   a. scarifier
   b. V-plow blade
   c. serrator
   d. ripper

SECTION TWO

2.0.0 SAFETY, INSPECTION, AND MAINTENANCE

Objective

Identify and describe safety, inspection, and service guidelines associated with a motor grader.
  a. Describe guidelines associated with motor grader safety.
  b. Describe prestart inspection procedures.
  c. Describe preventive maintenance requirements.

Performance Task 1

Demonstrate a proper prestart inspection and maintenance on a motor grader.

Most companies have policies and guidelines about safety, pre-startup procedures, and machine maintenance. It is important to be familiar with those policies and always follow them. This section covers many of the basic guidelines and procedures that typically apply to motor graders.

2.1.0 Safety Guidelines

Safety can be divided into three areas: safety of the operator, safety of others, and safety of the equipment. You, as the operator, are responsible for performing the work safely, protecting the public and co-workers from harm, and protecting the equipment from damage. Knowing the equipment and concentrating on safety are the easiest ways to keep a job site safe.

> **WARNING!**
>
> Motor graders are complex and powerful machines that require complete attention for proper control during operation.

2.1.1 Operator Safety

Nobody wants to have an accident or be hurt. There are a number of things you can do to protect yourself and those around you from getting hurt on the job. Mainly, be alert and avoid accidents.

Know and follow your employer's safety rules. Your employer or supervisor will provide the requirements for proper dress and safety equipment. The following are recommended safety procedures for all situations:

- Clean steps, grab irons, and the operator's compartment.
- Mount and dismount the equipment carefully using three-point contact and facing the machine.
- Wear PPE as required by company and site policy when operating the equipment.
- Wear the seat belt.
- Do not wear loose clothing or jewelry that could catch on controls or moving parts.
- Keep the windshield, windows, and mirrors clean at all times.
- Never operate equipment under the influence of alcohol or drugs.
- Never smoke while refueling or checking batteries or fluid.
- Never remove protective guards or panels while in operation. Replace guards that break or become damaged.
- Do not use hands to check for fluid leaks. Hydraulic and cooling systems operate at high pressure, and fluids under high pressure can cause serious injury. Use cardboard or some other solid material to check for hydraulic and cooling system leaks.
- Always lower the moldboard to the ground before performing any service, or when leaving the motor grader unattended.
- Use extreme caution when operating under low overhead obstructions, or when grading where there may be underground wires and pipes.

> **WARNING!**
>
> Getting in and out of equipment can be dangerous. Always face the machine and maintain three points of contact when you are mounting and dismounting. That means you should have three out of four of your hands and feet on the equipment. That can be two hands and one foot or one hand and two feet.

Since graders are often used to clear snow from roadways or other areas, additional precautions may be needed to protect yourself in cold, snowy weather. First, always dress suitably. Wear protective clothing, especially warm gloves, hats, and socks because fingers, ears, and toes are the first extremities to freeze. Wear moisture-repellant clothing and dress in layers. Layers of lightweight clothing will provide more warmth than a single heavyweight garment.

Lights are intensified when reflected from snow and can cause eye fatigue or temporary blindness (*Figure 38*). Wear good quality sunglasses to protect your eyes from the strong light. When sunglasses are not needed, remove them to maximize your vision.

2.1.2 Safety of Co-Workers and the Public

You are responsible for your own safety, as well as the safety of other individuals who may be working nearby (*Figure 39*). When working in areas that are very close to pedestrians or motor vehicles, be aware of people around the machine and avoid any potential accidents.

REPRINTED COURTESY OF CATERPILLAR INC.

22305-14_F38.EPS

Figure 38 Prepare for reflected light in snow conditions.

Keeping Warm

Any time you are working in frigid temperatures, you need to be aware of the symptoms of exposure and frostbite before it is too late. Prolonged exposure to cold weather can cause numbness in the extremities, as well as drowsiness and confusion. Watch for these signs of hypothermia:

- Uncontrollable shivering
- Weakness, loss of coordination
- Confusion, lack of judgment
- Pale and cold skin
- Extreme drowsiness
- Slowed breathing or heart rate

Frostbite typically begins in the extremities, the ears, nose, hands, and feet. Watch for progressive numbness and discoloration of the skin. To avoid problems in cold weather, take breaks in a warm area when necessary. If you cannot get to a warm area, exercise such as jogging in place will help to warm you. Keep warm liquids handy, but avoid drinking alcohol.

REPRINTED COURTESY OF CATERPILLAR INC.

22305-14_SA03.EPS

The main safety points when working around other people include the following:

- Walk around the equipment to make sure that everyone is clear of the equipment before starting and moving it.
- Before beginning work in a new area, take a walk to locate any cliffs, steep banks, holes, power or gas lines, or other obstacles that may be hazardous.
- Maintain a clear view in all directions. Do not carry any equipment or materials that obstruct the view.
- Always look before changing directions.
- Never allow riders on the motor grader.
- If you are working in traffic, find out what warning devices are required. Make sure you know the rules and the meaning of all flags, hand signals, signs, and markers used.
- Use a spotter or flagger when working close to traffic.

2.1.3 Equipment Safety

The motor grader has been designed with certain safety features to protect the operator, as well as the equipment. For example, it has guards, canopies, shields, roll-over protection, and seat belts. Know the equipment's safety devices and be sure they are in working order.

Use the following guidelines to keep the equipment in good working order:

- Perform pre-start inspection and lubrication daily.
- Test the service and parking brakes before operating the equipment.
- Look and listen to make sure the equipment is functioning normally. Stop if it is malfunctioning. Correct or report trouble immediately.
- Check hinge pins of all components and their securing devices often.
- Always travel with the circle rail and moldboard raised to allow for sufficient clearance between the ground and the blade.
- Never exceed the manufacturer's limits for speed or for operating on inclines.
- Never park on an incline.
- Always turn the engine off and secure the controls before leaving the machine. Lower all operating surfaces to the ground.
- When loading, transporting, and unloading equipment, know and follow the manufacturer's recommendations for loading, unloading, and tie-down.

REPRINTED COURTESY OF CATERPILLAR INC.

22305-14_F39.EPS

Figure 39 Be aware of other workers and equipment on the job site.

- Make sure clearance flags, lights, and other required warnings are on the equipment when transporting or roading it.

The basic rule is know the equipment. Learn the purpose and uses of all gauges and controls, as well as the equipment's limitations. Never operate the machine if it is not in good working order. Some basic safety rules of operation include the following:

- Do not operate the grader from any position other than the operator's seat.
- Do not coast. Neutral is for standing only.
- Maintain control when going downhill. Do not shift gears.
- Whenever possible, avoid obstacles such as rocks, fallen trees, curbs, or ditches.
- If you must cross over an obstacle, reduce speed and approach at an angle to reduce the impact on the equipment and yourself.

2.2.0 Prestart Inspection Procedures

The first thing done each day before beginning work is conducting the daily inspection using a checklist. This should be done before starting the engine. The inspection will identify any potential problems that could cause a breakdown and will indicate if the machine can be operated. The equipment should be inspected before, during, and after operation.

The daily inspection is often called a walkaround. The operator should walk completely around the machine checking various items. This general inspection of the equipment should include checking for the following:

> **NOTE**
> If a leak is observed, locate the source of the leak and fix it. If leaks are suspected, check fluid levels more frequently.

> **WARNING!**
> Do not check for hydraulic leaks with bare hands. Use cardboard, wood, or other solid material. Pressurized fluids can cause severe injuries to unprotected skin. Long-term exposure can cause cancer or other chronic diseases.

- Leaks (oil, fuel, hydraulic, and coolant)
- Worn or cut hoses
- Loose or missing bolts
- Trash or dirt buildup
- Cut or damaged tires
- Broken or missing parts
- Damage to gauges or indicators
- Damaged hydraulic lines
- Dull, worn, or damaged cutting edge
- General wear and tear

Some manufacturers require that daily maintenance be performed on specific parts. These parts are usually those that are the most exposed to dirt or dust and may malfunction if not cleaned or serviced. For example, the service manual may recommend lubricating specific bearings every 10 hours of operation, or always cleaning the air filter before starting the engine. *Figure 40* shows some of the locations of points to be inspected on a grader.

Before beginning operation, check the fluid levels and top off any that are low. Check all the machine's major functions including the following components:

- *Air cleaner* – If the machine is equipped with an air cleaner service indicator, observe the indicator. If it shows red, the air filter and intake chamber need to be cleaned. If the machine does not have a service indicator attachment, you must remove the air cleaner cover and inspect the filter (*Figure 41*). Clean out any dirt at the bottom of the bowl. In dusty conditions, pay close attention to cleaning the air filter.
- *Battery* – Check the battery cable connections.
- *Blade and attachments* – Check this hardware to make sure there is no damage that would create unsafe operating conditions or cause an equipment breakdown. Make sure the blade is not cracked or broken. Check the teeth on the ripper or scarifier.
- *Circle rail and supports* – Check this hardware to make sure there is no damage that would create unsafe operating conditions or cause equipment breakdown: circle rail crank, circle rail shift cylinder, circle rail top face, clamp plate, guide plate bearing surfaces, upper slide rail, and slide shaft bar. Check the lubricant level and fill as necessary.
- *Cooling system* – Check the coolant level and make sure it is at the level specified in the operating manual. Also check radiators and coolers to make sure there are no obstructions or damage.

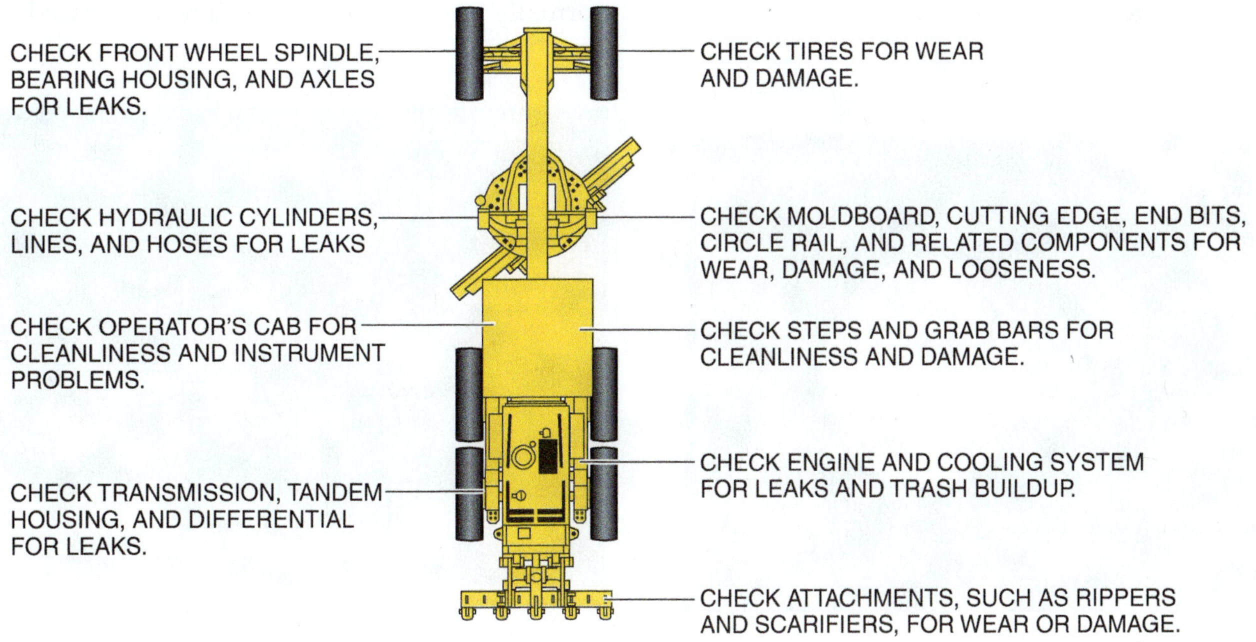

CHECK FRONT WHEEL SPINDLE, BEARING HOUSING, AND AXLES FOR LEAKS.
CHECK HYDRAULIC CYLINDERS, LINES, AND HOSES FOR LEAKS
CHECK OPERATOR'S CAB FOR CLEANLINESS AND INSTRUMENT PROBLEMS.
CHECK TRANSMISSION, TANDEM HOUSING, AND DIFFERENTIAL FOR LEAKS.
CHECK TIRES FOR WEAR AND DAMAGE.
CHECK MOLDBOARD, CUTTING EDGE, END BITS, CIRCLE RAIL, AND RELATED COMPONENTS FOR WEAR, DAMAGE, AND LOOSENESS.
CHECK STEPS AND GRAB BARS FOR CLEANLINESS AND DAMAGE.
CHECK ENGINE AND COOLING SYSTEM FOR LEAKS AND TRASH BUILDUP.
CHECK ATTACHMENTS, SUCH AS RIPPERS AND SCARIFIERS, FOR WEAR OR DAMAGE.
REPRINTED COURTESY OF CATERPILLAR INC.
22305-14_F40.EPS

Figure 40 Daily inspection spots for motor graders.

- *Drive belts* – Check the condition and tightness of engine drive belts.
- *Engine oil* – Check the engine oil level and make sure it is in the safe operating range
- *Environmental controls* – The machine may be equipped with lights and windshield wipers. If operating under conditions where these accessories are needed, make sure they work properly.
- *Fuel level* – Check the level in the fuel tank(s). Do this manually with the aid of the fuel dip stick or marking vial. Do not rely only on the fuel gauge at this point. Check the fuel pump sediment bowl if one is fitted on the machine.
- *Hydraulic fluid* – Check the hydraulic fluid level in the reservoir.
- *Pivot points* – Clean and lubricate all pivot points.

> **NOTE**
>
> In cold weather it is sometimes better to lubricate pivot points at the end of a work shift when the grease is warm. Warm the grease gun before using it for better grease penetration.

- *Transmission fluid* – Measure the level of the transmission fluid to make sure it is in the operating range.

The operator's manual usually has detailed instructions for performing periodic maintenance. If there are any problems with the machine that you are not authorized to fix, inform the foreman or field mechanic (*Figure 42*). Get the problem fixed before beginning operations.

2.3.0 Preventive Maintenance

Preventive maintenance involves an organized effort to regularly perform periodic lubrication and other service work in order to avoid poor performance and breakdowns at critical times. Performing preventive maintenance on the grader will keep it operating efficiently and safely, thereby avoiding the possibility of costly failures in the future.

Preventive maintenance of equipment is essential and not that difficult, if you have the right tools and equipment. The leading cause of premature equipment failure is putting things off. Preventive maintenance should become a habit, performed on a regular basis.

Maintenance intervals for most machines are established by the Society of Automotive Engineers (SAE) based on hours of run time, and have been adopted by most equipment manufacturers. Instructions for preventive maintenance and specified run time intervals are usually found in the operator's manual for each piece of equipment. Common time intervals are: 10 hours (daily), 50 hours (weekly), 100 hours, 250 hours, 500 hours, and 1,000 hours. The operator's manual will also include lists of inspections and servicing activities required for each time interval. The *Appendix* contains an example of typical periodic maintenance requirements.

When servicing a motor grader, follow the manufacturer's recommendations and service chart. Any special servicing for a particular piece of equipment will be highlighted in the manual. As previously mentioned, the service chart will normally recommend specific intervals, based on

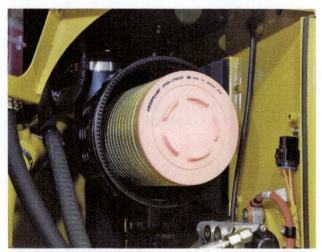

REPRINTED COURTESY OF CATERPILLAR INC.

22305-14_F41.EPS

Figure 41 Check the air filter.

REPRINTED COURTESY OF CATERPILLAR INC.

22305-14_F42.EPS

Figure 42 Report maintenance problems to foreman or field mechanic.

hours of run time, for such things as changing oil, filters, and coolant.

Hydraulic fluids should be changed whenever they become dirty or break down due to overheating. Continuous and hard operation of the hydraulic system can heat the hydraulic fluid to the boiling point and cause it to break down. Filters should also be replaced during regular servicing.

Before performing maintenance procedures, always complete the following steps:

Step 1 Park the machine on a level surface to ensure that fluid levels are indicated correctly.

Step 2 Lower all equipment to the ground. Operate the controls to relieve hydraulic pressure.

Step 3 Engage the parking brake.

Step 4 Lock the transmission in neutral.

Step 5 Turn the engine start switch to the Off position and remove the key. Place a lock-out/tagout notice on the switch.

Step 6 Turn the battery disconnect switch key to the Off position and allow the engine to cool before performing maintenance.

Preventive Maintenance Records

Accurate, up-to-date maintenance records are essential for tracing the history of a piece of equipment. Each machine should have a record that describes any inspection or service that is performed and the corresponding time intervals. Typically, an operator's manual and some sort of inspection sheet are kept with the equipment at all times. Operators should update maintenance records as part of their daily routine.

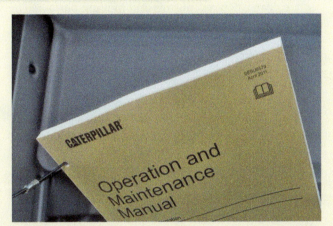

REPRINTED COURTESY OF CATERPILLAR INC.

22305-14_SA04.EPS

Additional Resources

Basic Equipment Operator, NAVEDTRA 14081, 1994 Edition. Morris, John T. (preparer), Naval Education and Training Professional Development and Technology Center.

Operator's Manual Grader, Heavy Road, Motorized Caterpillar Model 130G, (NSN 3805-01-150-4795), TM 5-3805-261-10, 1992. Washington, DC: Department of the Army.

2.0.0 Section Review

1. While operating a motor grader, avoid wearing loose clothing and jewelry because they _____.

   a. are fire and electrical hazards
   b. do not look neat and professional
   c. can catch on controls and moving parts
   d. cannot transfer heat away from the body

2. Some motor grader manufacturers require daily maintenance to be performed on certain parts that are the most _____.

   a. likely to be backordered from the manufacturer
   b. difficult to access during scheduled maintenance
   c. expensive to replace in the event of a breakdown
   d. exposed to dirt or dust that could cause them to malfunction

3. Before performing any maintenance procedures on a motor grader, first make sure to _____.

   a. raise the blade and any attachments
   b. park the grader on a level service
   c. lock the transmission in forward or reverse
   d. maintain fluid pressure in all systems

SECTION THREE

3.0.0 BASIC STARTUP AND OPERATIONS

Objective

Describe basic startup and operating procedures for a motor grader.
  a. Describe startup, warm-up, and shutdown procedures.
  b. Describe basic maneuvers and operations.
  c. Describe common work activities.

Performance Tasks 2, 3, and 4

Perform proper startup, warm-up, and shutdown procedures.

Perform basic maneuvers on a motor grader, including moving forward, moving backward, and turning.

Demonstrate basic motor grader operation by:

• Grading a rough grade by following grade stakes placed along a 300-foot (91-meter) section that is at least double the width of the machine.

• Demonstrating rotation of the blade for high-bank grading.

• Cutting a V-ditch with a 3-to-1 slope.

Trade Terms

Heel of the blade: The following end of the blade, regardless of which side of the machine the blade is located on.

Pitch: The angle of the moldboard in relation to a vertical plane.

Side-casting: Placing material in a windrow by pushing the load off the trailing end of the mold-board.

Toe of the blade: The leading end of the blade, regardless of which side of the machine the blade is located on.

A well-maintained motor grader that is operated properly can perform many different tasks. Before operation begins, operators must understand the importance of proper startup, warm-up, and shutdown procedures.

3.1.0 Startup, Warm-Up, and Shutdown

The startup, warm-up, and shutdown of an engine are very important. Proper startup lengthens the life of the engine and other components. A slow warm-up is essential for proper operation of the machine under load. Similarly, the shutdown of the machine is critical because of all the hot fluids circulating through the system.

3.1.1 Startup

There may be specific startup procedures for the piece of equipment you are operating. Check the operator's manual for a specific procedure. In general, the startup procedure should follow this sequence:

Step 1 Be sure all controls are in neutral. Blade and attachment controls should be in the hold position.

Step 2 Engage the parking brake.

> **NOTE**
> When the parking brake is engaged, an indicator light on the dashboard will light up or flash. If it does not, stop and correct the problem before operating the equipment.

Step 3 Make sure there are no people close to the equipment.

Step 4 Engage the transmission control lock lever safety (if equipped).

Step 5 Place the ignition switch in the On position. The engine oil pressure light should light up.

Step 6 Move the hand throttle to low idle. Press the starter button or turn the engine key switch to the start position until the engine starts.

> **CAUTION**
> Never operate the starter for more than 30 seconds at a time. If the engine fails to start, refer to the operator's manual before activating the starter again.

Step 7 Warm up the engine for at least five minutes. A longer period will be required in cold temperatures.

Step 8 Check all the gauges and instruments to make sure they are working properly. Many machines perform an automatic self-test when first started.

Step 9 Shift the gears to low range.

Step 10 Release the parking brake and depress the service brakes.

Step 11 Check all the controls for proper operation.

Step 12 Check the service brakes for proper operation.

Step 13 Check the steering for proper operation.

Step 14 Manipulate the controls to be sure all components are operating properly.

> **CAUTION**
>
> Do not operate the moldboard if the hydraulics are sluggish. Allow the machine to warm up until the hydraulics function normally. The hydraulics can fail if not warmed up completely.

Step 15 Shift the gears to neutral and lock.

Step 16 Reset the brake.

Step 17 Make a final check for leaks, unusual noises, or vibrations.

Some machines have special procedures for starting the engine in cold temperatures. Some units are equipped with glow plugs or starting aids, which may be reviewed in the operator's manual.

> **WARNING!**
>
> Ether is used in some older equipment to start cold engines. Ether is a highly flammable gas and should only be used under strict supervision in accordance with the manufacturer's instructions.

3.1.2 Warm-Up

A slow warm-up is essential for proper operation of the machine under load. As soon as the engine starts, release the starter switch and keep the engine speed low until the oil pressure registers. The oil pressure light should come on briefly and then go out. If the oil pressure light does not turn off within 10 seconds, stop the engine, investigate, and correct the problem. Many machines do a self-test upon first starting. Make sure that all systems are operating normally.

Let the engine warm up to operating temperature before moving the grader. A longer warm-up period is necessary in cold weather. If the temperature is at or slightly above 32°F (0°C), let the engine warm up for 15 minutes. If the temperature is between 32°F (0°C) and 0°F (–18°C), warm the engine for 30 minutes. If the temperature is less

than 0°F (–18°C) or hydraulic operations are sluggish, additional time is needed. Follow the manufacturer's procedures for cold starting.

Check the other gauges and indicators to see that the engine is operating normally. Check that the water temperature and oil pressure indicators are in the normal range. If there are any problems, shut down the machine and investigate or get a mechanic to look at the problem.

3.1.3 Shutdown

The shutdown of the machine is critical because of all the hot fluids circulating through the system. These fluids must cool so that they can cool the metal parts before the engine is switched off. Proper shutdown will reduce engine wear and possible damage to the machine.

Shutdown should follow a specific sequence, such as the following:

Step 1 Find a dry, level spot to park the grader. Stop the grader by decreasing the engine speed and placing the direction lever in neutral. Depress the service brakes and bring the machine to a full stop.

Step 2 Place the transmission in neutral and engage the parking brake. Lock out the controls if the machine has a control lock feature.

Step 3 Lower the moldboard so that it rests on the ground. If there is a ripper or any other attachment, be sure it is also lowered.

Step 4 Place the speed control in low idle and let the engine run for approximately five minutes.

> **CAUTION**
>
> Failure to allow the machine to cool down can cause excessive temperatures in the turbocharger.

Step 5 Turn the engine key switch to the Off position and remove the key.

Step 6 Release the hydraulic pressure by moving the control levers until all movement stops.

Some machines have disconnect switches, guards, or locks for added security. Check the machine and secure any additional features. Close the battery disconnect switch and fuel shutoff valve. Close and lock vandal guards. Lock the parking brake and transmission. These controls provide additional safety and deter unauthorized users. Always engage any additional security systems when leaving the grader unattended.

3.2.0 Basic Maneuvers and Operations

More than skill is involved in operating a motor grader. Effective operation requires constant attention to the controls and a good sense of placement of materials. The operator who can think ahead and decide what adjustments must be made to keep the blade at the correct elevation, *pitch*, and angle will be successful in performing any activity with the motor grader.

Do not take risks. If there is doubt about the capability of the machine to do some work, stop the equipment and investigate the situation. Discuss it with the competent person. Whether it is a slope that may be too steep or an area that looks too unstable, know the limitations of the equipment. Decide how to do the job before starting. Once you start working, it may be too late, causing extra work for others or unsafe conditions.

Preparing to work involves getting organized in the cab, fastening the seat belt, and starting the machine. Mount the equipment using the grab rails and foot rests. Adjust the seat so you can see clearly and reach all the controls.

> **WARNING!**
> OSHA requires that approved seat belts and ROPS be installed on heavy equipment. Old equipment must be retrofitted. Do not use heavy equipment that is not equipped with these safety devices.

Much of the work done with motor graders is repetitive, meaning that several basic maneuvers are performed many times with slight variations in the blade height, angle, and pitch. As a result, there are several basic steps that should be followed no matter what activity is being performed. They are usually carried out in the same sequence.

Follow these six basic steps for effective operation:

Step 1 Study the area. Look at the site plan to see if there is anything that could cause a problem. Walk around the site and check the grade stakes. Look for things that are not on the plans, such as holes, washouts, signs, underground utilities, overhead obstructions, rocks, or anything else that may interfere with the operation.

> **NOTE**
> Contact the local One-Call number to identify any underground utilities before digging.

Step 2 Position the motor grader directly in front of the work area. Check the grade and determine where the material will flow. Check all around the motor grader for obstructions and make sure other workers are clear before moving forward.

Step 3 Set the blade to the proper position for the work. The height of the blade will depend on the type of material and the finish required: for mixing, the blade must be above the hard surface; for finishing, it must be on the surface; and for rough grading, the blade must be on, or cutting slightly into the surface.

> **CAUTION**
> Do not articulate the grader while it is standing still. Do not leave the grader articulated on hot asphalt on a hot day. This could damage the asphalt.

The angle of the blade also depends on the job: cutting and moving material requires a sharp angle; windrowing requires a smaller angle. The tilt of the blade should be forward for finish grading and slightly backwards for moving material.

Step 4 Set the gear. Do not stop to change gears in the middle of a pass, or you may cause the wheels to dig into the surface and alter the finish. Low gears are used for cutting; higher gears are used for light cutting, or for driving the motor grader from one job location to the next.

Step 5 Drive straight ahead. If the front end of the motor grader tends to veer out of line, lean the wheels in the opposite direction to correct the motion.

Step 6 Make few blade changes. Changing the blade angle or shape will cause a change in the elevation of the grade. Make as few blade changes and passes as possible to ensure a uniform finish.

Always be sure the equipment is in good operating order before beginning any activity. Make sure the daily check has been performed and that proper startup procedures have been carried out.

Basic maneuvering of a motor grader involves moving forward, moving backward, and turning. To perform grading operations with the machine, you must be able to operate the controls effectively while the machine is in motion.

The controls consist of levers, switches, foot pedals, and a steering wheel grouped in different configurations according to the manufacturer's design.

On most models, the transmission control and engine speed console is to the right of the operator's seat. It has a gear shift lever that controls the forward and reverse gears. A hand throttle allows the operator to maintain a constant speed in any gear without keeping a foot on the accelerator pedal. Most graders have an electronically activated hydraulic steering wheel mounted on a pedestal in the operator's cab.

Levers that control the movement of the moldboard, circle rail, and other attachments are located to the right and left of the steering pedestal. The arrangement of these levers will vary based on the model and manufacturer.

3.2.1 Moving Forward

The first basic maneuver is learning to drive forward. To put the motor grader in forward motion, follow these steps:

Step 1 Raise the blade and other attachments high enough to be clear of obstructions. Check the controls for proper operation.

Step 2 Push down on the service brake pedal to keep the machine from moving.

Step 3 Disengage the parking brake and unlock the transmission lever.

Step 4 Move the hand throttle to the desired engine speed.

Step 5 Move the transmission lever to the forward position. Select the desired engine speed to move the machine.

Step 6 Release the brake pedal.

Step 7 Depress the accelerator to move forward and obtain the desired speed. Accelerate, shifting one gear at a time, increasing the engine speed as necessary.

Step 8 When slowing down, downshift one gear at a time. When downshifting under a load, increase the engine speed to match the ratio of the lower gear speed.

3.2.2 Reverse and Backing

To change the direction of travel, the machine should be stopped using the service brake. Move the transmission control lever to neutral, and then move it to the reverse direction and desired gear. Press the accelerator pedal to begin moving in reverse. Use the grader's mirrors when backing up.

When a motor grader has to make a number of passes over a distance of less than 1,000 feet (about 305 meters), it is usually more efficient to back the motor grader the entire distance to the starting point than to turn around and return from the far end. If the passes cover a distance of 1,000 feet (about 305 meters) or more, as in snow removal, it is more efficient to turn the motor grader around and start blading from the far end back to the starting point. The combined maneuvering advantages of leaning wheels and articulated frame are a big help in turning the machine around.

3.2.3 Steering and Turning

When the grader is being turned, the amount of turning that takes place depends on whether the machine is configured with the frame straight, articulated, or offset. Straight-frame steering is normally used for long passes, such as road maintenance.

Articulating the frame shortens turns in confined areas and lets the operator counteract any side thrust during normal grading. To counteract side thrust, position the front of the rear wheels toward the **heel of the blade** or moldboard. Offset steering helps the operator keep the rear wheels on solid ground in operations such as ditch cleaning, and places most of the machine's weight behind the load in heavy grading.

When a motor grader is pushing a load with the blade set at an angle, the movement of the load tends to swing the front of the motor grader toward the **toe of the blade**. To counteract this side thrust, lean the top of the front wheels toward the discharge side of the blade, or make the cut more shallow.

Because the motor grader is a large and long piece of equipment, it is difficult to turn around or change direction sharply. Leaning the wheels and using articulation will make turning around easier. When turning around in a tight area, it may be necessary to move forward and backward several times while changing the position of the front wheels and the articulation.

Where there is sufficient space to make a U-turn with the equipment, use articulation to help reduce the required turning radius. Using this procedure allows operators to quickly line up the

---

motor grader to make the next pass without having to back up.

3.2.4 Operating the Blade

The moldboard of the blade is shaped so that graded material will roll and mix as it is being cut. The cutting edge protects the moldboard edges from the abrasive action of the grading material and is meant to be replaced often. Before operating the grader, check the serviceability of the cutting edge and replace it if necessary.

A motor grader blade can be adjusted to many different positions. The final position depends on the job and the materials being graded. The blade can be raised and lowered as little as a fraction of an inch to adjust to the grade being finished. Movements of the blade are typically controlled from operating levers in the cab, but some new models have joystick controls for the blade and attachments (*Figure 43*). The levers or joystick will operate the hydraulic cylinders and gears that connect the blade to the circle rail and the circle rail to the frame of the motor grader. The controls connecting the blade to the circle rail allow the blade to be shifted to the right and left or pitched forward and backward. The blade can also be rotated around the circle rail, which allows positioning at different angles relative to the direction of travel. The circle rail is controlled by a set of hydraulic cylinders that allow it to be moved horizontally and vertically in different configurations.

Typical operating procedures for controlling the blade are as follows:

- *Right hand blade lift lever* – Pulling this lever raises the right hand side of the blade. Pushing this lever lowers the right hand side of the blade.
- *Left hand blade lift lever* – Pulling this lever raises the left hand side of the blade. Pushing this lever lowers the left hand side of the blade.
- *Circle rail shift lever* – Pulling this lever shifts the circle rail drawbar to the right. Pushing this lever shifts the circle rail to the left.

REPRINTED COURTESY OF CATERPILLAR INC.

22305-14_F43.EPS

Figure 43 Joystick controls for the blade and attachments.

- *Circle rail rotation lever* – Pulling this lever rotates the circle rail clockwise. Pushing this lever rotates the circle rail counterclockwise.
- *Wheel lean lever* – Pulling this lever leans the front wheels to the right. Pushing this lever leans the wheels to the left.
- *Blade shift lever* – Pulling this lever shifts the blade to the right. Pushing this lever shifts the blade to the left.
- *Blade tilt (pitch) lever* – Pulling this lever tilts the blade toward the rear of the motor grader, so the bottom of the blade is more forward than the top. Pushing this lever tilts the top of the

Three-Point Turns

Even though a motor grader can dwarf a typical automobile, you can still use a conventional three-point turn method. To make a three-point turn on a roadway, use judgment for the amount of area the motor grader requires to turn around. Travel forward as far as possible. Lean and turn the wheels in the opposite direction before starting to back up. Once you have backed up as far as possible, lean and turn the wheels to the new direction of travel. Pull forward and straighten the wheels after the turn is completed. When you leave the wheels on grade and lean in the direction of the turn, the motor grader makes the turn with ease. Always back across the road or ditch and leave the front wheels on the roadway.

22305-14 **Motor Graders**

blade forward, so the top of the blade is more forward than the bottom.

- *Scarifier lever* – Pulling this lever raises the scarifier. Pushing this lever lowers the scarifier.

On older models of motor graders, there are two basic configurations for the controls. The two-hand blade lift configuration places the right and left blade lift levers on opposite sides of the row of levers. The one-hand blade lift configuration places the right and left blade lift levers together at the outside right hand position of the row of levers. Regardless of the type of control on the grader, they are all designed to control the position of the circle and blade. *Figure 44* shows some common blade operating positions.

> **NOTE**
>
> The blade can be positioned so that material is cast between the wheels or to the side of a wheel. It is very important to position the blade so that material is not cast into the path of the wheels.

Figure 45 shows the top view of a blade and circle rail arrangement. The drawing on the left (A) has the blade perpendicular to the frame, which is designated as 0 degrees. The drawing on the right (B) shows the blade at a 45-degree angle to the first position. Whenever the blade is positioned at any angle other than 0 degrees, the leading edge of the blade is called the toe, and the trailing edge is called the heel.

As the angle of the blade increases, the amount of material spilling off the heel also increases. As the angle decreases, a greater amount of the load is directed straight ahead. With the blade straight across the circle rail (0 degrees), the effect is the same as bulldozing or pushing the material forward. Remember that the blade is not designed to push a great deal of material forward. If a lot of material needs to be moved a long distance, consider using a dozer.

Raising or lowering the blade determines the depth of the cut. Raising the blade also determines whether the material is cast into windrows or spread evenly. When the heel of the blade is raised, material will spill out underneath and be spread along the surface by the forward motion of the blade. Lowering the heel adjusts the flow so that very little or none of the material spills out from under the blade; instead, it spills off the heel to form a windrow. Note that when one end of the blade is raised, the other end is lowered, like a board on a pivot point. Raising the heel too far will drive the toe into the ground.

Pitch is the vertical angle of the blade. Pitch adjustment produces either a cutting or dragging action. *Figure 46* shows four common pitch adjustments. For normal grading operations, the blade is kept near the center of pitch adjustment (A), so that the top of the blade is directly over the cutting edge. For greater cutting action, the blade is pitched backward, as shown in (B). This way, the cutting edge slices into the ground and the material rolls up on the moldboard.

For mixing and laying operations, the blade is pitched slightly forward as shown in drawing (C). This pitch is used for making only light, rapid cuts because the cutting ability of the blade is decreased.

For spreading or maintaining surface material, and for snow removal, the blade is pitched farther forward to the position shown in drawing (D). In this position, the blade tends to ride over hard material rather than cutting it. The blade will not be damaged by riding over obstructions. This pitch creates an uneven surface, especially when it is made at higher speeds.

For normal grading operations, use the straight pitch position of the blade so that the top edge is vertical or slightly ahead of the blade. This puts the cutting edge at its best angle for most normal grading. When greater cutting action is needed, use the backward pitch on the blade. This causes the cutting edge to dig in and plow material up and over the moldboard. Dirt can build up and flow over the top of the blade into the circle rail. When this happens, adjust the pitch to decrease the flow of the material over the top.

For the best mixing and rolling action, pitch the blade slightly ahead of the vertical position. When spreading material or maintaining roads, the blade should be tipped forward. This tends to compact material and fill low spots. Use only as much forward tip as needed for compacting or filling. Excessive use of this position wears the cutting edge bevel faster than other methods.

3.2.5 Calibrating the Laser Setup

If the motor grader is equipped with a laser leveling system for grading work, the receiving unit and controls must be calibrated before use. This must be done after the laser plane or laser sending unit has been set up and the reference height for the unit established.

> **NOTE**
>
> Calibration of a laser instrument should be performed by a qualified person in accordance with the manufacturer's instructions.

22305-14_F44.EPS

Figure 44 Blade operating positions.

3.3.0 Work Activities

The motor grader is a versatile piece of equipment. It can do many tasks well, but it was never designed to pick up material and transport it over long distances. However, with a skilled operator at the controls, a motor grader can move and spread large amounts of material with great accuracy. The main job of the motor grader is grading. This includes both rough and finish grading

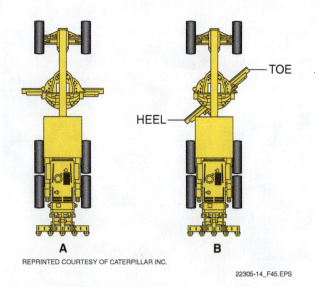

REPRINTED COURTESY OF CATERPILLAR INC.

22305-14_F45.EPS

Figure 45 Angle of the blade.

of roads, streets, building pads, parking lots, and any other areas requiring a smooth or level surface.

Motor graders are often used with other, less precise, heavy equipment. For example, a bulldozer can travel ahead of a grader to push large amounts of material away, while a scraper can follow a grader to pick up any material left in a windrow.

Because turning the motor grader takes time and increases the machine's cycle time, whenever possible, plan the work so you have ample turning room. Reverse blading (turning the blade 180 degrees and operating the grader in reverse while grading) is a good time saver, but not for an inexperienced operator. Do not attempt reverse blading until you are experienced and confident in your ability to operate the grader while moving forward.

3.3.1 Windrowing

Windrowing is the process of creating a long narrow pile of material on the ground. A motor grader creates a windrow by scraping material from the ground with the blade, rolling and mixing the material with the moldboard. It then casts the material off the blade in a continuous row so it can be moved to another location, such as the crown of a road. Windrows are also created for easier pickup of material by other equipment, such as a scraper, or for the purpose of mixing and blending materials together.

To make a windrow of material on the right side of the motor grader, use the following procedure:

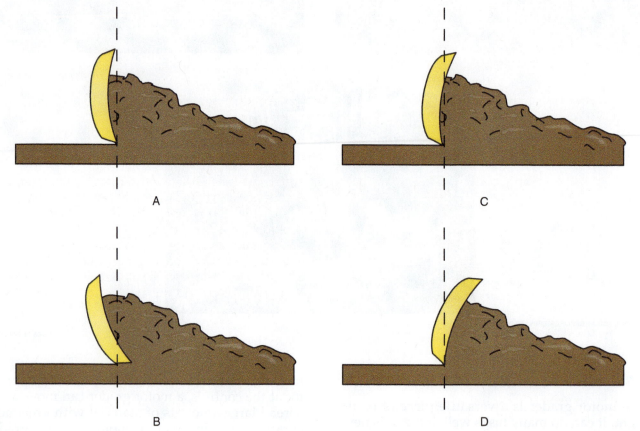

22305-14_F46.EPS

Figure 46 Common blade pitch adjustments.

NCCER – *Heavy Equipment Operations Level Three*   22305-14

Step 1 Circle the blade right to position the toe of the blade behind the left front tire.

Step 2 To avoid driving over the windrow, slide the blade shift so the heel is clear of the right rear wheels, as shown in *Figure 47*.

Step 3 Lower the toe and heel of the blade to begin cutting the material. As the desired depth is reached, material should flow freely off the heel of the blade and form the windrow outside the wheel line.

Step 4 Continue moving forward, making the windrow until the end of the cut is reached.

Step 5 Raise the blade clear of the ground and stop. Select Reverse and back up until you return to the starting point.

Step 6 Stop and select the first forward gear.

Step 7 Move the machine to the left and position the left front tire to the right of the windrow. In this position, you should be straddling the windrow with the motor grader.

Step 8 Lower the blade to the ground to make a light grading pass. The material should flow off the heel of the blade on the right side.

Step 9 As the windrow becomes larger, adjust the motor grader's front tires to lean toward the windrow to counteract side drifts.

Step 10 Continue this process until the windrow reaches the desired size.

To make a windrow for pickup or spreading, begin by running the grader over the material with the blade set at a sharp angle, so the material rolls off to one side. Then make another pass using the low speed range with the blade set in the opposite direction. The material now rolls off to the other side. This creates a long pile of material with shaped sides and a uniform height. This material can now be spread again or picked up by a front-end loader or scraper.

3.3.2 Blending and Spreading Material

Blending material with the motor grader is done by mixing the windrows of two or more materials together. Once the materials have been piled close together, they can be blended by making another pass on one side of a windrow and folding the material from that windrow into the one next to it. Once the two windrows have been combined, they can be further blended by making additional passes back and forth. Repeat this process until the material looks uniform.

After obtaining the required mix, position the motor grader at one end of the windrow of mixed material. Angle the blade at 20 to 30 degrees, lift the heel, and spread the material over the surface by blading back and forth across the area.

3.3.3 Cutting Ditches

Cutting ditches is one of the main work activities for motor graders in a public works department or highway maintenance division. Ditches are normally cut alongside rural roads and highways to channel storm water runoff away from the road. In subdivision and commercial work, shallow ditches are cut to drain runoff away from building pads and other structures. The two basic types of ditches constructed with a motor grader are shown in *Figure 48*.

The V-ditch (*Figure 48A*) is the basic profile for situations where the expected runoff is minimal and space is limited. The flat bottom ditch profile (*Figure 48B*) is more difficult to construct, but should be used when a great deal of runoff is expected, such as a large area or a steep grade.

There are three basic parts to both types of ditches. The inslope is the slope between the bottom of the ditch and the area being drained. The backslope is on the back side of the ditch and usually has a steeper slope. The bottom of the ditch can either be flat or V-shaped. The grade line of the ditch bottom determines which way the water flows.

Typically, cutting a ditch involves making several passes in one or both directions to cut the inslope, backslope, and bottom, and then clean up any excess material. The steps for making a V-shape ditch are as follows:

22305-14_F47.EPS

Figure 47 Adjust the blade to avoid driving over the windrow.

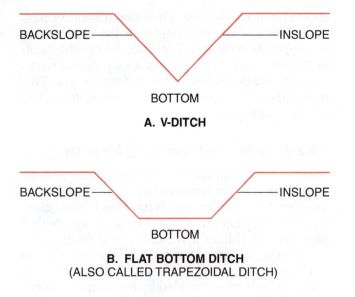

A. V-DITCH

B. FLAT BOTTOM DITCH
(ALSO CALLED TRAPEZOIDAL DITCH)

Note: The grade line at the bottom determines the direction of flow.

22305-14_F48.EPS

Figure 48 Ditch profiles.

Step 1 Look for the marked center line of the ditch. Position the motor grader with the right front wheel approximately 3 feet (just under 1 meter) to the left of the marks or stakes. Rotate the circle rail so the blade is approximately 2.5 feet (about 0.76 meter) behind the right front wheel.

Step 2 Position the blade so the toe is lined up to dig the ditch bottom, with the heel positioned so that material will be cast off in a windrow on the inside of the left rear wheel. Roll the moldboard to the rear and use the right-hand blade lift lever to lower the toe until it is several inches into the surface. If the ground is dry or hard, the motor grader may lift a little, but once it moves forward, the blade will dig in. The heel of the blade should be above the surface.

Step 3 When beginning the cut, do not try to cut too much on the first pass. Keep the speed control in the low range. Make as many passes as necessary to reach the depth and slope called for by the stakes or specifications. If the job has been done properly, a windrow will have formed on the shoulder of the embankment with material cut from the ditch.

Step 4 Backslopes can be cut with either a shallow or a steep slope. Most shallow cuts have a 4:1 slope; deep cuts have a 2:1 slope. There are two ways to cut the backslope: the first method places the spoil on the inslope or shoulder; the second method places the spoil on the top of the backslope.

The first method involves completing the inslope and turning the motor grader around so that it is facing the opposite direction. Doing this eliminates the need to make other adjustments. Some of the material should roll off the right side of the blade and land in the ditch bottom and on the inslope. After completing the backslope, back up, reposition the motor grader, and remove the material from the inslope, rolling it onto the shoulder.

For the second method, do not turn the motor grader around. Instead, position it so that the right wheel is on the center line of the ditch. However, this time shift the circle rail right of center. Adjust the circle rail so that the left side of the blade is forward (that is, opposite from where it was for the inslope). Now the toe is the left hand end of the blade. It should be approximately 3 feet (just under 1 meter) from the left front wheel. The heel should extend past the intermediate right rear wheel.

Step 5 Use the blade shift to place the toe over the center line of the ditch. Lower the toe and the heel to the desired depth. With the speed control again in the low range, make a pass. Material from the bottom and inslope should move from left to right and land at the top of the backslope.

Step 6 After the ditch is complete, make as many passes as necessary to level the windrow to a proper grade between the ditch and the surrounding area. If necessary, the extra material coming out of the ditch can be used to level up any irregularities.

When constructing flat bottom ditches, several steps are required in addition to those described for construction of a V-bottom ditch. After the flow line has been established, reposition the motor grader at the beginning of the ditch. Rotate the blade to the desired angle so the blade will cut the desired width of the ditch bottom. Material cut by the blade should be cast onto the inslope for removal from the ditch. After cutting the bottom to the desired profile, move the motor grader over to the inslope and grade the excess material up the inslope and onto the shoulder. Always be careful not to let the blade touch the front tire when turning the grader. This can damage the tire or the blade.

3.3.4 Ditching in Wet Areas

Ditches are sometimes soft and will not support the full weight of the equipment. The motor grader can get stuck in the ditch if the drive wheels are allowed to enter soft areas. If the blade can be moved far enough to the side while keeping the tandem wheels on stable ground, it will take only a couple of passes and clean the inslope and the ditch bottom.

Assuming that the ditch to be cleared is to the left of the motor grader, position it so the front wheel closest to the center line of the ditch is offset two to three feet (about 0.6 to 0.9 meters) to the right. Shift the circle rail as far left as possible, and rotate the circle rail so the blade heel is under the center of the motor grader. If the ditch is wet, offset the tandem wheels to allow them to ride on the shoulder of the road.

Lift material from the ditch and dump it under the motor grader. By articulating the tandem wheels on the road and the front wheels into the ditch, the motor grader can be backed up until the toe of the blade is at the end of the culvert. Then start the cut and windrow the material onto the shoulder or roadbed. When filling washouts on existing backslopes, the blade can be angled back and side-shifted enough so that material can be rolled up the slope to fill the washed-out areas.

3.3.5 Ripping and Scarifying

There are two attachments for the motor grader that can be used to loosen, break up, or scarify hard material: the ripper and the scarifier.

The shanks on the ripper (*Figure 49*) can be lowered and raised to different levels. For tough surfaces, use fewer teeth, so that maximum force will be concentrated on the points where the teeth enter the ground. As the material is loosened, additional teeth may be added. Whenever the ripper is being used, keep the frame of the motor grader straight, and rip in a straight line. Come to a complete stop before lifting the teeth out of the ground, then turn the motor grader around to begin the next pass.

The scarifier is usually attached to the front of the motor grader between the front wheels. The scarifier attachment also has individual shanks and can be raised or lowered to meet scarifying requirements. Use all the shanks in light material and fewer shanks in heavier material. *Figure 50* shows a scarifier attachment in use.

The procedure for operating the scarifier is as follows:

Step 1 Position the motor grader about 20 feet (6 meters) from the starting point and in line with the area to be scarified.

Step 2 Lower the scarifier bar until the teeth are about 2 inches (5 centimeters) above the ground.

REPRINTED COURTESY OF CATERPILLAR INC.

22305-14_F49.EPS

Figure 49 Ripper attached to the rear of a motor grader.

Effective Ditch Cutting

Ditch cutting is done with the blade angled between the wheels. The front wheels are usually leaned in the direction of the side where the material will be cast. In flat-bottom ditch construction, the blade is at a sharp angle, so the width of the cut is between the two ends of the blade.

A high-bank cut is made by moving the circle rail to an extreme vertical position on one side of the motor grader. The circle rail can be moved up or down. In a wide side reach configuration, the blade is extended as far as possible to one side of the circle rail.

Reverse blading involves turning the blade completely around by rotating the circle rail 180 degrees. In this configuration, the motor grader moves backwards and scrapes material in a reverse direction. Keep in mind, though, that reverse blading is difficult and should not be performed by an inexperienced operator.

REPRINTED COURTESY OF CATERPILLAR INC.

22305-14_F50.EPS

Figure 50 Scarifier attached to the front of a motor grader.

Step 3 Shift the transmission into low range and move forward. When operating speed is reached, lower the scarifying teeth to the desired cutting depth at the starting point. If the engine begins to lug, shift to a lower gear. If it stalls, there are two options that may improve its performance. The first is to reduce the depth of the cut. The second involves removing some of the shanks from the operating position. When removing shanks, always take the same number from each end of the bar.

Step 4 Continue to scarify in a straight line.

Step 5 When the end of the section has been reached, raise the shanks 4 to 6 inches (about 10 to 15 centimeters) above the surface.

Step 6 Turn the motor grader around and continue to perform the work until all material in the section has been scarified.

3.3.6 Grading Unpaved Roads

One of the main activities of a motor grader in state and county highway departments is grading unpaved roads. This maintenance process is performed to smooth out the surfaces that have become rutted or washboarded, or to fill in potholes and depressions that are formed by moisture and traffic.

The ideal time for grading an unpaved road is after the road surface has received some moisture and has dried out just enough so that the binding material will stick to the aggregate, cementing it together.

Figure 51 shows a motor grader smoothing the top surface of a rural road, leveling humps and ridges, and filling in low spots and potholes with material being carried forward on the blade. Note that the blade is pitched slightly forward to force the material into low spots. The material is being moved from the edge of the road to the center so that it can be used to create a crown.

To perform grading of an unpaved road, use the following procedure:

Step 1 After normal checks of the equipment, make a final check of the area for clearance.

Step 2 Make sure the traffic control devices are in place.

Step 3 Downshift to the working speed.

REPRINTED COURTESY OF CATERPILLAR INC.

22305-14_F51.EPS

Figure 51 Grading a rural road.

Step 4 Position the blade at a 20- to 30-degree angle, with the toe of the blade behind the right front wheel, and so the heel casts material outside of the left rear wheel.

Step 5 Position the motor grader to grade from the right ditch or shoulder toward the center of the road on the first pass.

Step 6 Lower the blade to the desired depth of cut when the motor grader is in position. Normally, you would start on the right side of the road with the toe of the blade at the bottom of the ditch or at the edge of the shoulder.

Step 7 Begin grading by moving forward.

Step 8 At the end of the graded section, raise the blade and turn the motor grader around.

Step 9 Position the blade at a 20- to 30-degree angle with the toe of the blade just outside of the left front wheel.

Step 10 Adjust the elevation of the blade to maintain a crown in the road. This is important, because if the road is left flat or low in the center, water will not drain to the sides. Poor drainage creates conditions that cause potholes and ruts to form.

Step 11 Lower the blade and begin grading one side of the road. Continue back to the starting point and make additional passes, adjusting the blade as needed until the proper shape is obtained.

Step 12 Before leaving the work area, remove any debris and boulders that were loosened by the grading operation.

3.3.7 Grading Around an Object

It will occasionally be necessary to grade around an obstruction such as a large boulder, tree, wall, or pier. Moving the motor grader over, or steering around the object, will divert the pass and cause problems later. If possible, it is better to move the blade sideways to clear the object while keeping the motor grader's wheels in a straight line. Experienced operators can use this technique and still grade very close to the object; inexperienced operators should be very careful not to damage the blade.

3.3.8 Grading Slopes

Slope work requires that constant attention be given to the direction of the motor grader and the position of the blade. If possible, make the first pass from above the slope to establish the proper angle. Side-shift the circle rail and blade toward the slope. Then, set the blade angle slightly toward the rear of the machine with a slight forward tilt. Reach the blade as far down the slope as possible and make the first pass. The motor grader can be in the straight frame position or it can be articulated for extra reach.

On the remaining passes, lean the front wheels toward the slope and position the upper front wheel above the windrow. This helps prevent the front end from drifting down the slope. Articulate the tandem wheels down the slope to increase the stability of the machine. Position the toe of the blade forward, and angled, so the material spills off the heel of the blade slightly outside or under the tandems. Set the blade pitch ahead of the vertical plane and side-shift the circle rail toward the uphill side. Repeat the procedure until all excess material is deposited at the bottom of the slope. The excess material can be windrowed for pickup by other equipment.

> **WARNING!**
>
> When working on steep slopes, the differential may be locked or unlocked at the operator's preference. A rollover with the motor grader articulated is very unlikely; however, always use extreme caution when working on slopes that are steeper than 3:1. Never exceed the manufacturer's recommendation for working on side slopes.

3.3.9 Finish Grading

Finish grading involves making the final passes over an area to bring the grade to the proper elevation within the specified tolerance, and cleaning up any irregularities that may have formed on the surface. Finish grading can have several different meanings, depending on the type of construction being done.

Finish grading of a ditch may mean the cleaning up of any spoils that were cast into the bottom of the ditch. Finish or final grading of a subgrade or base would require the smoothing and trimming of the surface to the specified elevation.

Finishing of a roadbed or a pad to the required elevation requires references. These references are usually finish grade stakes called blue tops that are placed every 50 feet (15.24 meters) along the road. They are also placed at intervals across the roadbed if there is to be a change in the slope at specific points, such as at the crown or at the inside edge of the pavement. After all the rough grading has been done on the roadbed, and the shoulders have been built and cut off vertically on the roadside, surveyors will set the blue tops to guide the final grading. The tops of the stakes are placed at the required elevation and are sprayed blue for easy identification. *Figure 52* shows a simple profile of where finish subgrade stakes would be placed on a two-lane road with a crown in the middle. The blue tops should be placed at 50 foot (15.24 meter) intervals (or closer) along the entire length of the project to guide the grading operation.

Using a motor grader to perform finish grading often requires the assistance of a grade checker who works with the motor grader operator. The grade checker lets the operator know when to raise or lower the blade to meet the height of the top of the stakes. Taking care to stay in plain sight of the operator, the checker signals the operator about how to adjust the motor grader for the proper grade. Sometimes the grade checker will use spray paint and mark directly on the ground to show where and how much to cut or fill.

> **WARNING!**
>
> Operators must always keep grade checkers and other ground workers in sight to avoid the risk of hitting or running over them with the equipment.

When an automated guidance system is used for finish grading, the blade is controlled automatically. In this situation, the checker does not need to spot check the elevation until after the operator has made a pass. Several passes are generally needed to complete a finish grade.

The work is done with the blade tilted slightly forward. All passes must be slow so that the wheels do not slip and damage the surface. Grade the entire length of the area in one pass; stopping and starting will cause unevenness. Because very little excess material will be accumulated, do not worry about side-casting large piles of dirt and creating big windrows. Most of the excess material will fill in the depressions and low spots that need to be evened out along the way.

For less accurate grading, or coarse finishing, move at approximately five to seven miles per hour (8 to 11 kilometers per hour). When grading dirt or gravel roads on level ground, operate at five miles per hour (about 8 kilometers per hour) or less. For the most accurate grading, or fine finishing, a speed of no more than two to three miles per hour (about three to five kilometers per hour) is needed. It may be difficult to go this slow at first, but practice and patience are required to produce an accurately finished grade with the best possible appearance.

3.3.10 Snow Plowing

In many areas, motor graders are used for snow plowing. Because of their roading ability and power, graders are good machines for this purpose.

Operating a snowplow should be done with extreme caution. Working in poor environmental conditions means that you need to make every effort to protect yourself and the equipment from accident and injury. Snow presents an additional hazard because it covers many obstacles. Some obstructions can dislocate or damage the plow. Always use extreme caution when plowing around the following:

- Bridge and pavement expansion joints
- Headwalls of culverts
- Cattle guards
- Signposts and guardrails
- Hard-packed snow or ice
- Road shoulders
- Raised pavement markers, curbs, and islands
- Fire hydrants

Only experience will allow you to develop a feel for plowing snow under different conditions.

Before you begin plowing, perform the normal pre-operational inspection and then check that the plow is operational. Set the blade angle

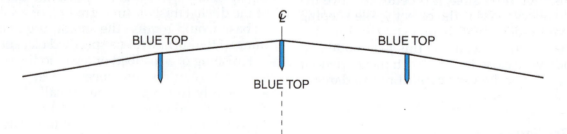

Blue top stakes show finished subgrade elevation.

22305-14_F52.EPS

Figure 52 Finish grade stakes.

as shallow as conditions allow. The shallower the angle, the wider the plow lane. A light, dry snow requires a shallow angle, but wet, heavy snow requires an acute angle.

> **NOTE**
>
> Be sure that the plow angle is not set so that it fails to clear a path for the grader wheels.

When plowing a roadway, plow in the direction of traffic flow so that the snow is removed to the shoulder of the road. For two-lane roads, plow from the centerline out to shoulders and make sure the blade overlaps the centerline slightly on the first pass. Plow to the low side of ramps or curves when possible. If plowed to the high side, the snow will melt across the highway and ice over. Be alert for vehicles approaching from behind. Rear-end collisions are common during snow removal operations. Plow away from wind whenever possible and practical. Clear snow beyond intersections before making a turnaround and raise the blade before making a turn. Use caution when making the turnaround.

Plowing can be performed in several different ways. One method uses the regular blade of the motor grader. Another method requires specially shaped snow plow blades (for instance, the V-plow blade or the one-way blade) that attach to the front of the motor grader or to the side of the regular blade.

For plowing snow with the regular blade, use the following procedure:

Step 1 Downshift to the working speed and tip the blade forward.

Step 2 Rotate the circle rail to sidecast material off the road.

Step 3 Line up the left front wheel of the motor grader on the center of the road.

Step 4 Lower the blade to the prescribed level. If no level has been prescribed, adjust it so it approximately 1 inch (2.5 centimeters) above the pavement.

> **CAUTION**
>
> Never adjust the blade level so that it digs into the pavement. This will damage the road surface and the blade.

Step 5 Lean the wheels in the direction of the side-cast snow.

Step 6 Continue moving the snow until the end of the pass is reached.

Step 7 Raise the blade and turn the machine around.

Step 8 Reposition the motor grader on the opposite side of the road for a return pass.

Step 9 Line up the left side, placing the left front wheel on the previously plowed area.

Step 10 Lower the blade to the pavement and continue the snow removal while making any adjustments necessary to avoid damaging the surface of the road. When the area is cleared of snow, raise the blade and move to the next area to be cleared.

The V-plow blade and the one-way blade are special attachments to the motor grader. Their attachment and use depend on the manufacturer's recommendations. Before using these attachments, read and understand the directions contained in the operator's manual.

3.3.11 *Roading a Motor Grader*

Because the motor grader is a rubber-tired machine, it can be driven on public highways with the proper safety precautions (*Figure 53*). The motor grader can be driven to various work sites at reasonable speeds over distances of several miles. This is common in rural areas where motor graders are used for county road maintenance and where traffic congestion is not a problem. Roading is not recommended on interstate highways or in heavily populated areas.

REPRINTED COURTESY OF CATERPILLAR INC.

22305-14_F53.EPS

Figure 53 Roading a motor grader.

Prepare the machine for roading by completing the following steps:

Step 1 Check all fluid levels and top off as required.

Step 2 Check the tire inflation pressure and inspect the tires for damage. Clean the windshield, side glass, and rear glass.

Step 3 Clean all lights and reflectors.

Step 4 Make sure the headlights, taillights, turn signals, and hazard warning lights function properly.

Step 5 Make sure the steering functions properly.

Step 6 Make sure the brakes function properly.

Step 7 Make sure all controls, indicators, and gauges function properly.

Step 8 Make sure there are flares, safety reflectors, and a fire extinguisher on the machine.

When traveling on public highways, obey all traffic regulations. Be alert and watch out for other drivers getting too close to the equipment. Ideally, there should be pilot vehicles driving in front and behind for added protection. Coordinate movements with these vehicles through the use of the mobile radio. Do not speed with the motor grader; it has a tendency to bounce at higher speeds, which could cause loss of control.

When it is necessary to leave the machine unattended, always do the following:

Step 1 Lower the blade and other attachments.

Step 2 Move the transmission control lever to neutral.

Step 3 Engage the parking brake.

Step 4 Pull the accelerator up past the detent to stop the engine.

Step 5 Turn the key start switch to the Off position and remove the key.

Step 6 Secure the equipment by locking the dashboard cover plate and cab. Chock the wheels.

If the machine is to be moved by truck, make sure the wheels of the motor grader fit on the bed of the transporter and do not hang over the edge. Drive the motor grader onto the trailer using the appropriate procedure for the type of trailer being used. Once in position, set the parking brake and put the transmission in neutral. The motor grader should be chained down using cross-chaining attached at the points specified in the operator's manual. Put wooden blocks underneath the blade and lower it to the bed of the transporter. Chock the wheels and turn off the main electrical switch.

Additional Resources

Basic Equipment Operator, NAVEDTRA 14081, 1994 Edition. Morris, John T. (preparer), Naval Education and Training Professional Development and Technology Center.

Operator's Manual Grader, Heavy Road, Motorized Caterpillar Model 130G, (NSN 3805-01-150-4795), TM 5-3805-261-10, 1992. Washington, DC: Department of the Army.

3.0.0 Section Review

1. During the shutdown of a motor grader, it is important to place the speed control in low idle and let the engine run for approximately five minutes because it _____.

   a. gives the operator time to check all the components
   b. helps clean water and debris out of the fuel lines
   c. enables the hour meter to synchronize and reset
   d. allows the engine fluids and parts to cool gradually

2. When a motor grader is pushing a load with the blade set at an angle, the movement of the load tends to swing the front of the motor grader toward the _____.

   a. downhill edge
   b. toe of the blade
   c. uphill edge
   d. front of the blade

3. A scraper is most likely to follow a motor grader so that it can _____.

   a. pick up material that has been windrowed by the grader
   b. make a more precise cut into the grade
   c. function as a ripper to remove rocks and debris
   d. cut the inslope and backslope of the ditch in one pass

SUMMARY

The motor grader's main purpose is to place and smooth material that is on the ground or other surfaces. Modern motor graders have an articulated frame that enables them to function in several configurations and perform numerous difficult tasks.

Like all heavy equipment, the motor grader is a large machine that takes practice and skill to operate well. Safety should be one of the primary concerns of the operator at all times. You must consider the safety of other workers and personnel at the job site, the equipment, and yourself. Some jobs will require working in and around traffic. Operators must also be alert to traffic and pedestrians in the general area. Do not operate the motor grader outside a work zone without proper flaggers or spotters.

Motor graders require maintenance on a regular basis. Perform a daily walk-around check of the equipment before beginning operations. During that time, carry out certain maintenance activities such as topping off fluids and cleaning and greasing fittings. Make sure that other maintenance is done periodically according to the schedules in the operator's manual.

There are six basic steps that should be followed when beginning to operate a motor grader:

- Look over the plan, check any grade stakes, and study the area by walking the job site.
- Position the motor grader directly in front of the work.
- Set the blade for the work to be done.
- Set the proper gear for the work.
- Drive straight ahead.
- Make as few blade changes as possible.

Uses of the motor grader include rough and finish grading, mixing and windrowing, spreading and dozing material, ditch cutting, ripping, and scarifying. Motor graders are also used for snow plowing in some parts of the country. When using the motor grader to perform finish grading, be sure to locate the stakes. Always keep the grade checker in view and follow the signals given for cutting and spreading material. This will reduce the chance of error and the possibility of an accident.

When roading the motor grader, obey all traffic signs and markings. Use pilot vehicles for additional safety whenever possible.

1. Which of the following motor grader activities allows other equipment, such as a scraper or a front-end loader, to pick up the graded material without having to do all of the preparation work?

   a. Straight dozing
   b. Finish grading
   c. Sidehill cutting
   d. Windrowing

2. Laser leveling or GPS equipment is most likely to be used with a grader that is doing _____.

   a. ditch cleaning
   b. finish grading
   c. rough grading
   d. snow plowing

3. Most modern motor graders have a(n) _____.

   a. rigid frame
   b. blade winch
   c. articulated frame
   d. scarified main blade

4. The component that is bolted onto the bottom of a moldboard to provide the primary digging action for the blade is called the _____.

   a. cutting edge
   b. drawbar
   c. scarifier
   d. lower end bit

5. The drawbar and circle rail allow an operator to _____.

   a. turn in a tight radius
   b. maneuver the blade
   c. reverse direction quickly
   d. make the front wheels lean

6. An instrument in the operator's cab that is most likely to display information about engine overspeed is the _____.

   a. voltmeter
   b. fuel level gauge
   c. engine temperature gauge
   d. tachometer

7. A control that is used to move the rear portion of a motor grader from the center position to the right or left of center is the _____.

   a. wheel lean control
   b. drawbar control
   c. articulation control
   d. circle rail control

8. A grader blade with a cutting edge that works much like a large rake is called a(n) _____.

   a. aggregate blade
   b. serrated blade
   c. wing blade
   d. straight blade

9. Reflected light that can cause eye fatigue or temporary blindness is especially hazardous to operators during _____.

   a. snow removal operations
   b. summertime activities
   c. any finish grading job
   d. early morning or late afternoon work

10. During a walk-around inspection, the operator notices that the engine fan belt is frayed. What is the most appropriate action to take?

    a. If the weather is cold, proceed with the job since the engine is unlikely to overheat.
    b. Loosen the belt slightly, spray the pulley groove with belt treatment, and continue until the belt will no longer function.
    c. Inform the foreman or field mechanic to fix the problem before beginning operations.
    d. Contact the belt manufacturer and ask for instructions on how to tighten the belt to prevent slippage.

11. Preventive maintenance on a motor grader is done primarily to _____.

    a. avoid having to perform daily walk-around inspections
    b. save operator time by not requiring them to check fluids
    c. meet federal requirements for maintaining service records
    d. avoid poor performance and breakdowns at critical times

12. Turning the battery disconnect switch to the OFF position and closing the fuel shutoff valve would most likely occur _____.

    a. during a prestart inspection check
    b. after the motor grader has been shut down
    c. at the end of the warm-up period
    d. just prior to starting up the grader

13. When a motor grader has to make a number of passes over a distance of less than 1,000 feet (about 305 meters), it is usually more efficient to _____.

    a. back the grader the entire distance to the starting point
    b. have a scarifier follow the grader to avoid overlap
    c. turn the grader around and proceed in the opposite direction
    d. articulate the grader to make numerous passes unnecessary

14. If a motor grader's blade is positioned perpendicular to the frame, the greatest amount of material will be _____.

    a. spilling off the heel of the blade
    b. left for the ripper to grade
    c. directed straight ahead
    d. spilling off the toe of the blade

15. To grade around an object that is in the way, it is best to _____.

    a. keep the blade stationary and steer the grader around the object
    b. articulate the front of the grader, but keep the blade and the rear wheels straight
    c. lift the toe of the blade to a high sloping angle to clear the object
    d. keep the grader's wheels in a straight line and move the blade sideways

Trade Terms Introduced in This Module

Articulated-frame motor grader: Frames that have a hinge or joint in the middle and are able to pivot for better traction and handling.

Blade: The blade comprises the moldboard, cutting edge, and end bits.

Blending: The mixing of two or more materials together until they are uniformly distributed throughout the soil mass.

Cutting edge: The sharp steel bar attached to the bottom of the moldboard used for cutting into the ground.

Heel of the blade: The following end of the blade, regardless of which side of the machine the blade is located on.

Moldboard: The long concave strip of metal that is used to push the dirt. The cutting edge is fastened to the bottom of the moldboard.

Pitch: The angle of the moldboard in relation to a vertical plane.

Rigid-frame motor grader: A single metal frame that extends from the front to the back of a motor grader.

Scarifying: To loosen the top surface of material using a set of metal shanks (teeth).

Side-casting: Placing material in a windrow by pushing the load off the trailing end of the moldboard.

Toe of the blade: The leading end of the blade, regardless of which side of the machine the blade is located on.

DAILY CHECKS AND LUBRICATION

Check the crankcase oil level:

- Check the oil level on the dipstick. The oil level must be between the marks. Some dipsticks have a cold check side and a hot check side. Be sure to read the cold side when performing a prestart inspection or checking a cold engine.
- If oil is to be added, remove the filler cap and add the required amount of oil to the crankcase. Replace the filler cap.
- Replace the dipstick when the oil level is correct.

Check the coolant level:

- Remove the radiator filler cap. Coolant level must be within 3 inches of the bottom of the filler neck.
- If coolant is low, add sufficient coolant to restore to the correct level.
- Replace the filler cap.

Drain the fuel filter sediment:

- Loosen the drain screw and drain the fuel for several seconds. Retighten the drain screw.

Check the transmission/hydraulic oil level:

- Be sure all equipment is lowered.
- A hydraulic check may be done with a dipstick or sight glass on the tank.
- Remove the dipstick. Oil must be in the crosshatched area on the dipstick. If necessary, remove the filler cap and add required oil.
- Replace the dipstick.

Lubricate the lift arms and pivot points:

- Lubricate the lift arm cylinders and bucket pivot points with two shots of grease each.
- Lubricate the steering cylinder, drive shaft, and pivot frame bearings.

Check the tires:

- Check the pressure.
- Check for cuts and tread wear.
- Check behind the tires for any obstruction around the axle.

Check the service and parking brakes to ensure that they are working properly.

Additional Resources

This module presents thorough resources for task training. The following resource material is suggested for further study.

Basic Equipment Operator, NAVEDTRA 14081, 1994 Edition. Morris, John T. (preparer), Naval Education and Training Professional Development and Technology Center.

Operator's Manual Grader, Heavy Road, Motorized Caterpillar Model 130G, (NSN 3805-01-150-4795), TM 5-3805-261-10, 1992. Washington, DC: Department of the Army.

Figure Credits

Deere & Company, Module opener, Figures 1–5, 7, 10, 21, 23, 27, 31, 32, 47

Reprinted Courtesy of Caterpillar Inc., Figures 6, 8, 11–18, 20, 22, 24–26, 28–30, 33–35, 38–45, 49–51, 53, SA01–SA04

Komatsu America Corp., Figure 9

Tim Davis/NCCER, Figures 19, 36, 37

Section Review Answers

Answer	Section Reference	Objective
Section One		
1.c	1.1.1	1a
2.a	1.2.0	1b
3.d	1.3.0	1c
4.b	1.4.1	1d
5.a	1.5.4	1e
Section Two		
1.c	2.1.1	2a
2.d	2.2.0	2b
3.b	2.3.0	2c
Section Three		
1.d	3.1.3	3a
2.b	3.2.3	3b
3.a	3.3.1	3c

NCCER CURRICULA — USER UPDATE

NCCER makes every effort to keep its textbooks up-to-date and free of technical errors. We appreciate your help in this process. If you find an error, a typographical mistake, or an inaccuracy in NCCER's curricula, please fill out this form (or a photocopy), or complete the online form at **www.nccer.org/olf**. Be sure to include the exact module ID number, page number, a detailed description, and your recommended correction. Your input will be brought to the attention of the Authoring Team. Thank you for your assistance.

Instructors – If you have an idea for improving this textbook, or have found that additional materials were necessary to teach this module effectively, please let us know so that we may present your suggestions to the Authoring Team.

NCCER Product Development and Revision
13614 Progress Blvd., Alachua, FL 32615

Email: curriculum@nccer.org
Online: www.nccer.org/olf

❏ Trainee Guide   ❏ Lesson Plans   ❏ Exam   ❏ PowerPoints   Other _____

Craft / Level: _____ Copyright Date: _____

Module ID Number / Title: _____

Section Number(s): _____

Description: _____

_____

_____

_____

_____

Recommended Correction: _____

_____

_____

_____

_____

Your Name: _____

Address: _____

_____

Email: _____ Phone: _____

Glossary

Articulated-frame dump truck: A type of off-road dump truck that has a permanent pivoting point In the frame that allows the front of the truck to turn so that all of the wheels follow the same path.

Articulated-frame motor grader: Frames that have a hinge or joint in the middle and are able to pivot for better traction and handling.

Automatic retarder control system: A system that works with the engine brake and the traction control system to electronically slow the vehicle during downhill travel.

Bank state: A term indicating that material, such as soil, is in its natural state. Also referred to as virgin ground.

Blade: The blade is comprised of the moldboard, cutting edge, and end bits; the main attachment on the front end of the machine that is used for moving material.

Blade float: When the blade is allowed to float over the surface for a smooth finish.

Blade pitch: The angle of the blade from the vertical.

Blade tilt: The angle of the blade from the horizontal.

Blending: The mixing of two or more materials together until they are uniformly distributed throughout the soil mass.

Blue tops: Stakes that are used to identify the final elevation of the sub-base. The tops are either painted with blue paint or covered with blue plastic tassels.

Boom: A part of the digging component (hoe) of an excavator. The boom is attached to the upper carriage and serves as the main lifting stick of the machine.

Breakout force: The amount of digging force that can be exerted by the bucket on material in its natural state.

Bucket: A U-shaped closed-end scoop that is attached to the backhoe.

Canopy: A section of an off-road truck body that extends above the operator's cab to help protect the operator from falling material.

Charging hoppers: The process of filling a hopper or temporary storage container with a material, such as coal, that will be dispersed later.

Circle rail: A ring-shaped component on the motor grader that controls the horizontal position and elevation of the moldboard.

Compaction: Using an engineered process, such as rolling, tamping, or soaking, to reduce the bulk and increase the density of soil.

Counterweight: Dead or nonworking load that is attached to one end or side of a machine to help improve the balance.

Crowd: The process of forcing the stick into digging, or moving the stick closer to the machine.

Crown: A slightly built-up section of the roadway, usually along the center line. The crown provides a slope to the pavement so that water will drain to either side.

Curl: Rotate the bucket.

Cutting edge: The sharp steel bar attached to the bottom of the moldboard used for cutting into the ground.

Demolition: The tearing down of buildings and other structures.

Density: The ratio of the weight of a substance to its volume.

Detention pond: A pond that will temporarily hold storm water until it can drain elsewhere.

Dozing: Using a blade to scrape or excavate material and move it to another place.

Draw: Move the stick back toward the operator.

Engine retarder: An alternate braking system activated from the cab that slows down the vehicle by reducing engine power.

Exhaust body heating system: A system that diverts some of the exhaust gas from a truck's exhaust pipe(s) to conduits in the body to help prevent material in the body from freezing.

Extend: Move the extendable stick outward.

Foot: In tamping rollers, one of a number of projections from a cylindrical drum that contact the ground.

Geotextile: A synthetic material used to filter water and soil, reinforce soil, or separate to unlike materials.

GLONASS: A Russian-owned Global Navigation Satellite System operated by the Ministry of Defense of the Russian Federation.

GNSS: An acronym for Global Navigation Satellite System. GNSS is the generic term used to describe a locating or grade control system that uses signals from either GPS or GLONASS satellites.

Gradient: The change of elevation per unit length; the slope along a specific line of a road surface, channel, or pipe.

Grapple: An excavator attachment that is used for lifting. The grapple has two or more hooks that can be closed on each other to pick up solid material, such as logs, by grabbing onto them.

Ground contact pressure (GCP): The weight of the machine divided by the area in square inches of the ground directly supporting it.

Grouser: A ridge or cleat across a track shoe that improves the track's grip on the ground.

Grubbing: Digging out roots and other buried material.

Haul truck: A name that is sometimes used to describe a rigid-frame dump truck or a mining truck.

Haunches: The portions of a pipe extending from the bottom to the springline, which is line of maximum diameter.

Heel of the blade: The following end of the blade, regardless of which side the blade is pointing.

Hoe: The component of an excavator that is made up of the boom, stick, and bucket. Also, a term used to refer to any type of excavator that digs by pulling the bucket from front to back.

Hoist: The mechanism used to raise and lower the dump body, typically consisting of two hydraulic cylinders.

Hubs: Surveying stakes set for reference purposes. Hubs are usually at the edge of, or outside of, the work area.

Hydraulic breaker (hammer): A hydraulic attachment for an excavator that is used for breaking boulders and other solid objects.

Infiltration system: Storm water runoff control system that filters contaminants from the runoff before the runoff is allowed to infiltrate soil.

Inslope (foreslope): The slope of a drainage ditch that is between the shoulder of the road and the flow line of the ditch.

Invert: The flow line of a pipe. This would be the bottommost visible surface on the inside of a pipe.

Joystick: A control mechanism that pivots about a fixed point and is used to control the mo-tion of an object. Joystick controls are used in an excavator to control the movement of the boom, stick, bucket, and upper carriage.

KG blade: A type of dozer-mounted blade used in forestry and land clearing operations. On some blades, a single spike called a stinger splits and shears stumps at the base.

Knuckle boom: A term sometimes used for a gooseneck boom and stick combination that resembles a knuckle at the pivot point of the boom and stick.

Lift: A layer of material that is a specific thickness; the depth of material that is being rolled by the roller.

Mat: Asphalt as it comes out of a spreader box or paving machine in a smooth, flat form.

Moldboard: The long concave strip of metal that is used to push the dirt. The cutting edge is fastened to the bottom of the moldboard.

Pad: On a segmented or sheepsfoot roller, the part of the roller that contacts the ground; also called the foot.

Pawl: Pivoted lever which has a free end to engage with the teeth of a cogwheel or ratchet so that the wheel or ratchet can only move one way.

Pitch: The angle of the moldboard in relation to a vertical plane.

Platen: A flat plate that is used to support another structure, and that can be rotated around a pivot point. The platen is the supporting base of a turntable mechanism.

Pozzolan: Originally used for volcanic ash, but now applies to any similar material that re-acts with lime and water to form cement.

Puddling: A process in which water is added to the soil until it is semi-liquid; the soil is then allowed to dry before being vibrated.

Quarries: Excavations or pits in which gravel, stone, and other material are mined.

Reach: Extend the stick away from the cab.

Red tops: Stakes that are used to identify the final elevation of the crushed aggregate base course.

Retention pond: A pond that will permanently hold the same level of water.

Retract: Move the extendable stick in.

Rigid-frame dump truck: A type of off-road truck in which the cab and the body are mounted on a common, non-pivoting, frame. Rigid-frame dump trucks are commonly referred to as haul trucks or mining trucks.

Rigid-frame motor grader: A single metal frame that extends from the front to the back of a motor grader.

Ripping: Loosening hard soil, concrete, asphalt, or soft rock.

Riprap: Broken stone, in pieces weighing from 15 to 150 pounds (6 to 68 kilograms) each, placed on the ground for protection against the action of water.

Robotic total station: A machine control system that uses infrared signals reflected by a target mounted on the grading equipment to provide data to the equipment's on-board computer about how to adjust the blade.

Rubble: Fragments of stone, brick, or rock that have broken apart from larger pieces.

Scarifying: To loosen the top surface of material using a set of metal shanks (teeth).

Settling: The natural wetting and drying process whereby soil particles become more compact and denser.

Sheepsfoot: A tamping roller with feet expanded at the outer tips.

Shooting-boom excavator: A term sometimes used to describe a telescoping-boom ex-cavator.

Side-casting: Placing material in a windrow by pushing the load off the trailing end of the moldboard.

Slope: Ground that forms a natural or artificial incline.

Slot dozing: A method that creates a trench for controlling spillage from the blade ends.

Spoil: Soil, rock, or other material that is removed from a ditch or other excavation.

Spoils: Excavated material from a digging operation.

Stick: The part of the digging component of an excavator between the end of the boom and the bucket. The bucket is attached to the end of the stick.

Stockpile: Material that is dug and piled for future use.

Superelevation: The increased elevation of one side of a curved roadway that allows for banking of the pavement to the inside.

Tamping roller: One or more steel drums fitted with projecting feet and towed with a box frame.

Tassel (whisker): A small plastic colored tag affixed to grade stakes for identification purposes.

Toe of the blade: The leading end of the blade, regardless of which side of the machine the blade is located on.

Traction control system: A computerized system that diverts torque from a spinning wheel to one or more of the other wheels to improve traction.

Trenching: Using a backhoe to dig a long, straight excavation with vertical walls.

Undercarriage: The lower frame of an excavator that supports the turntable and has the tracks or wheels attached.

Undercutting: The digging of material from beneath the machine. Also, the digging of material from a bank or vertical face of a trench below the grade.

Underground mining dump truck: A type of articulated truck in which the vehicle height is reduced by moving the cab forward of the front axle, thus enabling the truck to be used in underground mining applications.

Upper carriage: The upper frame of an excavator that includes the turntable, engine, cab, controls, and counterweights.

Upper structure: The revolving turntable, swing mechanism, counterweight, boom, and cab of a telescoping-boom excavator.

Vibratory roller: A compacting device that mechanically vibrates the soil while it rolls. It can be self-propelled or towed.

Virtual reference station (VRS): An imaginary reference station that is established by a computer based on data from numerous GPS receivers and the grading equipment that serves as the source of data for grading control.

Winch: An attachment commonly mounted on the rear of a dozer that uses an electric motor and wire rope to pull equipment, trees, and other objects.

Index

A

ABS. *See* Anti-lock braking systems (ABS)
Air brakes, (22310):12
Air pollution
    diesel engine exhaust emissions, (22310):6, 19
    dust controls, (22307):17
Angling blade, (22302):10
Anti-lock braking systems (ABS), (22310):13
Appian Way, (22307):2
Armrest controls, (22305):11
Articulated-frame motor grader, (22305):1, 4, 6, 47
Articulated-frame off-road dump trucks
    components
        axles, (22310):5, 6
        basics, (22310):5
        body and body liners, (22310):6–7
        engine compartment, (22310):6
        exhaust body heating system, (22310):6–7
        extensions, (22310):6–7
        instrument panel/alert indicators, (22310):9, 10, 11
        operator's cab, (22310):6
        spill guard, (22310):7
        tailgate, (22310):7
        typical, (22310):7
    defined, (22310):1, 32
    rigid-frame off-road trucks vs., (22310):6
    tractor trailer vs., (22310):5
    uses, (22310):6
Asphalt
    compacting, (22203):4, 31
    grinding to grade, (22303):14, 16
    ripping, (22305):16
Asphaltic binders, (22307):15–16
Augers, (22303):13
Auto-leveling alignment lasers, (22307):4
Automatic guidance systems
    for backhoes, (22307):30
    for dozers, (22302):11, (22307):2, 3
    for excavators, (22304):41–42, (22307):30
    GPS-based, (22305):3, 17–19, (22307):9–12
    laser-based, (22304):41–42, (22305):17–18, 32, (22307):2, 3, 30
    for motor graders, (22305):11, 16–19, 32, 40
    robotic total station, (22307):1, 8, 9, 10, 12, 36
    satellite-based, (22307):8–9
Automatic retarder control system, (22310):1, 13, 32
Axles
    articulated-frame dump trucks, (22310):5, 6
    rigid-frame dump trucks, (22310):2, 3, 4

B

Backfilling, (22302):30–31, (22303):34
Backhoe/loader, (22303):1, 5
Backhoes
    advantages of, (22303):1
    attachments, types and uses
        augers, (22303):13

        buckets, (22303):14, 15
        cold planer, (22303):14, 16
        compactor, (22303):14
        hydraulic breaker, (22303):14, 15, 32–33
        rippers, (22303):13
        street pads, (22303):13–14, 15
    components
        boom cylinder, (22303):4
        control panel, (22303):5–7
        engines, (22303):5
        excavating arm, (22303):1, 3, 4
        extendable stick, (22303):2
        instrument panel/alert indicators, (22303):8–9
        operator's cab, (22303):4, 5
        stabilizers, (22303):4, 13–14, 27–28, 38
        stick cylinder, (22303):4
        typical, (22303):4
    configurations, (22303):1–2
    controls
        boom, (22303):9
        boom swing, (22303):9
        bucket, (22303):9
        four-lever, (22303):11–12
        hydraulics, (22303):4
        normal operations, (22303):9
        stabilizers, (22303):9
        stick, (22303):9
        two-lever, (22303):10–11
        vehicle movement, (22303):12–13
    gauges
        engine coolant temperature, (22303):9
        fuel level, (22303):8
        prestart inspection, (22303):25
        tachometer, (22303):9
        transmission oil temperature, (22303):8
        voltmeter, (22303):8
    maintenance records, (22303):21
    operations
        attachments, connecting/disconnecting, (22303):13
        automatic guidance systems, (22307):30
        efficiency in, (22303):23–24
        gauge and indicator checks, (22303):25
        maintenance, (22303):20–21, 47–49
        maneuvering, (22303):6, 26–29
        preparations, (22303):24
        prestart inspections, (22303):18–20
        on slopes, (22303):28
        startup and shutdown procedures, (22303):24–25
        in unstable soil, (22303):28, 38–39
    range of motion, (22303):4
    roading, (22303):39–40
    safe operation
        co-worker safety, (22303):18
        equipment, (22303):18
        operator safety, (22303):17
    sizing for trenching, (22303):30

Backhoes (*continued*)
  transporting, (22303):40
  types of, (22303):1–2
  work activities
      backfilling, (22303):34
      common, (22303):1
      confined areas, (22303):36–37
      demolition, (22303):32–33
      excavation, footings and foundations, (22303):35
      pipe, setting, (22303):33–34
      stockpiles, loading from, (22303):30–31
      trenching and loading, (22303):30, 31–32, 37–38
Bank state, (22302):23, 27, 44
Base course finish grading, (22307):26–27
Bedding material, placing, (22304):37
Binders, (22307):14–16, 17
Bitumens, (22307):15–16
Blade, (22302):1, 44, (22305):1, 47. *See also* Dozer blades;
      Motor grader blades
Blade angle, (22203):14
Blade float, (22302):23, 27–28, 32, 44
Blade lift, (22203):14
Blade pitch, (22203):14, (22302):23, 27, 44
Blade tilt, (22203):14, (22302):23, 27, 44
Blading, (22304):40
Blending, (22305):1, 35, 47
Blue tops, (22305):39–40, (22307):22, 36
Body and body liners
  articulated-frame off-road trucks, (22310):6–7
  rigid-frame dump trucks, (22310):3, 4, 5
Bomag steel-wheel rollers, (22203):7
Boom
  backhoe controls, (22303):9
  defined, (22304):1, 47
  track-mounted hydraulic excavator, (22304):8–9
Boom cylinder, (22303):4
Boom extensions, (22304):16
Boom swing, (22303):9, (22304):29, 30
Brake oil temperature gauge, (22310):8
Braking system
  compaction equipment, (22203):17
  dozers, (22302):7, 8
  off-road dump trucks, (22310):12–13, 25, 28
  telescoping-boom excavators, (22304):11–13
Breakout force, (22304):1, 13, 47
Bucket operations, (22303):9, (22304):27–29
Buckets, types and uses
  backhoe, (22303):14, 15
  defined, (22303):1, 46
  ditch cleaning, (22303):15, (22304):6, 15, (22307):4
  dredging, (22304):14
  formed ditch bucket, (22304):15
  mass excavation, (22304):13
  multi-purpose, (22303):15
  pavement removal, (22304):7, 15
  ripper bucket, (22304):7
  rock bucket, (22304):14
  side tip bucket, (22303):15
  telescoping-boom excavators, (22304):2, 7
  trenching, (22304):13
  utility bucket, (22304):13

C

Calcium chloride, (22307):16
Call Before You Dig number, (22304):19
Canopy, (22310):1, 32
Caterpillar 815F Series 2, (22203):7–8

Caterpillar 825G series, (22203):9
Cement, (22307):15
Charging hoppers, (22302):1, 11, 44
Circle rail, (22305):7, 8, 13, (22307):22, 28, 36
Coal blade, (22302):13
Cold planer, (22303):14, 16
Combination rigid-frame off-road body, (22310):5
Commercial driver's license, (22310):1
Compaction
  asphalt, (22203):31–32
  backfilling process, (22203):30
  backhoes for, (22303):14
  cement, (22203):31
  defined, (22203):1, 37
  dozer attachment for, (22302):14, 15, 36
  and leveling process, (22203):30
  method selection variables, (22203):28–29
  quality testing, (22203):29–30
Compaction equipment
  attachments
      dozer blades, (22203):14, 30
      interchangeable rollers, (22203):13
      water spray units, (22203):13–14, 17
  components
      instrument panel/alert indicators, (22203):9–10, 25
      operator's cab, (22203):8
  controls
      disconnect switches, (22203):11
      engine start switch, (22203):12–13
      seat adjustment, (22203):11–12
      steering wheel adjustment, (22203):11–12
      vehicle movement, (22203):13
  historically, (22203):1
  operations
      gauge and indicator checks, (22203):25
      inspection, (22203):18
      maintenance, (22203):18–20
      maneuvering basics, (22203):25–26
      preparations, (22203):23
      of scrapers, (22203):17
      startup and shutdown procedures, (22203):24–25
  operator's manual, (22203):7
  safe operation
      basic guidelines, (22203):16–17
      braking, (22203):17
      responsibility for, (22203):16
      on slopes, (22203):17
      tires and, (22203):17–18, 21
  selection of, (22203):27
  servicing rollers, (22203):19–20
Compass satellite system, (22307):9
Confined areas backhoe operation, (22303):36–37
Counterweight, (22304):1, 9, 47
Co-worker safety, ensuring around
  backhoes, (22303):18
  dozers, (22302):18
  excavators, (22304):18–19
  motor graders, (22305):21–22
  off-road dump trucks, (22310):18
Crawler dozers, (22302):1
Crowd, (22303):4, 9, 46
Crowns, (22307):2, 14, 16, 24–25, 36
Curbs, finish grading, (22307):31
Curl, (22303):4, 9, 46
Cushion blade, (22302):11–12
Cutter head, (22304):15
Cutting edge, (22305):1, 7, 47

D

Demolition, (22304):1, 5, 7–8, 39–40, 47. *See also* Hydraulic breaker (hammer)
Density, (22203):1, 37
Detention pond, (22307):14, 18, 36
Diesel/electric powertrain, rigid-frame dump truck, (22310):2, 3
Diesel engines exhaust emissions, (22310):6, 19
Dipper stick, (22304):1–2, 8–9
Disc brakes, (22310):13
Discing, (22302):36, 38
Disconnect switches, (22203):11
Discs, (22302):14, 15
Distance meters, laser, (22307):4
Ditch buckets, (22303):15, (22304):15, (22307):4
Ditches
   cutting, shaping, cleaning (ditching)
      dozers, (22302):29–30
      motor graders, (22305):3, 35–37
      telescoping-boom excavators, (22304):6
   finish grading, (22307):27–29
Dozer blades
   controls, (22203):14, (22302):9
   maneuvering, (22302):27–28
   selecting, (22302):9–10, 27–28
   types of
      coal, (22302):13
      cushion, (22302):11–12
      general-purpose, (22302):10
      KG blade, (22302):1, 12, 13, 44
      landfill, (22302):12
      production, (22302):11
      push plate, (22302):11
      rake, (22302):12, 13
      V-tree cutter, (22302):12–13, 29–30
      wood chip, (22302):13
Dozers
   alternate names for, (22302):1
   attachments, types and uses
      discs, (22302):14, 15, 36, 38
      rippers, (22302):13–14
      rollers, (22302):14, 36–38
      side booms, (22302):14–16, 38, 39
      tree shear, (22302):14
      winches, (22302):1, 2, 14, 15, 44
   components
      instrument panel/alert indicators, (22302):4–8, 26
      operator's cab, (22302):2–4, 7
      typical, (22302):2, 3
   controls
      blade controls, (22203):14, (22302):9
      GPS-based, (22302):11
      rippers, (22302):9
      seat adjustment, (22302):4, 25
      vehicle movement, (22302):2, 7–8
   gauges
      engine coolant temperature, (22302):5–6
      fuel level, (22302):6
      hydraulic oil temperature, (22302):6–7
      torque converter oil temperature, (22302):6
   large vs. small, (22302):1–2
   operations
      automatic guidance systems, (22302):11, (22307):2, 3
      blade operations, (22302):27–28
      efficiency in, (22302):25
      inspections, (22302):19–20, 45–46
      maintenance, (22302):20–21
      maneuvering basics, (22302):25–27
      startup and shutdown procedures, (22302):23–25
      in unstable soil, (22302):33
      warm-up procedure, (22302):24
   performance variables, (22302):27–28
   safe operation
      attachments, (22302):13, 15
      co-worker safety, (22302):18
      equipment, (22302):18–19
      operator safety, (22302):17
      public safety, (22302):18
      in rocky areas, (22302):36
      seat adjustment, (22302):25
      windshield wipers, (22302):4
   transporting, (22302):38–39
   types of, (22302):1–2, 3
   uses, (22302):1
   work activities
      backfilling, (22302):30–31
      compaction, (22302):36–38
      confined areas, (22302):32–33
      discing, (22302):38
      ditching, (22302):29–30
      dozing, (22302):28–29
      finishing, (22302):32, (22307):2
      heavy loads, hoisting and carrying, (22302):38, 39
      land clearing, (22302):30
      large objects, moving, (22302):31–32
      pushing scrapers, (22302):33–34
      ripping, (22302):35
      sidehill cutting, (22302):31
      slopes, cutting and building, (22302):29–30
      stockpiles, building, (22302):29–30
      towed attachments used in, (22302):36–38
      winching, (22302):34–35
Dozing, (22302):1, 12, 28–29, 44
Draw, (22303):30, 46
Drawbar, (22305):7, 8
Dredging bucket, (22304):14
Drifting. *See* Dozing
Driver's license, commercial, (22310):1
Dropping-weight compactor, (22203):5–6
Dual slope rigid-frame off-road body, (22310):5
Dump trucks. *See also* Off-road dump trucks
   functions, (22310):1
   on-road vs. off-road, (22310):1
Dust controls, (22307):16, 17

E

EG. *See* Electrical density gauge (EG)
Electrical density gauge (EG), (22203):29–30
Electrical shock, (22304):20
Emergency shutoffs, (22310):15
Enclosed rollover protective structures (EROPS), (22303):4
Engine body heating system, (22310):27
Engine coolant temperature gauge, (22203):9, (22302):5–6, (22303):9, (22310):8
Engine key switch, (22302):5
Engine overspeed indicator, (22310):8
Engine retarder, (22310):1, 13, 32
Engines
   articulated-frame dump trucks, (22310):6
   backhoes, (22303):5
   rigid-frame dump trucks, (22310):2, 3, 4
Engine start switch, (22203):12–13
Equipment safety
   dozers, (22302):18–19
   excavators, (22304):19
   motor graders, (22305):22

Equipment safety (*continued*)
   off-road dump trucks, (22310):18–19
EROPS. *See* Enclosed rollover protective structures (EROPS)
Erosion control, (22307):18–20
E-stick, (22303):2
Excavating arm, backhoes, (22303):1, 3, 4
Excavators. *See also* Telescoping-boom excavators
   advantages of, (22304):30
   attachments, types and uses
      blades, (22304):15, 40
      boom extensions, (22304):16
      buckets, (22304):1, 6, 7, 13–15
      cutter head, (22304):15
      grapple, (22304):1, 8, 15, 41, 47
      hydraulic breaker (hammer), (22304):8, 16, 39–40, 47
      magnets, (22304):8, 41
      pile driver, (22304):15, 16
      rippers, (22304):15
      ripper tooth, (22304):8
      shears, (22304):8, 15, 16
      thumb, (22304):15
   components, (22304):1–2
   lift capacity, (22304):35–36, 48
   maintenance records, (22304):23
   maneuvering
      basics, (22304):25–27
      hoe and bucket operations, (22304):27–29
      safety features, (22304):28–29
      stability guidelines, (22304):30
   operations
      automatic guidance systems, (22304):41–42, (22307):30
      efficiency in, (22304):30–31
      maintenance, (22304):21–23
      prestart inspection, (22304):20–21
      in standing water, (22304):30
      startup and shutdown procedures, (22304):24–25
      in unstable soil, (22304):30, 38–39
      warm-up procedure, (22304):25
   safe operation
      co-worker safety, (22304):18–19
      equipment, (22304):19
      general guidelines, (22304):30–31
      operator safety, (22304):18
      overhead power lines, (22304):19–20
      public safety, (22304):18–19
      for stability, (22304):30
      swing momentum, (22304):29, 30
      underground utilities, (22304):19–20
      when transporting, (22304):29
   track-mounted, (22304):2–3, 4, 8–12
   transporting, (22304):29, 42–43
   types of, (22304):1–2
   wheel-mounted, (22304):2–3, 4
   work activities
      bedding material, placing, (22304):37
      blading, (22304):40
      demolition, (22304):7–8, 39–40
      digging, (22304):5–6, 31–33, 40–41
      foundations, (22304):33, 34
      lifting objects, (22304):7, 35–36
      loading trucks, (22304):33–34, 35, 40–41
      logging, (22304):41
      material handling, (22304):41
      pipe, lifting, laying, and setting, (22304):7, 36–38
      scrap-handling, (22304):8, 41
      trenching, (22304):5–6, 36–37
Exhaust body heating system, (22310):1, 6–7, 32
Exhaust emissions, diesel engines, (22310):6, 19

Extend, (22303):1, 2, 46
Extendable stick backhoes, (22303):2
Extenda-dig option, (22303):2

F

Falling object protective structure (FOPS), (22203):8, (22310):3
Filtration, geotextile, (22307):19
Fingertip armrest controls, (22305):11
Finish grade stakes, (22305):39–40, (22307):22, 24
Finish grading
   base course, (22307):26–27
   crowns, (22307):2, 14, 16, 24–25, 36
   curbs, (22307):31
   ditches, (22307):27–29
   parking lots, (22307):31
   sidewalks, (22307):31
   slopes, (22304):6–7, 15, (22305):39, (22307):25–26
   specifications, (22307):22–25
   subgrade, (22307):26–27
   superelevation, (22307):25
   trenches, (22307):29–30
Finish grading equipment
   dozers, (22302):32, (22307):2
   motor graders, (22305):2–3, 39–40, (22307):1–2
   scrapers, (22307):2–3
   telescoping-boom excavators, (22304):6–7, (22307):3–4
Finish grading instrumentation
   automatic grade control systems
      GPS-based, (22305):3, (22307):9–12
      laser-based, (22307):7–8
      robotic total station, (22307):8, 9, 10
      satellite-based, (22307):8–9
      spot checking with, (22305):40
      virtual reference station (VRS), (22307):1, 9, 36
   laser instruments
      benefits of using, (22307):4
      calibration, (22307):7–8
      for dozers, (22307):2, 3
      maintenance, (22307):7
      operation of, (22307):6
      receivers, (22307):5–6
      safety, (22307):6–7
      setup, (22307):6
      transmitters, (22307):4–5
Fixed-beam laser, (22307):4
Fly ash, (22307):15
Foot, (22203):16, 37
Footings excavation, (22303):35
FOPS. *See* Falling object protective structure (FOPS)
Foreslope (inslope), (22304):1, 6, 47
Forestry, (22302):13–14, (22304):41
Formed ditch bucket, (22304):15
Foundation excavations, (22303):35, (22304):33, 34
Four-lever backhoe controls, (22303):11–12
Fuel level gauge, (22203):10, (22302):6, (22303):8, (22305):10, (22310):8
Fuel shutoff switch, (22203):11

G

Galileo satellite system, (22307):9
Gateless coal, (22310):5
GCP. *See* Ground contact pressure (GCP)
Geotextiles, (22307):14, 18–20, 36
G-force, (22307):26
Global Navigation Satellite System (GNSS), (22307):1, 9, 36
Global Positioning System (GPS), (22307):8–9, 12

Global Positioning System (GPS)-based guidance systems
    dozers, (22302):11, (22307):3
    excavators, (22304):42
    grade control, (22307):3, 9–12
    motor graders, (22305):3, 17–19
GLONASS, (22307):1, 9, 36
GNSS. *See* Global Navigation Satellite System (GNSS)
Gooseneck boom, (22304):1–2
GPS. *See* Global Positioning System (GPS)
Grade checker, (22305):40
Grade control systems, automatic. *See* Automatic guidance
    systems
Gradient, (22307):22, 27, 36
Grading blade, (22304):15
Grapple, (22304):1, 8, 15, 47
Ground contact pressure (GCP), (22203):1, 2, 37
Grouser, (22302):17, 20, 44
Grubbing, (22302):1, 12, 23, 44

H

Hammer. *See* Hydraulic breaker (hammer)
Haul truck, (22310):1, 2, 32
Haunches, (22304):24, 38, 47
Heavy-duty U blades, (22302):11, 12
Heel of the blade, (22305):27, 30, 47
High-speed dozers, (22302):1–2, 3
High-track dozers, (22302):1, 3
Hoe, (22304):1, 2, 8, 47
Hoist, (22310):1, 32
Hoist controls, (22310):13–14
Hoisting and carrying activities, (22302):38, 39
Hour meter, (22203):10, (22310):8
Hubs, (22307):22, 36
Hydrated lime, (22307):15
Hydraulic brakes, (22310):12–13
Hydraulic breaker (hammer)
    backhoes, (22303):14, 15, 32–33
    defined, (22304):1, 47
    excavators, (22304):8, 16, 39–40
Hydraulic excavator. *See* Excavators
Hydraulic oil temperature gauge, (22203):10, (22302):6–7,
    (22305):10

I

Idle control systems, (22310):10
Infiltration system, (22307):14, 18, 36
Ingersoll-Rand DD-138 steel-wheel vibratory roller,
    (22203):8
Inslope (foreslope), (22304):1, 6, 47
Instrument panel/alert indicators
    articulated-frame off-road trucks, (22310):9, 10, 11
    backhoes, (22303):8–9
    compaction equipment, (22203):9–10, 25
    dozers, (22302):4–8, 26
    motor graders, (22305):8–10
    off-road dump trucks, (22310):7–9
    rigid-frame dump trucks, (22310):8–9, 11
    telescoping-boom excavators, (22304):11
Invert, (22307):22, 29, 36

J

John Deere Grade Pro (GP) fingertip armrest controls,
    (22305):11
Joystick, (22304):1, 10, 47
Joystick controls
    backhoes, (22303):12–13
    dozers, (22302):2, 7–8

excavator hoe and bucket operations, (22304):27–29
motor graders, (22305):7, 10–12

K

KG blade, (22302):1, 12, 13, 44
Knuckle boom, (22304):1, 2, 47

L

Land clearing, (22302):30
Landfill blade, (22302):12
Laser-based guidance systems for grade control
    backhoes, (22307):30
    dozers, (22307):2, 3
    excavators, (22304):41–42, (22307):30
    motor graders, (22305):17–18
Laser instruments
    benefits of using, (22307):4
    calibrating, (22307):7–8
    maintenance, (22307):7
    operation of, (22307):6
    receivers, (22307):5–6
    safety, (22307):6–7
    setup, (22307):6
    transmitters, (22307):4–5
Lift, (22203):27, 28, 37
Lift capacity, excavators, (22304):35–36, 48
Lifting activities
    dozers, (22302):38, 39
    excavators, (22304):7, 35–36
Lime, (22307):15
Load capacity
    off-road dump trucks, (22310):5, 18, 26
    rigid-frame dump trucks, (22310):2
Loading activities, (22304):33–34, 35, 40–41
Logging, (22304):41
Low-track dozers, (22302):1, 3

M

Magnets, (22304):8, 41
Mass excavation bucket, (22304):13
Mat, (22203):27, 32, 37
Mine Safety and Health Administration (MSHA), (22310):1
Mine specific design (MSD II), (22310):5
Moldboard, (22305):1, 3, 7, 47
Motor grader blades
    components
        circle rail, (22305):7, 8, 13, (22307):22, 28, 36
        cutting edge, (22305):1, 7, 47
        drawbar, (22305):7, 8
        moldboard, (22305):1, 3, 7, 47
    controls, (22305):13–14, 31
    functions, primary, (22305):1
    heel, (22305):27, 30, 47
    maneuvering, (22305):31–32
    pitch adjustments, (22305):34
    toe, (22305):27, 30, 47
    types and uses
        serrated, (22305):15–16
        snow plow, (22305):3–4, 14–15, 41
        straight, (22305):16
Motor graders
    attachments, types and uses
        rippers, (22305):16, 37–38
        scarifiers, (22305):16, 37–38
        snow blowers, (22305):3
    components
        instrument panel/alert indicators, (22305):8–10

Motor graders
  components (*continued*)
    operator's cab, (22305):6–8
    typical, (22305):4, 6–7
    wheels, (22305):12
  configurations, (22305):4
  controls
    blade controls, (22305):13–14, 31
    braking controls, (22305):12
    common arrangements of, (22305):10–11
    gauges, (22305):9–10
    transmission controls, (22305):12
    vehicle movement, (22305):7, 10–12
  function, primary, (22305):33–34
  maintenance records, (22305):25
  operations
    automatic guidance systems, (22305):11, 16–19, 32, 40
    blade operations, (22305):31–32
    inspections, prestart, (22305):22–24, 48
    laser calibration, prestart, (22305):32
    maintenance, (22305):24–25, 48
    maneuvering basics, (22305):29–31, 34
    startup and shutdown procedures, (22305):27–29
    in unstable soil, (22305):37
    warm-up procedure, (22305):28
  power of, (22305):5
  roading, (22305):41–42
  safe operation
    basics, (22305):22
    co-worker safety, (22305):21–22
    equipment, (22305):22
    mirrors, (22305):9
    operator safety, (22305):20–21
    public safety, (22305):21–22
    seat adjustment, (22305):9
    windshields, (22305):8
  size of, (22305):5
  transporting, (22305):42
  types of
    articulated-frame, (22305):1, 4, 6, 47
    rigid-frame, (22305):1, 4, 47
  work activities
    blending, (22305):35
    common, (22305):1
    ditch cutting and cleaning, (22305):3, 35–37
    finish grading, (22305):2–3, 39–40, (22307):1–2
    grading around objects, (22305):39
    mixing, (22305):1
    ripping, (22305):37–38
    rough grading, (22305):1, 2
    scarifying, (22305):37–38
    slopes, grading, (22305):39
    snow plowing, (22305):3–4, 40–41
    spreading, (22305):1–2, 35
    unpaved roads, grading, (22305):38–39
    windrowing, (22305):1, 34–35
Multi-purpose bucket, (22303):15

N

Nuclear testing, (22203):29–30

O

Object detection system, (22310):14–15
Occupational Safety and Health Administration (OSHA)
    regulations
      overhead power lines, operating around, (22304):19–20
      trench wall slope, (22304):37

Occupational Safety and Health Administration (OSHA)
    Standards *29 CFR*
      1926 (laser safety precautions), (22307):6
      1926.601 (off-road trucks), (22310):1
      1926.601(b)(1) (braking controls), (22310):12
      1926.601(b)(10) (rigid-frame off-road body), (22310):5
    rollover protective structure (ROPS), (22203):23
    seatbelts, (22203):23
Odometer, (22310):8
Off-road dump trucks. *See also* Articulated-frame off-road
      dump trucks; Rigid-frame off-road dump trucks
  controls
    braking control, (22310):12–13, 25, 28
    electrical center, (22310):15
    emergency shutoffs, (22310):15
    engine body heating system, (22310):27
    engine control, (22310):10
    exterior lighting, (22310):15
    ground level controls, (22310):15
    hoist control, (22310):13–14
    instrument panel/alert indicators, (22310):7–9
    object detection system, (22310):14–15
    traction control, (22310):1, 11–12, 25, 32
    transmission control, (22310):10–11, 12
  functions, (22310):1
  operations
    curves, (22310):25–26
    dumping, (22310):26–28
    inspection procedures, prestart, (22310):19
    loading, (22310):26
    maintenance, (22310):20–21
    maneuvering basics, (22310):24–25
    operator responsibilities, (22310):1, 17–18
    startup and shutdown procedures, (22310):22–23
    warm-up procedure, (22310):22–23
  operator's manual, (22310):8, 18, 21
  safe operation
    backward tipovers, (22310):27–28
    co-worker safety, (22310):18
    emergency shutoffs, (22310):15
    equipment, (22310):18–19
    hills, (22310):25
    load capacity, (22310):5, 18, 26
    object detection systems, (22310):14–15
    operator guidelines, (22310):17–18, 23–24, 28
    public safety, (22310):18
    runaway vehicles, (22310):27, 28
    signal persons, (22310):23
    skids, (22310):28
    training in, (22310):23
  size of, (22310):12
One Call Center, (22304):19
One-tooth ripper, (22303):13
One-way blade, (22305):15, 41
On-road dump truck operators, (22310):1
On-road dump trucks, (22310):1
Operator cab
  articulated-frame dump trucks, (22310):6
  backhoes, (22303):4, 5
  compaction equipment, (22203):8
  dozers, (22302):2–4, 7
  motor graders, (22305):6–8, 7–8
  rigid-frame dump trucks, (22310):3, 4, 9
Operator safety guidelines
  backhoes, (22303):17
  dozers, (22302):17
  excavators, (22304):18
  motor graders, (22305):20–21

off-road dump trucks, (22310):17–18, 28
OSHA. *See* Occupational Safety and Health Administration (OSHA)

P

Pads, (22203):1, 5, 37, (22303):13–14
Parking lots, finish grading, (22307):31
PAT blade, (22302):10
Pavement removal bucket, (22304):7, 15
Pawl, (22302):23, 38, 44
Pile driver, (22304):15, 16
Pipe, lifting, laying, and setting
    backhoes, (22303):33–34
    excavators, (22304):7, 36–38
Pipe booms, (22302):14
Pipe laser, (22307):30
Pitch, (22305):27, 29, 47
Platen, (22304):1, 10, 47
Plate tamper, (22303):14, 15
Pneumatic tire roller, (22203):2–3, 4, 27
Polymers, (22307):16
Pozzolan, (22307):14, 15, 36
Production blades, (22302):11
Public safety, ensuring around
    dozers, (22302):18
    excavators, (22304):18–19
    motor graders, (22305):21–22
    off-road dump trucks, (22310):18
Puddling, (22203):1, 37
Push plate blade, (22302):11
Push-type rehandling bucket, (22304):14

Q

Quarries, (22310):1, 32
Quicklime, (22307):15

R

Rake blade, (22302):12
Reach, (22303):4, 9, 46
Red tops, (22307):22, 36
Retarder control systems, automatic, (22310):1, 13, 32
Retention pond, (22307):14, 18, 36
Retract, (22303):30, 46
Reverse float, (22302):32
Reverse neutralizer, (22310):10–11
Rigid-frame motor grader, (22305):1, 4, 47
Rigid-frame off-road dump trucks
    articulated-frame dump trucks vs., (22310):6
    basics, (22310):1–2
    capacity, (22310):2
    components
        axles and wheels, (22310):2, 3, 4
        body, (22310):3, 4, 5
        engines and motors, (22310):2, 3, 4
        instrument panel/alert indicators, (22310):8–9, 11
        operator's cab, (22310):3, 4, 9
        typical, (22310):4
    defined, (22310):1, 32
    functions, (22310):2
    size of, (22310):3
    typical, (22310):2
Ripper bucket, (22304):7
Ripper controls, (22302):9
Rippers
    backhoes, (22303):13
    dozers, (22302):9, 13–14
    excavators, (22304):15

motor graders, (22305):16, 37–38
Ripper tooth, (22304):8
Ripping, (22302):23, 35, 44, (22305):16, 37–38
Riprap, (22307):1, 4, 36
Riprap placement, (22304):7
Roading
    backhoes, (22303):39–40
    motor graders, (22305):41–42
    telescoping-boom excavators, (22304):4
Roadways
    ancient, (22307):2
    asphalt
        compacting, (22203):4, 31
        grinding to grade, (22303):14, 16
        ripping, (22305):16
    concrete, (22303):14, 16
    crowns, (22307):2, 14, 16, 24–25, 36
    grading
        crowns, (22307):24–25
        finish grading, (22305):39–40, (22307):24–25
        superelevation grading, (22307):25
        unpaved, (22305):39–40
    grinding to grade, (22303):14, 16
    resurfacing, (22307):14, 15–16
    ripping, (22304):7, 15
    sidehill cutting, (22302):31
    snow plowing, (22305):41
Robotic total station, (22307):1, 8, 9, 10, 12, 36
Rock bucket, (22304):14
Rollover protective structure (ROPS), (22203):8, (22303):4, (22310):3
ROPS. *See* Rollover protective structure (ROPS)
Rotating-beam lasers, (22307):5
Rough grading, (22305):1, 2
Rubble, (22303):30, 31, 46
Runaway truck ramps, (22310):27, 28
Runaway vehicles, (22310):27, 28
Runoff control, (22307):18

S

SAE. *See* Society of Automotive Engineers (SAE) maintenance schedules
Safe operation
    of backhoe attachments, (22303):13
    boom swing and, (22304):29, 30
    of compaction equipment, (22203):16–19, 21
    of dozer attachments, (22302):13, 15
    of dozers, (22302):4, 25, 36
    emergency shutoffs, (22310):15
    of excavators, (22304):19–20, 29, 30
    laser instruments, (22307):6–7
    load capacity and, (22310):5, 18, 26
    loading, (22310):26
    mirrors for, (22305):9
    of motor graders, (22305):8, 9
    object detection systems, (22310):14–15
    of off-road dump trucks, (22310):5, 14–15, 18, 23, 25–28
    overhead power lines, (22304):19–20
    in rocky areas, (22302):36
    runaway vehicles, (22310):27, 28
    of scrapers, (22203):17
    seat adjustments for, (22302):25, (22305):9
    signal persons, (22310):23
    skids, (22310):28
    on slopes, (22203):17, (22310):25
    tires and, (22203):17–18, 21
    underground utilities, (22304):19–20

...re operation (*continued*)
  in unstable soils, (22302):33, (22303):28, 38–39, (22305):37
  when transporting, (22304):29
  windshield wipers, (22302):4
Safety
  of co-workers, ensuring around
    backhoes, (22303):18
    dozers, (22302):18
    excavators, (22304):18–19
    motor graders, (22305):21–22
    off-road dump trucks, (22310):18
  equipment
    dozers, (22302):18–19
    excavators, (22304):19
    motor graders, (22305):22
    off-road dump trucks, (22310):18–19
  of operators, when operating
    backhoes, (22303):17
    dozers, (22302):17
    excavators, (22304):18
    motor graders, (22305):20–21
    off-road dump trucks, (22310):17–18, 28
  of the public, ensuring around
    dozers, (22302):18
    excavators, (22304):18–19
    motor graders, (22305):21–22
    off-road dump trucks, (22310):18
Sand cone test, (22203):29
S blade, (22302):10
Scarifier/ripper combinations, (22305):15, 16
Scarifiers, (22305):16, 37–38
Scarifying, (22305):1, 16, 37–38, 47
Scraper blade, (22305):14–15
Scrapers
  finish grading, (22307):2–3
  pushing, dozers for, (22302):33–34
  safe operation, (22203):17
Scrap handling, (22304):8, 41
Seat adjustment controls, (22203):11–12, (22302):4, 25, (22305):9
Security panels, (22203):11
Segmented-pad rollers, (22203):5, 27
Self-leveling lasers, (22307):5
Serrated motor grader blades, (22305):15–16
Settling, (22203):1, 37
Shears, (22304):8, 15, 16
Sheepsfoot, (22203):1, 37
Sheepsfoot roller, (22203):5–6, (22302):14, 15, 36
Shooting-boom excavator, (22304):1, 2, 47
Shovel excavator, (22304):40–41
Side booms, (22302):14–16
Side-casting, (22305):27, 40, 47
Sidehill cutting, (22302):31
Side sloping, (22304):6–7
Side tip bucket, (22303):15
Sidewalks, (22307):31
Silt fencing, (22307):19
Single beam laser, (22307):4
Slopes
  backhoe operations on, (22303):28
  cutting and building, (22302):29–31
  defined, (22304):1, 47
  finish grading, (22304):6–7, 15, (22305):39, (22307):25–26
  safe operation on, (22203):17
Slot dozing, (22302):23, 29, 44
Snow blowers, (22305):3
Snow plow blades, (22305):3–4, 14–15
Snow plowing, (22305):3–4, 40–41

Society of Automotive Engineers (SAE) maintenance
  schedules, (22302):20, (22303):19, (22304):21, (22305):24, (22310):20
Soil
  density testing, (22203):29–30
  dozer performance variable, (22302):27–28
  moisture content, (22302):28
  particle size, (22302):27
  unstable
    backhoe operations on, (22303):28, 38–39
    dozer work in, (22302):33
    grading on, (22305):37
    motor graders on, (22305):37
  voids, (22302):28
Soil compaction, (22203):1, 37
Soil modification, (22307):14–16
Soil over-compaction, (22203):23
Soil stabilization
  benefits, (22307):14
  binders, (22307):14–16, 17
  erosion control, (22307):18–20
  planning for, (22307):16
  procedures, (22307):14
  roadway resurfacing, (22307):14
  subgrade preparation, (22307):16–17
Speedometer, (22203):10, (22310):8
Spill guard, (22310):7
Spoils, (22303):1, 46, (22304):24, 32, 47
Spreading activities, (22305):1–2, 35
Stability
  backhoes, (22303):1
  excavators, (22304):30
Stabilizers, backhoe, (22303):4, 9, 27–28, 38
Steel-wheel roller, (22203):3–4, 13, 14, 27
Steering wheel adjustment, (22203):11–12
Stick, (22303):9, (22304):1, 47
Stick cylinder backhoe, (22303):4
Stockpiles
  building, (22302):29–30
  defined, (22302):1, 11, 44
  loading, (22303):30–31
Storm water runoff, (22307):18
Straight blade, (22302):10
Straight dozing, (22302):28–29
Straight motor grader blades, (22305):16
Straw wattles, (22307):20
Street pads, (22303):13–14, 15
Subgrade finish grading, (22307):26–27
SU blade, (22302):11
Superelevation, (22307):22, 25, 36
Superelevation grading, (22307):25
Switches
  disconnect switches, (22203):11
  engine key switch, (22302):5
  engine start switch, (22203):12–13
  fuel shutoff switch, (22203):11

T
Tachometer, (22203):10, (22303):9, (22305):10, (22310):9
Tailgate, articulated-frame dump trucks, (22310):7
Tamping roller, (22203):1, 5, 37
Tassel (whisker), (22307):22, 36
Telescoping-boom excavators
  advantages of, (22304):6, (22307):3–4
  attachments, types and uses, (22307):3–4
  components
    braking system, (22304):11–13
    instrument panel/alert indicators, (22304):11

functions, (22304):2, 3, (22307):3–4
pavement removal bucket, (22304):7
roading, (22304):4
straight boom with pivoting bucket, (22304):2
track-mounted, (22304):4–5
truck-mounted, (22304):4, 10
wheel-mounted, (22304):4–5
work activities
    ditch cutting, shaping, cleaning, (22304):6
    excavation, general, (22304):6
    finish grading, (22304):7
    ripping pavement, (22304):7
    riprap placement, (22304):7
    side sloping, (22304):6–7
    trenching, (22304):6
Temperature gauges
brake oil, (22310):8
engine coolant, (22203):9, (22302):5–6, (22303):9, (22310):8
hydraulic oil, (22203):10, (22302):6–7, (22305):10
torque converter oil, (22302):6
transmission oil, (22203):9, (22303):8
Throttle locks, (22310):10, 12
Thumb, (22304):15
Tipovers, (22310):27–28
Tires
for equipment safe operation, (22203):17–18, 21
pneumatic tire roller, (22203):2–3, 4
Toe of the blade, (22305):27, 30, 47
Torque converter oil temperature gauge, (22302):6
Track-mounted excavators, (22304):2–3, 4, 8–12
Track-mounted telescoping-boom excavators, (22304):4–5
Traction control system
defined, (22310):1, 11, 32
off-road dump trucks, (22310):11–12, 25
Transmission controls
compaction equipment, (22203):13
dozers, (22302):8
motor graders, (22305):12
off-road dump trucks, (22310):10–11, 12
Transmission oil temperature gauge, (22203):9, (22303):8
Transporting
backhoes, (22303):40
dozers, (22302):38–39
excavators, (22304):29, 42–43
motor graders, (22305):42
safety guidelines, (22304):29
Tree shear attachment, (22302):14
Trenches
backfilling, (22302):30–31
finish grading, (22307):29–30
wall slope regulations, (22304):37
Trenching, (22303):1, 46
Trenching bucket, (22304):13
Trenching equipment
backhoes, (22303):30, 31–32, 37–38
excavators, (22304):5–6, 36–37
telescoping-boom excavators, (22304):6
Truck-mounted telescoping-boom excavators, (22304):4, 10
Two-lever backhoe controls, (22303):10–11

U

U blade, heavy-duty, (22302):11, 12
Undercarriage, (22304):1, 2, 47
Undercutting, (22304):24, 30, 47
Underground mining dump truck, (22310):1, 7, 8, 32
Upper carriage, (22304):1, 11, 47
Upper structure, (22304):1, 8, 47
Utility bucket, (22304):13

V

Vandal guards, (22203):11
Variable-radius blade, (22302):10
Vehicle movement controls
backhoes, (22303):12–13
compaction equipment, (22203):13
dozers, (22302):2, 7–8
motor graders, (22305):7, 10–12
Vibratory compactor, (22203):4–5
Vibratory plate tamper, (22303):14, 15
Vibratory roller, (22203):1, 4–5, 37
Vibromax single-drum rollers, (22203):7
Virtual reference station (VRS), (22307):1, 9, 36
Voltmeter, (22303):8, (22305):9
V-plow blade, (22305):3–4, 14–15, 41
VRS. *See* Virtual reference station (VRS)
V-tree cutter blade, (22302):12–13, 29–30

W

Water
availability for soil stabilization, (22307):16
non-potable, use of, (22307):17
Water spray units, (22203):13–14, 17
Weather, soil stabilization and, (22307):16
Wheel dozers, (22302):1–2, 3, 4
Wheel-mounted excavators, (22304):2–3, 4
Wheel-mounted telescoping-boom excavators, (22304):4–5
Wheels
motor graders, (22305):12
rigid-frame dump trucks, (22310):2, 3, 4
tilt front, motor graders, (22305):12
Whisker (tassel), (22307):22, 36
Winch
defined, (22302):1, 44
dozer attachment, (22302):2, 14, 15
Winching procedure, (22302):34–35
Windrowing, (22305):1, 34–35
Windshield wipers, (22302):4
Wood chip blade, (22302):13

X

X-body, (22310):5